AF744099

Min Xu, Xiaomin An, Wei Kang, Guangning Li

Modern Computational Aeroelasticity

Also of interest

Computational Chemistry Methods, Applications
Ramasami, Ponnadurai (Ed.), 2020
ISBN 978-3-11-062906-4, e-ISBN (PDF) 978-3-11-063162-3,
e-ISBN (EPUB) 978-3-11-062992-7

Computational Physics: With Worked Out Examples in FORTRAN and MATLAB
Michael Bestehorn, 2018
ISBN 978-3-11-051513-8, e-ISBN (PDF) 978-3-11-051514-5,
e-ISBN (EPUB) 978-3-11-051521-3

Non-Newtonian Fluids, A Dynamical Systems Approach
Boling Guo, Chunxiao Guo, Yaqing Liu, Qiaoxin Li, 2018
ISBN 978-3-11-054923-2, e-ISBN (PDF) 978-3-11-054961-4
e-ISBN (EPUB) 978-3-11-054940-9

Optical Measurement Mechanics
Kaifu Wang, 2018
ISBN 978-3-11-057304-6, e-ISBN (PDF) 978-3-11-057305-3
e-ISBN (EPUB) 978-3-11-057320-6

De Gruyter Frontiers in Computational Intelligence
ISSN: 2512-8868, eISSN 2512-8876

Min Xu, Xiaomin An, Wei Kang, Guangning Li

Modern Computational Aeroelasticity

DE GRUYTER

国防工業出版社
National Defense Industry Press

Authors
Min Xu
Shaanxi Aerospace Flight Vehicle Design Key Laboratory,
School of Astronautics, Northwestern Polytechnic University,
Xi'an 710072, China
xumin@nwpu.edu.cn

Xiaomin An
Shaanxi Aerospace Flight Vehicle Design Key Laboratory,
School of Astronautics, Northwestern Polytechnic University,
Xi'an 710072, China
frank805@nwpu.edu.cn

Wei Kang
Shaanxi Aerospace Flight Vehicle Design Key Laboratory,
School of Astronautics, Northwestern Polytechnic University,
Xi'an 710072, China
wkang@nwpu.edu.cn

Guangning Li
Shaanxi Aerospace Flight Vehicle Design Key Laboratory,
School of Astronautics, Northwestern Polytechnic University,
Xi'an 710072, China
lgning@nwpu.edu.cn

ISBN 978-3-11-057647-4
e-ISBN (PDF) 978-3-11-057668-9
e-ISBN (EPUB) 978-3-11-057662-7

Library of Congress Control Number: 2020935449

Bibliographic information published by the Deutsche Nationalbibliothek
The Deutsche Nationalbibliothek lists this publication in the Deutsche Nationalbibliografie; detailed bibliographic data are available on the Internet at http://dnb.dnb.de.

© 2021 National Defense Industry Press and Walter de Gruyter GmbH, Berlin/Boston
Cover image: Devrimb/iStock/Getty Images Plus
Typesetting: Integra Software Services Pvt. Ltd.
Printing and binding: CPI books GmbH, Leck

www.degruyter.com

Preface

High flight speed and lightweight structure have increasing significance in commercial and military aircraft design, whereby the nonlinear aeroelastic computation is the key ingredient for aerodynamic load analysis. This book, which provides the state of the art and recent progress in the numerical method for nonlinear aeroelastic computation and load analysis, is written to be used as a postgraduate textbook in aerospace engineering. It is consciously written for students to read, understand, be inspired and challenged.

With the ever-growing capability of computing hardware and software, computational fluid dynamics (CFD)/computational structure dynamics (CSD) coupling technology emerges for analyzing the characteristics of nonlinear aeroelastic system and remains the state of the art in the past decades. Furthermore, the CFD/CSD coupling technology has been widely used in the research, where high-fidelity computation is necessary, such as aerosturcture optimization, limit cycle oscillation computation and active flow control.

The material covered throughout this book was compiled from the prominent research output of our group in the area of computational aeroelasticity since the past decades in the Northwestern Polytechnical University. Moreover, the book is featured as follows:

1. **Chapter 1:** An introduction to the methods for nonlinear aeroelastic system analysis (written by Xu Min).
2. **Chapter 2:** This chapter is devoted to describe the state-of-the-art CFD technologies for unsteady aerodynamics (written by Xu Min and Li Guangning).
3. **Chapter 3:** The purpose of this chapter is to describe the numerical methods to solve the structure system with geometric nonlinearity (written by An Xiaomin and Kang Wei).
4. **Chapter 4:** The chapter is devoted to obtain an interpolation between CFD/CSD coupling interface followed by the CFD mesh deformation techniques (written by An Xiaomin an Xu Min).
5. **Chapter 5:** This chapter provides numerical methods to solve the CFD/CSD coupling system and performs accuracy analysis (written by An Xiaomin and Xu Min).
6. **Chapter 6:** This chapter is to construct the reduced order models for rapid aeroelastic computation (written by Xu Min and Kang Wei).
7. **Chapter 7**: Software development for computational aeroelasticity (written by Xu Min and Xie Liang).

I want to thank my students and colleagues for their help in the development of this book. I would like to especially extend my gratitude to Dr. Yao Weigang, Dr. Cai Tianxing, Dr. Dou Yibin, Dr. Zhang Zijian, Dr. Xie Liang, Dr. Chen Hao, Dr. Chen Tao,

https://doi.org/10.1515/9783110576689-202

Dr. Zhang Bin, Dr. Xie Dan and Mr. Wang Yabin. This version is translated by An Xiaomin, Kang Wei, Xie Dan, Li Guangning and Yao Weigang.

Many research works presented in this book were funded by the National Natural Science Foundation of China and the Youth Foundation Project (Grant Nos. 11402212, 11202165, 10272090, 90405002, 90816008). Chinese Academy of Science academician Chen Shilu and Yu Benshui kindly provided me with their expertise and helped in editing the text. Also, special thanks go to Niu Xudong for his excellent help in producing this book.

Modern Computational Aeroelasticity was designed to convey the art of mathematics and beauty of the great minds in the subject of aerodynamics and aeroelasticity. I hope that this book can bring readers into the world of aeroelastic design where great challenges remain yet to be overcome.

Xi'an China
Xu Min

Contents

Chapter 1
Introduction

Aeroelasticity is a discipline which studies the mutual interaction among aerodynamic, elastic and inertial forces [10, 7]. The flexible structure deforms under aerodynamic load, and the deformation in turn affects the aerodynamic load. This interaction between deformation and load leads to a variety of aeroelastic phenomena such as flutter onset, limit cycle oscillation (LCO) and vortex-induced vibration (VIV). In general, the aircraft structure can experience fatigue, damage and collapse amid the growing transient structural vibration if the flight speed exceeds the threshold velocity or flutter onset speed before it reaches the static aeroelastic stability [21, 27]. From the application perspective, aeroelasticity can be broadly categorized into two main branches, namely civil and aerospace engineering. Similar to aircraft structure, aeroelasticity is also applicable to civil engineering, for example, flow around high-rise buildings, suspension bridges, electricity transmission lines and offshore pipelines. This book only concentrates the aeroelasticity regarding the aerospace engineering, however, the theatrical and numerical methods herein are transferable and can be easily applied for civil engineering.

Aeroelasticity can be broadly divided into two main categories: static and dynamic aeroelasticity. Static aeroelasticity mainly studies structural deformation subject to steady aerodynamic force, such as divergence and control surface reversal, whereas the topics of the dynamic aeroelasticity cover dynamic stability of a linear aeroelastic system (flutter), transient load analysis (e.g., Gusts) [14, 28], VIV and LCO.

Aircraft can experience a range of static and dynamic aerodynamic loads arising from various maneuvers, such as take-off and landing, as well as different flow conditions such as gust and turbulence. These loads are critical and impose significant constraints on the structure design, for example, structural sizing, flutter clearance and topology optimization. Therefore, how to compute these loads efficiently and accurately is of significant interest in the aeroelastic community as the load features is explicitly linked with aeroelastic effects. Aircraft can also be viewed as an automatic control system which is coupled with the structural vibration to maintain stable flight. The lightweight structural is commonly adopted in aircraft structure design, which tends to make the structural natural frequency within the normal operating frequency band of the flight control system. Therefore, the control input is coupled with structural vibration and makes the aeroelastic effects more pronounced [6] and is able to excite the structural degree of freedom. The feedback control system can be affected by the elastic deformation, whereby the observer detects not only the change of the rigid motion of the aircraft but also the structural deformation, which is viewed as an additional feedback input introduced into the flight control system. The technology dealing with interaction between aeroelasticity and control system is known as aeroservoelasticity [1].

https://doi.org/10.1515/9783110576689-001

The conventional method for aeroelastic analysis is mainly developed in the frequency domain and contains many assumptions, such as sufficiently small structural deformation for linearization. Therefore, it is difficult, if possible, to deal with nonlinear aerodynamic problems arising from transonic speed and high angle of attack [26, 31]. With the development of computational fluid dynamics (CFD) and computational structural dynamics (CSD) technology, the foundation for the time domain simulation of aeroelastic system with CFD/CSD coupling technology is laid. The numerical analysis of aeroelasticity directly in the time domain can obtain the transient structural response, variation law of the internal/external load and stability characteristics, especially in the transonic region, the CFD/CSD coupling methodology can capture the influences of shock/shock, shock/vortex, vortex/vortex and shock/boundary layer interaction for physical analysis of aeroelastic system [17]. Furthermore, the CFD/CSD coupling method, which is a cross-disciplinary technology for fluid–structure interaction, not only has academic impact, but also has significant industrial impact in aeronautical engineering. However, the nature of the method inevitably leads to large computational scale, expensive computational cost and long engineering aeroelastic design period [24]. Therefore, it is necessary to develop fast yet accurate unsteady aerodynamic models in lieu of CFD.

The aeroelastic problems such as flutter onset also compromise the flight performance and ride comfort of aircraft. For severe cases, it may damage aircraft structure integrity, and therefore, jeopardize the safety of passengers. Furthermore, the improvement of flight speed and reduction of the weight and rigidity of aircraft structure render the aeroelastic problem more pronounced in the aircraft design, especially in recent years, the aerospace industry in China has been growing vigorously, and the aeroelasticity receives unprecedented attention, which promotes innovative research and development significantly in aeroelasticity for future aircraft structure design in China.

1.1 Development history of aeroelasticity

The aeroelastic problems emerge in the early history of aviation technology and become more pronounced as the aviation industry further grows. The first structural failure attributed to aeroelasticity occurred on December 8, 1903, nine days before the successful flight of the Wright Brothers' heavier-than-air-powered aircraft at Kitty Hawk. During the flight test of the tandem monoplane built by Professor Samuel P. Langley of the Smithsonian Institute, the torsional divergence was observed in the front wing as result of insufficient stiffness and the Langley's aerodrome plunged into Potomac River. It was believed that the failure was a result of typical static aeroelastic divergence, whereby the elastic twist of the wing grows to infinity exponentially.

Shortly after the Langley's failure, tail surface flutter was found on a British Handley Page twin-engine bomber as a result of insufficient torsional stiffness in 1916, which is the first recorded flutter incidence. The similar tail flutter was experienced only a year later on the DH-9 aircraft. The first major development of aircraft flutter analysis was pioneered by the preeminent British engineer and scientist F. W. Lanchester.

Flight testing of the Bristol Bagshot aircraft, which was built by Bristol Aeroplane Company and first flown in 1927, revealed that the aileron efficiency reduced to zero and became negative progressively as the flight speed increased. This adverse effect is known as aileron reversal and significantly compromises the stability and controllability of aircraft.

Aviation industry experienced rapid growth since World War II (1939), and the flight speed of aircraft increased to the transonic regime, which raised new and challenging aeroelastic problems. When a National Aeronautics and Space Administration pilot flew the new P-80 aircraft for high-speed flight tests in 1944, a severe aileron vibration appeared, which was viewed as a single degree of freedom flutter caused by the coupling of the aileron deflection and the chordwise motion of the shock on the wing, and this phenomenon is called "aileron buzz."

The first supersonic flight was achieved by US Air Force Captain Charles Yeager in the Bell X-1 demonstrator in 1947. At supersonic speed, aeroelastic instability of thin plate or panel flutter occurs frequently resulting in structural fatigue, excessive noise and functional failure of component.

In 1946, the British engineer and scientist Arthur Roderick Collar proposed "a triangle of forces" or Collar's triangle as shown in Fig. 1.1 describing the mutual interrelation between inertial (I), elastic (E) and aerodynamic (A) forces for aeroelasticity. Each vertex in the triangle represents one of the three forces. Any of two vertices or forces in Collar's triangle can define a major discipline. For example, the

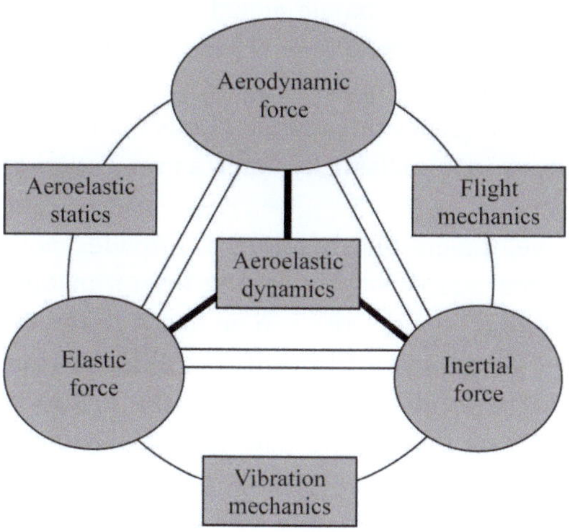

Fig. 1.1: Aeroelastic Collar's triangle of forces.

discipline of *vibration mechanics* can be defined by linking the elastic and inertial forces, and the connection between aerodynamic and inertial forces gives rise to *flight mechanics*. The linking between elastic and aerodynamic forces is in the realm of *static aeroelasticity*. However, it requires all the three forces to interact for the *dynamic aeroelasticity* to occur [9, 16, 19].

1.2 Computational aeroelasticity method

The aeroelasticity is multidisciplinary and challenged by difficulties from a variety of disciplines such as aerodynamics, structural dynamics and flight stability. Therefore, the aeroelastic problems of aircraft remain an attractive topic for aircraft designers.

The conventional computational aeroelasticity method is mainly developed in frequency domain, whereby airfoil flutter is assumed to be harmonic oscillation and aerodynamic force is only computed in frequency domain. For aeroelastic stability analysis (e.g., flutter onset), classical linear method such as Theodorson, slender body, piston theories and doublet lattice method (DLM) is valid and can be used to compute unsteady aerodynamic force with high efficiency. Therefore, the classical method is extensively adopted in aeronautical engineering for flutter onset prediction. However, the assumption inherent in the development of the methods (e.g., linear and inviscid), restricts their application for nonlinear aerodynamics, such as flow separation. Many types of software have been commercialized for aeroelasticity computation. For example, MSC.NASTRAN is widely used for structural dynamics and aeroelasticity. The FlightLoads module can solve static and dynamic aeroelastic problems by integrating the capabilities of MSC.NASTRAN and PATRAN. The internal unsteady aerodynamic model in FlightLoads is constructed by DLM, which cannot deal with the shape of aircraft and nonlinear aerodynamics.

As mentioned earlier, the frequency domain analysis requires linearization and contains many assumptions. The constructed aerodynamic model is based on the potential flow and sufficiently small angle of attack, which renders it difficult to deal with nonlinear aerodynamic problems, such as transonic speed and high angle of attack. In addition, the aforementioned frequency domain method is not feasible for the static–dynamic integration analysis in aeroelasticity, such as prediction of the subcritical or supercritical response.

Since the mid-to-late 1990s, the development of CFD and CSD provides more accurate solutions especially for nonlinear aeroelasticity arising from transonic speed and large elastic deformation [20]. Extensive work can be found in the literature regarding aeroelasticity computation using CFD/CSD coupling in time domain for time-accurate solution. The basic idea is to calculate the unsteady aerodynamic force on the elastic body using CFD and transfer the aerodynamic load to CSD system to update the elastic deformation for CFD system. Therefore, time-accurate solution such as structural response history can be produced by the coupling framework [15, 19, 23]. The CFD/CSD coupling method has the following advantages for solving aeroelastic problems.

(1) High accuracy. Unsteady aerodynamic force computation with high accuracy requirement for aeroelastic analysis is solved by CFD. The accuracy is much higher than the classical and empirical methods as CFD can deal with complex aircraft configuration and flow problems in transonic and hypersonic regime.
(2) Rich information. Time domain analysis can produce time-accurate load and structural response of aircraft, which can be analyzed extensively and provide insights on the mechanism of various aeroelastic phenomena.
(3) Strong scalability. Modular programming can be adopted to couple different levels of CFD and CSD systems. The up-to-date technologies can be conveniently implemented. Furthermore, the comprehensive analysis on mechanical environment can be performed by coupling with other disciplines.

As the CFD/CSD coupling requires two disciplines, namely CFD and CSD, the development of CFD/CSD coupling is closely associated with the evolution of CFD and CSD. The development and evolution of these two disciplines is elucidated in Fig. 1.2.

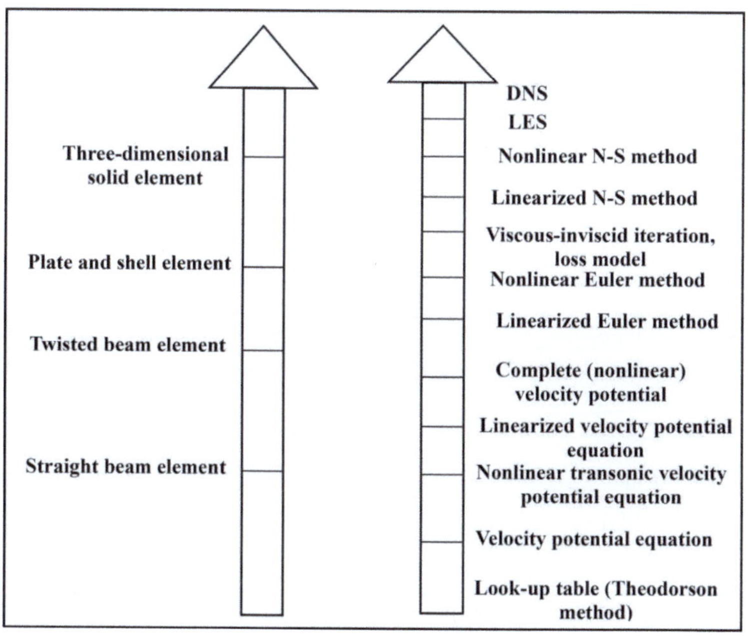

Fig. 1.2: Models and evolutions in CFD/CSD coupling.

The calculation of aerodynamic force experiences the Theodorson, small disturbance velocity potential method, transonic small disturbance nonlinear velocity potential method, linearized velocity potential method, (nonlinear) velocity potential method, linearized Euler method, nonlinear Euler method, linearized Navier–Stokes (NS) method, nonlinear NS method, large eddy simulation, direct numerical simulation and so on. The finite element model (FEM) of CSD has experienced beam-like elements (straight beams, twisted beams), plate-shell elements (thin shells, thick shells) and three-dimensional solid elements.

The aeroelastic model was developed from the early two-degree-of-freedom rigid wing shown in Fig. 1.3 to FEM, whereby the fluid system is three-dimensional unsteady viscous compressible flow, whereas the structure system is three-dimensional volume elements. Recently, aeroelasticity has been investigating nonlinear aerodynamics arising from shock wave, shock wave and boundary layer coupling, flow separation and so on, and nonlinear structure dynamics such as large deformation, contact, elastoplasticity and friction damping. In the past decades, CFD technology is developed extensively in many aspects such as numerical scheme, mesh automatic generation and turbulence modeling. Solving a linear CSD system is a mature technology and significant progress has been made in the method to solve nonlinear CSD system. In addition, speed, storage and architecture of computer have also been improved greatly. Therefore, nonlinear aeroelasticity is developed rapidly and has a broad range of applications in numerous fields such as nuclear engineering, pipelines, vessels, turbines, engine blade flow field simulation, the descending process of parachutes, bridges, dams, high-rise buildings, hydraulics, medical fluid mechanics and human organology [19, 20].

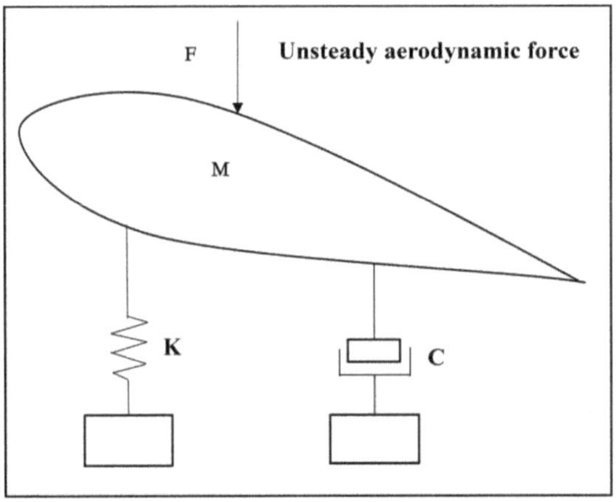

Fig. 1.3: Two-dimensional aeroelastic model.

CFD/CSD coupling method, which is a cross-disciplinary approach for fluid–solid interaction, has significant academic and industrial impact in understanding and solving complex mechanics in aeronautical engineering. Furthermore, CFD/CSD coupling offers an alternative method to solve nonlinear aeroelastic problems and captures aeroelastic effects accurately in aircraft design, which can be developed into a new design concept [8].

1.3 Computational aeroelasticity software

Aeroelastic analysis based on CFD/CSD coupling has been implemented in open-source and commercial software. For example, ZONA technology in the United States has developed aeroelasticity analysis software such as ZERO, ZONAIR and ASTROS. CFDRC also integrated the fluid–solid coupling (FSI) module into CFDRC-Fastran. Based on CFX's pioneering advantages in fluid mechanics analysis, ANSYS and CFX have jointly introduced a FSI simulation system. CFL3D V6.0 is an open-source CFD code and implemented the aeroelastic analysis module. Professor Charbel Farhat at Stanford University and the German Aerospace Center (DLR) developed numerical methods extensively for CFD/CSD coupling and applied the methods to solve engineering problems [8, 22]. At present, the mature aeroelastic analysis software emerges in an endless stream and can be divided into four categories as follows [4, 12, 13].

The first category: using the classical methods without graphical user interface (GUI) such as FLEXSTA and DYLOFLEX. Interface modeling function is not provided. Therefore, the model data is completely dependent on the user to specify, which introduces error easily, and the simplification of the model further reduces the reliability and accuracy of the result.

The second category: using the classic methods with a complete GUI interface such as Flds and ZARERO.

The third category: modern aeroelasticity software based on CFD/CSD coupling, such as STARS, ENSAERO and CFL3DV6, which however, has no GUI interface.

The fourth category: CFD software is foundation, the user interface of which is applied to complete the structure solver and the aeroelastic analysis in time domain such as Fluent and CFX.

So far, most of the aeroelasticity software used in Chinese research institutions belongs to the first and second categories, which is obviously difficult to satisfy the needs of aircraft design in the new situation. The aeroelasticity software based on CFD/CSD coupling, such as STARS, ENSAERO and CFL3DV6 still lacks a mature user interface, hence, the input of fluid data, structural information and the coupling interface specification must be completed with text files produced manually. Although CFD software, such as Fluent or CFX, has a friendly GUI, it has no interface for CFD/CSD coupling. It is cumbersome, if possible, for users to provide CSD solver and implement CFD/CSD coupling procedure using this type of software. Even if the exploitation is completed, still there is a problem of poor versatility and heavy repetitive workload.

Since the 1990s, many universities and research institutes in China have carried out the exploration of aeroelastic time domain simulation based on CFD/CSD coupling algorithm, and achieved many good achievements. These researches mainly focus on two aspects: application and self-developed software type.

(1) Application type: mainly relies on the direct purchase of foreign commercial softwares to carry out research, such as NASTRAN and ZARERO, or uses foreign independent academic software (such as fluid calculation using CFX, Fluent and structural calculation using Ansys) for secondary development. CFD and CSD are combined into a coupling form to obtain the coupling effect of fluid–structure calculation. This idea and process are widely used in engineering applications.
(2) Self-developed software type: it is the complete use of self-developed CFD, CSD and interface programs to form a fluid–structure coupled calculation system, which can be applied in engineering applications. The direct purchase of commercial fluid–structure coupling software has the disadvantages of less output information, oversimplified modeling, overlinearization and a large limitation of the range of flow speed. Both the secondary development of commercial softwares and the self-developed softwares have problems of computational robustness and versatility issues. In addition, the developed code has not been successfully softwareized and platformized, and the operation is cumbersome, learning interface is hard; therefore, it is difficult to satisfy the needs of the engineering institutions. Moreover, the domestic CFD and FEM software technologies started too late, and there is still a big gap compared with foreign excellent software, which is mainly reflected as follows: (1) Most software still adopt the Fortran language, which results in the poor versatility, scalability and maintainability, relatively backward architecture and difficult upgrading for. (2) The mode for software development needs to be improved, the means for project management is simple, and the software quality is difficult to guarantee. (3) The interface of the CFD software has no unified standard, which limits data sharing and communication between different software. (4) The interface design is relatively backward, and the human–computer interaction is simple and thus cannot meet the needs of engineers.

1.4 Key techniques for solving aeroelastic problems by CFD/CSD coupling method

In the nonlinear aeroelastic problem, two different physical fields of structure and aerodynamic interact on the coupling interface and affect each other. On the one hand, the fluid itself involves a large number of nonlinear phenomena, such as complex turbulent motion, shock waves induced by high-speed motion, boundary layer separation caused by shock waves, unsteady vortex shedding, motion and evolution as well as unsteady flow motion due to structural deformation or vibration. On the other hand, the structure involves nonlinear large geometric deformation, nonlinearity of elastoplastic materials and interface nonlinearities from indefinite contact surfaces. In addition, even if the respective physical fields are linear, the uncertain coupling on the common interface of CFD and CSD will cause new nonlinear problems. Therefore, solution of the coupling of CFD and CSD is not a simple superposition of these two problems. A CFD/CSD system coupled through continuity-compatible conditions on the interface is a highly nonlinear problem, and improper processing will lead to a failure in the calculation of the entire system. The key techniques for solving nonlinear aeroelastic problems based on CFD/CSD coupling method are discussed further.

1.4.1 Simulation of nonlinear characteristics of the fluid and solid domains

In view of the nonlinearity of the structure and fluid fields, foreign researchers in this area have made great progress, using many different solving models: in the fluid solver, from the simple potential flow model to the three-dimensional Reynolds average NS equation, and in the structural solver, from linear beam theory to nonlinear finite element theory. There are three combinations of research methods for solving the typical aeroelastic coupling in the time domain: (1) linear aerodynamic solver + linear structural solver; (2) nonlinear aerodynamic solver + linear structural solver and (3) nonlinear aerodynamic solver + nonlinear structural solver [30].

1.4.2 Motion strategy for moving grid

Moving-grid technology needs to be considered for solving unsteady aerodynamics based on the CFD method. There are two commonly used moving-grid methods: transfinite interpolation method for structural mesh and spring analog method for unstructured mesh. Other moving-grid methods are all based on the two methods with improvement of computational efficiency, accuracy and stability, and also satisfy the consistency and compatibility conditions between fluid grid motion and fluid integral solution [2, 3, 5, 11, 18, 25].

1.4.3 Continuity and compatibility conditions for fluid and structure interface

The first is the time marching synchronization technique of solving two independent domains of fluid and structure. Three coupling ways including fully coupling, tightly coupling and loosely coupling have been developed for CFD/CSD coupling system, which respectively correspond to different time marching methods, and ensuring the coordination of physical time and calculating time and reducing the coupling error are both the key points of the research [29]. Another research focus is on the information transformation of the interface to ensure the conservation of mass, momentum and energy on the coupled interface and the solution of the coupled equation. Currently, there are two types of mature methods: one is the surface assembly method, which uses a known point to obtain a surface spline function to interpolate unknown points, such as infinite-plate splines; another one is the surface tracking method, using the shape function of the local finite element to interpolate the information of unknown points, such as constant volume transformation method.

1.4.4 Calculation efficiency of fluid and structure coupling

One of the characteristics of coupled time domain calculation is that a large number of time marching steps are needed. The calculation is often time consuming if the number of discrete meshes in spatial domain is huge. Therefore, one of the current research interests is to simplify the entire coupling calculation using multiblock and parallelized technologies. The technology benefits from the development of communication technology for distributed processing. The algorithm design not only divides the two physical domains, but also parallelizes the coupling interface, moving-grid technology and coupling way.

In addition, the research focus of improving the aeroelastic calculation efficiency lies in the reduced-order model technology for unsteady aerodynamic, by which we can study the essential characteristics of the coupled system and efficiently perform system analysis for further application in the control system design and multidisciplinary optimization. The current developed technologies mainly include Volterra series, proper orthogonal decomposition technology, harmonic balance technology, and artificial neural network. These methods have their own advantages and shortcomings, so combining these methods together may perform better.

To sum up, the nonlinear aeroelastic study has both important theoretical and practical significance for aircraft design and analysis, and the CFD/CSD coupling solving technique is an important way to achieve nonlinear aeroelastic analysis. It can be seen that CFD/CSD coupling for solving nonlinear aeroelastic problems in the time domain is a complex nonlinear system technology, any link of which is worthy of in-depth research and analysis.

Bibliography

[1] Aeroelasticity committee of Chinese society of aerodynamics. Proceedings of the 9th national academic exchange conference on aeroelasticity, Szechuan, Chengdu, F, 2005.8. (in Chinese)

[2] Ali ABH, Soulaïmani A. A parallel-distributed approach for multi-physics problems with application to computational nonlinear aeroelasticity. Canadian Aeronautics and Space Journal. 2004, 4(50).

[3] BenElHajAli A, Soulaimani A, Feng ZA. distributed computing-based methodology for nonlinear aeroelasticity. AIAA Paper 2002–0868, 2002.

[4] Blair M, Noh T, Cerra JA. An aeroservoelastic analysis method for analog or digital systems. AIAA Paper 85-3090, 1985.

[5] Brown G, Geuzaine P, Farhat C. Three-field-based nonlinear aeroelastlic simulation technology: status and application to the flutter analysis of an f-16 configuration. AIAA Paper 2002-0870, 2002.

[6] Chae S, Hodges DH. Dynamics and aeroelastic analysis of missiles. AIAA Paper Paper 2003-1968, 2003.

[7] Chen G, et al. Aeroelastic design basis. Beijing: University of Aeronautics and Astronautics Press; 2004. 10. (in Chinese).

[8] Cotoi I, Botez R, Doin A, et al. Method validation for aeroservoelastic analysis. AIAA Paper 2002-1481, 2002.

[9] Edwards J, Dowell E. Strganac TW. Nonlinear aeroelasticity. AIAA Paper 2003–1816, 2003.

[10] Fu X, H.W. Principles of aeroelasticity. Shanghai: Science and Technical Literature Press; 1982. 2. (in Chinese).

[11] Geuzaine P, Farhat C, Brown G, et al. Nonlinear flutter analysis of an f-16 in stabilized, accelerated, and increased angle of attack flight conditions. AIAA Paper 2002–1490, 2002.

[12] Goodman CE, Pitt DM FAMUSS: a new aeroservoelastic modeling tool. AIAA Paper-92-2395-CP, 1992.

[13] Gupta KK. Stars – an integrated, multidisciplinary, finite-element, structural, fluids, aeroelastic, and aeroservoelastic analysis computer program. NASA Technical Memorandum 4795, 1997.

[14] Haddadpour H, Navazi HM. Aero-thermoelastic stability of functionally graded plates. Journal of Composite Structures. 2006, (10), 1016.

[15] Hodges H, Patil MJ. On the importance of aerodynamic and structural geometrical nonlinearities in aeroelastic behavior of high-aspect-ration wings. AIAA Paper 2000-1448, 2000.

[16] Kamakoti R. Fluid–structure interaction for aeroelastic applications. Progress in Aerospace Sciences. 2004, (40), 535–558.

[17] Kim DH. Transonic/supersonic aeroelastic instability of an all movable wing with structural nonlinearity. AIAA Paper 2003-1412, 2003.

[18] Lewis AP, Smith MJ. Extension of a Euler/Navier-Stokes aeroelastic analysis method for shell structures. AIAA Paper -98-2656, 1998.

[19] Liu DD, Schuster DM, Huttsell LJ. Computational aeroelasticity: success, progress, challenge. AIAA Paper 2003-1725, 2003.

[20] Luo H, Baum JD, Mestreau EL, et al. Recent developments of a coupled CFD/CSD methodology. AIAA Paper 01-31097, 2001.

[21] Mayuresh JP. Nonlinear aeroelastic analysis of joined-wing aircraft. AIAA Paper 2003-1487.

[22] Nick RD. The F-22 structural/aeroelastic design process with MDO examples. AIAA Paper 98-4732, 1998.

[23] Obayashi S, Yang GW, Nakanlieh J. Aileron buzz simulation using an implicit multiblock aeroelastic solver. Journal of Aircraft. 2003, 40(3), 580–589.
[24] Pendleton E A flight research program for active aeroelastic wing technology. AIAA Paper- 96-1574–CP.
[25] Potsdam MA, Guruswamy GP. A parallel multiblock mesh movement scheme for complex aeroelastic applications. AIAA Paper 2001-0716, 2001.
[26] Seigler TM, Jae-Sung B, Inman DJ. Aerodynamic and aeroelastic considerations of a variable-span morphing wing. AIAA Paper 2004-1726, 2004.
[27] Shkadov LM, Dmitriev VG, Denisov VE. The flying-wing concept – chances and risks. AIAA Paper 2003-2887.
[28] Tang D, Dowell E. Nonlinear aeroelasticity and unsteady aerodynamics. AIAA Paper Paper 2002-0003, 2002.
[29] Visbal MR, Gordnier RE. Development of a three-dimensional viscous aeroelastic solver for nonlinear panel flutter. AIAA paper 2000-2337, 2000.
[30] Xu M, AN X, Chen S. An overview of CFD/CSD coupled solution for nonlinear aeroelasticity. Advances in Mechanics. 2009, 39(03), 284–298. (in Chinese).
[31] Zink PS, Love MH, Stroud RL, et al. Impact of actuation concepts on morphing aircraft structures. AIAA Paper 2004-1724, 2004.

Chapter 2
Unsteady CFD technology

2.1 Introduction

In this chapter, the key techniques of steady and/or unsteady flow computation are discussed in the framework of finite volume method (FVM), including spatial discretization of Navier–Stokes (N-S) equations, that is, the numerical approximation of the convective and viscous fluxes, as well as of the source term, temporal discretization method, turbulence modeling and the so-called geometric conservation law (GCL) in unsteady cases where the relative motion of grid/mesh with fluid is considered. The CFD system is formulated using aforementioned techniques for aeroelasticity computation in time domain. At the end of this chapter, several steady and unsteady examples in engineering applications are presented, so that readers can have an intuitive understanding, and also the basic knowledge of CFD applications in numerical simulations of flow field.

2.2 Governing equations

In Cartesian coordinate system, the three-dimensional unsteady compressible N-S equations in conservation form are written as follows [1, 11–13, 16, 29, 34]:

$$\frac{\partial}{\partial t}\iiint_{\Omega} Q dV + \oiint_{\partial\Omega} F \cdot n dS = 0 \tag{2.1}$$

where $\boldsymbol{Q} = (\rho, \rho u, \rho v, \rho w, \rho e)^T$ is the so-called conservative solution vector, and $\partial\Omega$ is the boundary of a control volume $\boldsymbol{\Omega}$ in the physical flow domain. The vector $\boldsymbol{n}$ denotes the outer normal direction the boundary. The total fluxes vector $\boldsymbol{F}$ through a control volume boundary consists of two components, with one is the convective fluxes $\boldsymbol{F_c}$, and the other one is viscous fluxes $\boldsymbol{F_v}$, which can be written as follows:

$$\boldsymbol{F} = \boldsymbol{F}_c - \boldsymbol{F}_v \tag{2.2}$$

where

$$\boldsymbol{F_c} = \begin{bmatrix} \rho u\boldsymbol{i} + \rho v\boldsymbol{j} + \rho w\boldsymbol{k} \\ (\rho u^2 + p)\boldsymbol{i} + \rho uv\boldsymbol{j} + \rho uw\boldsymbol{k} \\ \rho uv\boldsymbol{i} + (\rho v^2 + p)\boldsymbol{j} + \rho vw\boldsymbol{k} \\ \rho uw\boldsymbol{i} + \rho vw\boldsymbol{j} + (\rho w^2 + p)\boldsymbol{k} \\ (\rho ue + up)\boldsymbol{i} + (\rho ve + vp)\boldsymbol{j} + (\rho we + wp)\boldsymbol{k} \end{bmatrix} \tag{2.3}$$

https://doi.org/10.1515/9783110576689-002

$$\boldsymbol{F}_v = \begin{bmatrix} 0 \\ \tau_{xx}\boldsymbol{i} + \tau_{xy}\boldsymbol{j} + \tau_{xz}\boldsymbol{k} \\ \tau_{yx}\boldsymbol{i} + \tau_{yy}\boldsymbol{j} + \tau_{yz}\boldsymbol{k} \\ \tau_{zx}\boldsymbol{i} + \tau_{zy}\boldsymbol{j} + \tau_{zz}\boldsymbol{k} \\ \Pi_x\boldsymbol{i} + \Pi_y\boldsymbol{j} + \Pi_z\boldsymbol{k} \end{bmatrix}. \tag{2.4}$$

In eq. (2.4),

$$\left.\begin{aligned} \Pi_x &= u\tau_{xx} + v\tau_{xy} + w\tau_{xz} - q_x \\ \Pi_y &= u\tau_{yx} + v\tau_{yy} + w\tau_{yz} - q_y \\ \Pi_z &= u\tau_{zx} + v\tau_{zy} + w\tau_{zz} - q_z \end{aligned}\right\} \tag{2.5}$$

where the viscous stresses are expressed as follows:

$$\left.\begin{aligned} \tau_{xx} &= 2\mu u_x - \tfrac{2}{3}\mu(u_x + v_y + w_z) \\ \tau_{yy} &= 2\mu v_y - \tfrac{2}{3}\mu(u_x + v_y + w_z) \\ \tau_{zz} &= 2\mu w_z - \tfrac{2}{3}\mu(u_x + v_y + w_z) \\ \tau_{xy} &= \tau_{yx} = \mu(u_y + v_x) \\ \tau_{xz} &= \tau_{zx} = \mu(u_z + w_x) \\ \tau_{yz} &= \tau_{zy} = \mu(v_z + w_y) \end{aligned}\right\} \tag{2.6}$$

q_x、q_y、q_z are the heat flux components in three dimensional directions, describing the work of the heat conduction in flow field, which follows the Fourier's law with respect to temperature

$$\left.\begin{aligned} q_x &= -k\tfrac{\partial T}{\partial x} \\ q_y &= -k\tfrac{\partial T}{\partial y} \\ q_z &= -k\tfrac{\partial T}{\partial z} \end{aligned}\right\}. \tag{2.7}$$

To close the entire system of N-S equations, two additional equations have to be supplied to describe the thermodynamic relations between the state variables. In pure aerodynamic problems, the assumption of calorically perfect gas is reasonable, which satisfies the equation of state. The internal energy or the enthalpy is determined by the pressure and the temperature. The abovementioned additional equations are as follows [31, 35]:

$$\left.\begin{aligned} p &= \rho RT \\ h &= c_p T \end{aligned}\right\} \tag{2.8}$$

Based on the hypothesis of Boussinesq [7, 8] for eddy viscosity, the dynamic viscosity in eq. (2.6) is expressed as the sum of a laminar and a turbulent component, that is, $\mu = \mu_l + \mu_t$. The laminar viscosity μ_l is computed based on the Sutherland function

$$\mu_l = \frac{1 + C_s}{T + C_s} T^{1.5}, \tag{2.9}$$

where $C_s = 117/T_\infty$.

The turbulent eddy viscosity μ_t is determined or computed from the numerical solutions of a turbulence model, which will be described in Section 2.7 in detail. k denotes the turbulent thermal conductivity coefficient, which is evaluated as

$$k = \frac{\gamma R}{\gamma - 1}\left(\frac{\mu_l}{Pr_l} + \frac{\mu_t}{Pr_t}\right), \tag{2.10}$$

where $\gamma = 1.4$ is the specific heat ratio for air at standards condition. R is the specific gas constant, which is a different value to different gas. For air at standard conditions, $R = 287.02J/(Kg \cdot K)$. The laminar Prandtl number and the turbulent Prandtl number shown in eq. (2.10) are generally assumed to be constant in the flow field ($Pr_l = 0.72$ and $Pr_t = 0.9$ for air).

2.3 Finite volume method

The FVM [40] is a spatial discretization method widely applied in CFD and has been adopted by many commercial software. The control volume is illustrated in Fig. 2.1. According to Gaussian's divergence theory, for a control volume Ω, which is assumed to be fixed in space, with the boundary denoted by $\partial\Omega$, the N-S equations can be discretized on the finite volume $\Omega_{i,j,k}$ as follows:

$$\frac{d}{dt}\int_{\Omega_{i,j,k}} \boldsymbol{Q} d\Omega + \int_{\partial\Omega_{i,j,k}} \boldsymbol{F_c} \cdot \boldsymbol{n} ds + \int_{\partial\Omega_{i,j,k}} \boldsymbol{F_v} \cdot \boldsymbol{n} ds = 0. \tag{2.11}$$

where the boundary $\partial\Omega_{i,j,k}$ consist of the following cell faces:

$$\partial\Omega_{i,j,k} = \boldsymbol{ds}_{i-\frac{1}{2},j,k} + \boldsymbol{ds}_{i+\frac{1}{2},j,k} + \boldsymbol{ds}_{i,j-\frac{1}{2},k} + \boldsymbol{ds}_{i,j+\frac{1}{2},k} + \boldsymbol{ds}_{i,j,k-\frac{1}{2}} + \boldsymbol{ds}_{i,j,k+\frac{1}{2}}. \tag{2.12}$$

On the control volume $\Omega_{i,j,k}$, the governing equations can be transformed to a semi-discretized form as follows, where the solution vector $Q_{i,j,k}$ is defined as averaged value with respect to the cell volume at the center, and $\mathrm{vol}_{i,j,k}$ is the volume of the control cell:

$$Q_{i,j,k} = \frac{1}{\mathrm{vol}_{i,j,k}} \int_{\Omega_{i,j,k}} \boldsymbol{Q} d\Omega. \tag{2.13}$$

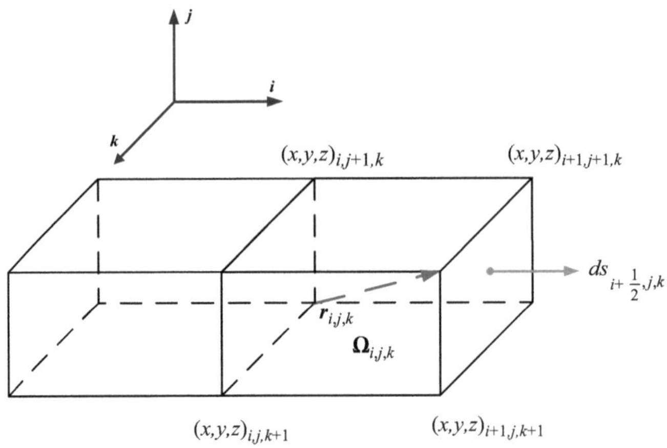

Fig. 2.1: The control volume of cell $\Omega_{i,j,k}$.

The control volume can be calculated with three cell face vectors and its main diagonal vector as follows:

$$\left.\begin{aligned} \boldsymbol{r}_{i,j,k} &= \left(x_{i+1,j+1,k+1} - x_{i,j,k}\right)\boldsymbol{i}_x + \left(y_{i+1,j+1,k+1} - y_{i,j,k}\right)\boldsymbol{i}_y + \left(z_{i+1,j+1,k+1} - z_{i,j,k}\right)\boldsymbol{i}_z \\ vol_{i,j,k} &= \frac{1}{3}r_{i,j,k}\cdot\left(\boldsymbol{ds}_{i-\frac{1}{2},j,k} + \boldsymbol{ds}_{i,j-\frac{1}{2},k} + \boldsymbol{ds}_{i,j,k-\frac{1}{2}}\right) \end{aligned}\right\}, \tag{2.14}$$

where $\boldsymbol{ds}$ is the cell face vector, $\boldsymbol{r}_{i,j,k}$ is the main diagonal vector of the control volume, and $\boldsymbol{ds}_{i-\frac{1}{2},j,k}$ can be written as

$$\boldsymbol{ds}_{i-\frac{1}{2},j,k} = \frac{1}{2}\left(\boldsymbol{a}_{i-\frac{1}{2},j,k} \times \boldsymbol{b}_{i-\frac{1}{2},j,k}\right), \tag{2.15}$$

where $\boldsymbol{a}$ and $\boldsymbol{b}$ are the two diagonal vectors on the face vector $ds_{i-\frac{1}{2},j,k}$ and can be formulated as the following:

$$\left.\begin{aligned} a_{i-\frac{1}{2},j,k} &= \left(x_{i,j+1,k+1} - x_{i,j,k}\right)\boldsymbol{i}_x + \left(y_{i,j+1,k+1} - y_{i,j,k}\right)\boldsymbol{i}_y + \left(z_{i,j+1,k+1} - z_{i,j,k}\right)\boldsymbol{i}_z \\ b_{i-\frac{1}{2},j,k} &= \left(x_{i,j,k+1} - x_{i,j+1,k}\right)\boldsymbol{i}_x + \left(y_{i,j,k+1} - y_{i,j+1,k}\right)\boldsymbol{i}_y + \left(z_{i,j,k+1} - z_{i,j+1,k}\right)\boldsymbol{i}_z \end{aligned}\right\}. \tag{2.16}$$

Then, the semidiscretized N-S equations can be written as

$$vol_{i,j,k}\frac{d}{dt}Q_{i,j,k} + W^c_{i,j,k} - W^v_{i,j,k} = 0, \tag{2.17}$$

where the convective fluxes $\boldsymbol{W}^c_{i,j,k}$ and the viscous fluxes $\boldsymbol{W}^v_{i,j,k}$ are expressed as

$$\begin{aligned}W^c_{i,j,k} = &\boldsymbol{F_c}\cdot \boldsymbol{ds}|_{i+\frac{1}{2},j,k} - \boldsymbol{F}_c\cdot \boldsymbol{ds}|_{i-\frac{1}{2},j,k} + \boldsymbol{F}_c\cdot \boldsymbol{ds}|_{i,j+\frac{1}{2},k} - \\ &\boldsymbol{F_c}\cdot \boldsymbol{ds}|_{i,j-\frac{1}{2},k} + \boldsymbol{F}_c\cdot \boldsymbol{ds}|_{i,j,k+\frac{1}{2}} + \boldsymbol{F}_c\cdot \boldsymbol{ds}|_{i,j,k-\frac{1}{2}} \\ W^v_{i,j,k} = &\boldsymbol{F_v}\cdot \boldsymbol{ds}|_{i+\frac{1}{2},j,k} - \boldsymbol{F}_v\cdot \boldsymbol{ds}|_{i-\frac{1}{2},j,k} + \boldsymbol{F}_v\cdot \boldsymbol{ds}|_{i,j+\frac{1}{2},k} - \\ &\boldsymbol{F_v}\cdot \boldsymbol{ds}|_{i,j-\frac{1}{2},k} + \boldsymbol{F}_v\cdot \boldsymbol{ds}|_{i,j,k+\frac{1}{2}} + \boldsymbol{F}_v\cdot \boldsymbol{ds}|_{i,j,k-\frac{1}{2}}\end{aligned} \tag{2.18}$$

2.4 Spatial discretization methodologies

The numerical solution of the N-S equation is generally divided into two parts, namely the convective term and the viscous term, which are processed separately. The viscous term has a linear form and shows an elliptical feature in mathematics and is generally discretized with the central difference method. The convective term is consistent with the corresponding Euler equations in form. Therefore, the methods of solving the Euler equations are also applicable to N-S equations. In this section, the main focus is on the numerical calculation of the convective fluxes [24, 38, 51, 57, 64]. At present, the spatial discretization methods are mature and applied widely, which can be generally divided into two categories: central difference and upwind schemes.

It is straightforward to construct a central difference scheme with a simple logical relationship. Based on the arithmetic average of the conservative variables, the convective flux at the cell face can be computed. As the central scheme does not have sufficient intrinsic dissipation to suppress the numerical oscillations, the artificial viscosity mechanism needs to be introduced to suppress such nonphysical oscillation, which is a blending of second- and fourth-order differences, and can make sure that the solving process of governing equations in shockwave-containing flow field is consistent with that in continuous regions. The central scheme was first implemented on structured grid by Jameson et al. [24]. The second-order difference term is nonlinear and introduces an entropy-like correction to avoid nonphysical numerical oscillations in the region of the shock, whereas the fourth-order difference term is basically linear and eliminates high-frequency errors to ensure the convergence in the numerical computation [58].

Based on eigenvalue analysis and propagation characteristics of Euler equations, a variety of upwind schemes is designed by considering the characteristics of the flow equations.

There are two main directions for the development of the upwind schemes: the Riemann problem and flux vector splitting (FVS) method. The Riemann problem is gradually developed into a flux difference splitting (FDS) scheme, which is the main direction of approximate Riemann method according to Godunov's idea. At the present, FVS scheme and FDS scheme are the two representative upwind schemes, where Van leer's FVS [64] scheme and Roe's FDS [51] are widely adopted in CFD community. Van leer's scheme decomposes the convective fluxes into two parts according to local Mach number normal to the surface of the control volume. Although Van leer's scheme is efficient and robust, the scheme is known to produce excessive dissipation when resolving boundary layer flow and predicts inaccurate stagnation temperature. For the contact discontinuity, there are still numerical fluxes which can smear the discontinuity, resulting in significant computation error in viscous region such as the boundary layer. In the framework of FDS scheme, the convective flux at a cell face of the control volume is evaluated with the left and the right state variables by solving the Riemann (shock tube) problem [6]. The idea was first introduced by Godunov [20]. Roe [51] and Osher and Solomon [43] developed approximate Riemann solvers to reduce the computational cost of the Godunov's scheme to approximate the solution of the Riemann problem. Roe's FDS scheme is applied widely because of high accuracy in resolving boundary layer flow and shock wave as compared to its FVS scheme counterpart [6]. However, when the Jacobian eigenvalue in flux evaluation is sufficiently small in the region of shock wave or discontinuities, Roe's FDS scheme can violate the entropy condition and produces unphysical solutions, therefore, the modification to the modulus of the eigenvalue needs to be introduce using the entropy correction ideas [28, 42]. Roe's FDS scheme has been well recognized as an upwind scheme, as the scheme resolves boundary layer flow and shock wave with high accuracy. Many entropy correction methods were proposed to further improve its accuracy and stability in the CFD community.

The advection upstream splitting method (AUSM) was proposed by Liou and Steffen [36, 38], which combines the advantage of Van leer's FVS and Roe's FDS scheme. AUSM scheme was subsequently evolved into different scheme, namely AUSMD/V [67] and AUSM+ [37, 39] respectively. The basic idea of the AUSM scheme is that the convective flux consists of two physically distinct parts, the convective and the pressure part. The first part is associated with the eigenvalue u and the second part is linked with the eigenvalues $u{+}c$ and $u{-}c$. AUSM scheme can be viewed as an improvement of Van leer's FVS scheme in terms of flux construction and a combination of FVS and FDS scheme from its dissipation properties viewpoint. AUSM scheme shows both the high discontinuity resolution of Roe's scheme and the high computational efficiency of Van Leer's scheme. Furthermore, numerical dissipation of the scheme is the rival of Roe's scheme and entropy correction is not necessary, which enables the ASUM scheme to predict shock wave and contact discontinuity with high accuracy and low computational cost. Therefore, the AUSM-type scheme has attracted more attention in the CFD community and evolved into many new variants, such as AUSM+, AUSMpw and AUSM+~up.

2.4.1 Central scheme with artificial dissipation

The following ordinary differential equation can be obtained by semidiscretizing the N-S equations spatially using the FVM:

$$\frac{d}{dt}\left(vol_{i,j,k}W_{i,j,k}\right)+\boldsymbol{Q}^{c}_{i,j,k}-\boldsymbol{Q}^{v}_{i,j,k}=0 \tag{2.19}$$

where $W_{i,j,k}=(\rho,\rho u,\rho v,\rho w,\rho e)^T_{i,j,k}$. $vol_{i,j,k}$ is the control volume, and $\boldsymbol{Q}^c_{i,j,k}$, $\boldsymbol{Q}^v_{i,j,k}$ are convective and viscous flux, respectively. The convective fluxes can be written as follows:

$$\begin{aligned}Q^{c}_{i,j,k}=\sum \boldsymbol{F}_c\cdot \boldsymbol{ds}=&\\ &\boldsymbol{F}^c_{i+\frac{1}{2},j,k}\boldsymbol{ds}_{i+\frac{1}{2},j,k}-\boldsymbol{F}^c_{i-\frac{1}{2},j,k}\boldsymbol{ds}_{i-\frac{1}{2},j,k}+\\ &\boldsymbol{F}^c_{i,j+\frac{1}{2},k}\boldsymbol{ds}_{i,j+\frac{1}{2},k}-\boldsymbol{F}^c_{i,j-\frac{1}{2},k}\boldsymbol{ds}_{i,j-\frac{1}{2},k}+\\ &\boldsymbol{F}^c_{i,j,k+\frac{1}{2}}\boldsymbol{ds}_{i,j,k+\frac{1}{2}}-\boldsymbol{F}^c_{i,j,k-\frac{1}{2}}\boldsymbol{ds}_{i,j,k-\frac{1}{2}}\end{aligned} \tag{2.20}$$

When Jameson's central finite difference method is adopted, $\boldsymbol{F_c}$ can be evaluated using the average of variables in the adjacent control volume,

$$\left.\begin{aligned}\boldsymbol{F}^c_{i\pm\frac{1}{2},j,k}&=\tfrac{1}{2}\left(\boldsymbol{F}^c_{i,j,k}+\boldsymbol{F}^c_{i\pm1,j,k}\right)\\ \boldsymbol{F}^c_{i,j\pm\frac{1}{2},k}&=\tfrac{1}{2}\left(\boldsymbol{F}^c_{i,j,k}+\boldsymbol{F}^c_{i,j\pm1,k}\right)\\ \boldsymbol{F}^c_{i,j,k\pm\frac{1}{2}}&=\tfrac{1}{2}\left(\boldsymbol{F}^c_{i,j,k}+\boldsymbol{F}^c_{i,j,k\pm1}\right)\end{aligned}\right\} \tag{2.21}$$

To suppress the numerical oscillations arising from the odd–even decoupling of the solution and the overshoots near shock wave, the artificial dissipation needs to be added for stability and convergence. Jameson's original artificial dissipation is isotropic with the form of a blend of second- and fourth-order differences as follows [24]:

$$\boldsymbol{D}_{i,j,k}=D\boldsymbol{W}_{i,j,k} \tag{2.22}$$

where D is the dissipation operator with the definition as follows:

$$\begin{aligned}D=\delta_i^-&\left[\alpha_{i,j,k}\left(\varepsilon^{2i}_{i,j,k}\delta_i^+-\varepsilon^{4i}_{i,j,k}\delta_i^{+3}\right)\right]\\ +\delta_j^-&\left[\alpha_{i,j,k}\left(\varepsilon^{2j}_{i,j,k}\delta_j^+-\varepsilon^{4j}_{i,j,k}\delta_j^{+3}\right)\right]\\ +\delta_k^-&\left[\alpha_{i,j,k}\left(\varepsilon^{2k}_{i,j,k}\delta_k^+-\varepsilon^{4k}_{i,j,k}\delta_k^{+3}\right)\right],\end{aligned} \tag{2.23}$$

δ^+, δ^- is the first-order forward difference and backward difference operators, respectively, and δ^{+3} is the third-order forward difference operator, that is,

$$\begin{aligned}
\delta_i^+ ()_{i,j,k} &= ()_{i+1,j,k} - ()_{i,j,k} \\
\delta_i^- ()_{i,j,k} &= ()_{i,j,k} - ()_{i-1,j,k} \\
\delta_i^{+3} ()_{i,j,k} &= ()_{i+2,j,k} - 3()_{i+1,j,k} + 3()_{i,j,k} - ()_{i-1,j,k}
\end{aligned} \tag{2.24}$$

$\alpha_{i,j,k}$ is the scaling factor of dissipative flux, that is,

$$\alpha_{i,j,k} = \frac{1}{2}\left(\frac{\mathrm{vol}_{i,j,k}}{\Delta t_{i,j,k}} + \frac{\mathrm{vol}_{i+1,j,k}}{\Delta t_{i+1,j,k}}\right) \tag{2.25}$$

where $\Delta t_{i,j,k}$ is the maximum local time step defined by the control volume $vol_{i,j,k}$.ε^{2i}, ε^{2j}, ε^{2k}. ε^{4i}, ε^{4j}, ε^{4k} are the second- and fourth-order adaptive coefficients associated with the flow variable gradients along the grid line in ***i,j,k*** direction, respectively:

$$\begin{aligned}
\varepsilon_{i,j,k}^{2i} &= \mu^{2i} \max\left(v_{i+1,j,k}, v_{i,j,k}\right) \\
\varepsilon_{i,j,k}^{4i} &= \max\left(0, \mu^{4i} - \varepsilon_{i,j,k}^{2i}\right),
\end{aligned} \tag{2.26}$$

where μ^{2i}, μ^{4i} are the second- and fourth-order artificial dissipation coefficients in i direction. $v_{i,j,k}$ is a pressure-based sensor used to switch off the fourth-order differences near shock wave to avoid strong oscillation of the solution [6]. The sensor can also be used to switch off the second-order differences in smooth region of the flow to reduce the dissipation introduced by artificial viscosity

$$v_{i,j,k} = \frac{\left|p_{i+1,j,k} - 2p_{i,j,k} + p_{i-1,j,k}\right|}{p_{i+1,j,k} + 2p_{i,j,k} + p_{i-1,j,k}}. \tag{2.27}$$

In general, the typical values of the parameters are $\mu^{2i} = \frac{1}{2}$,, $\mu^{4i} = \frac{1}{128} \sim \frac{1}{64}$.

In order to extend the implementation of Jameson's central difference scheme from subsonic and transonic flows to supersonic and hypersonic flows simulations, Turkel et al. [61] introduced a total variation diminishing (TVD)-like modification based on the flux limit concept, where the pressure-based sensor is modified as

$$v_{i,j,k} = \frac{\left|p_{i+1,j,k} - 2p_{i,j,k} + p_{i-1,j,k}\right|}{(1-\boldsymbol{\omega})(P_{\mathrm{TVD}})_{i,j,k} + \boldsymbol{\omega} p_{i,j,k}} \tag{2.28}$$

and

$$\left.\begin{aligned}
(p_{\mathrm{TVD}})_{i,j,k} &= \left|p_{i+1,j,k} - p_{i,j,k}\right| + \left|p_{i,j,k} - p_{i-1,j,k}\right| \\
p_{i,j,k} &= p_{i+1,j,k} + 2p_{i,j,k} + p_{i-1,j,k}
\end{aligned}\right\}. \tag{2.29}$$

The parameter $\boldsymbol{\omega}$ is set as $0 < \boldsymbol{\omega} \le 1.0$.

2.4.2 Van Leer's FVS scheme

Van Leer's FVS scheme [64] is derived from characteristic decomposition of the convective flux. The convective flux is split into a positive and a negative part, that is

$$\delta_\xi \boldsymbol{F} = \delta_\xi^- \boldsymbol{F}^+ + \delta_\xi^+ \boldsymbol{F}^- \tag{2.30}$$

where δ_ξ^- and δ_ξ^+ are the backward difference and forward difference operators, respectively. Ma_ξ is defined as the contravariant Mach number in the ξ direction.

$$Ma_\xi = \frac{\bar{u}}{c} \tag{2.31}$$

where

$$\bar{u} = \frac{U}{|\mathbf{grad}\ \xi|}$$

$U = u\xi_x + v\xi_y + w\xi_z + \xi_t$ is the contravariant velocity in the ξ direction

$$|\mathbf{grad}\ \xi| = \sqrt{\xi_x^2 + \xi_y^2 + \xi_z^2}.$$

For locally supersonic flow, where $|Ma_\xi| \geq 1$,

$$\left.\begin{aligned} F^+ &= F, F^- = 0(Ma_\xi \geq +1) \\ F^- &= F, F^+ = 0(Ma_\xi \leq -1) \end{aligned}\right\} \tag{2.32}$$

And for locally subsonic flow, where $|Ma_\xi| < 1$,

$$F^\pm = \frac{|\mathbf{grad}\xi|}{J} \begin{bmatrix} f_{\text{mass}}^\pm \\ f_{\text{mass}}^\pm \left[\hat{k}_x(-\bar{u} \pm 2c)/\gamma + u\right] \\ f_{\text{mass}}^\pm \left[\hat{k}_y(-\bar{u} \pm 2c)/\gamma + v\right] \\ f_{\text{mass}}^\pm \left[\hat{k}_z(-\bar{u} \pm 2c)/\gamma + w\right] \\ f_{\text{energy}}^\pm \end{bmatrix} \tag{2.33}$$

where

$$f_{\text{mass}}^\pm = \pm\rho c(Ma_\xi \pm 1)^2/4$$

$$f_{\text{energy}}^\pm = f_{\text{mass}}^\pm \left[\frac{(1-\gamma)\bar{u}^2 \pm 2(\gamma-1)\bar{u}c + 2c^2}{\gamma^2 - 1} + \frac{u^2 + v^2 + w^2}{2} - \frac{\hat{k}_t}{\gamma}(-\bar{u} \pm 2c)\right]$$

$$\begin{aligned}\hat{k}_x &= \xi_x/|\mathbf{grad}\xi|\\ \hat{k}_y &= \xi_y/|\mathbf{grad}\xi|\\ \hat{k}_z &= \xi_z/|\mathbf{grad}\xi|\\ \hat{k}_t &= \xi_t/|\mathbf{grad}\xi|\end{aligned}$$

2.4.3 Roe's FDS scheme

Roe's FDS scheme decomposes the convective flux over a face of the control volume into a sum of the wave contributions, while maintaining the conservation properties of the Euler equations [6]. On the grid cell face of $i+\frac{1}{2}$, the convective flux is expressed as

$$\boldsymbol{F}_{i+\frac{1}{2}} = \frac{1}{2}[\boldsymbol{F}(Q_l) + \boldsymbol{F}(Q_r) - |\boldsymbol{A}|(\boldsymbol{Q}_r - \boldsymbol{Q}_l)]_{i+\frac{1}{2}}, \tag{2.34}$$

Q_l and Q_r are the Roe-averaged flow variables on the left and right sides of the cell face $i+\frac{1}{2}$, where

$$|A|(Q_r - Q_l) = |A|\Delta Q = \begin{bmatrix} \alpha_4 \\ \tilde{u}\alpha_4 + \hat{k}_x\alpha_5 + \alpha_6 \\ \tilde{v}\alpha_4 + \hat{k}_y\alpha_5 + \alpha_7 \\ \tilde{w}\alpha_4 + \hat{k}_z\alpha_5 + \alpha_8 \\ \tilde{H}\alpha_4 + (\tilde{\bar{u}} - \hat{k}_t)\alpha_5 + \tilde{u}\alpha_6 + \tilde{v}\alpha_7 + \tilde{w}\alpha_8 - \left(\frac{\tilde{c}^2}{\gamma-1}\right)\alpha_1 \end{bmatrix} \tag{2.35}$$

The variables with superscript "~" denotes the so-called Roe-averaged variables, which are written as follows:

$$\left.\begin{aligned} \tilde{\rho} &= \sqrt{\rho_r\rho_l} \\ \tilde{u} &= \left(u_l + u_r\sqrt{\rho_r/\rho_l}\right)/\left(1 + \sqrt{\rho_r/\rho_l}\right) \\ \tilde{v} &= \left(v_l + v_r\sqrt{\rho_r/\rho_l}\right)/\left(1 + \sqrt{\rho_r/\rho_l}\right) \\ \tilde{w} &= \left(w_l + w_r\sqrt{\rho_r/\rho_l}\right)/\left(1 + \sqrt{\rho_r/\rho_l}\right) \\ \tilde{H} &= \left(H_l + H_r\sqrt{\rho_r/\rho_l}\right)/\left(1 + \sqrt{\rho_r/\rho_l}\right) \\ \tilde{c}^2 &= (\gamma - 1)\left[\tilde{H} - (\tilde{u}^2 + \tilde{v}^2 + \tilde{w}^2)/2\right] \end{aligned}\right\} \tag{2.36}$$

where $\hat{k}_x$, $\hat{k}_y$, $\hat{k}_z$ and $\bar{u}$ are defined as

$$\begin{gathered}\left(\hat{k}_x, \hat{k}_y, \hat{k}_z, \hat{k}_t\right) = -\left(\xi_x, \xi_y, \xi_z, \xi_t\right) / |\mathbf{grad}\ \xi|. \\ \tilde{\bar{u}} = \hat{k}_x \tilde{u} + \hat{k}_y \tilde{v} + \hat{k}_z \tilde{w} + \hat{k}_t\end{gathered} \tag{2.37}$$

The parameters $\alpha_1 \sim \alpha_8$ are expressed as

$$\left.\begin{aligned} a_1 &= \left|\frac{\mathbf{grad}\ \xi}{J}\right| \left|\tilde{\bar{u}}\right| \left(\Delta\rho - \frac{\Delta p}{\tilde{c}^2}\right) \\ a_2 &= \frac{1}{2\tilde{c}^2} \left|\frac{\mathbf{grad}\ \xi}{J}\right| \left|\tilde{\bar{u}} + \tilde{c}\right| \left(\Delta p + \tilde{\rho}\tilde{c}\Delta\bar{u}\right) \\ a_3 &= \frac{1}{2\tilde{c}^2} \left|\frac{\mathbf{grad}\ \xi}{J}\right| \left|\tilde{\bar{u}} - \tilde{c}\right| \left(\Delta p - \tilde{\rho}\tilde{c}\Delta\bar{u}\right) \\ a_4 &= a_1 + a_2 - a_3 \\ a_5 &= \tilde{c}\left(a_2 - a_3\right) \\ a_6 &= \left|\frac{\mathbf{grad}\ \xi}{J}\right| \left|\tilde{\bar{u}}\right| \left(\tilde{\rho}\Delta u - k_x \tilde{\rho}\Delta\bar{u}\right) \\ a_7 &= \left|\frac{\mathbf{grad}\ \xi}{J}\right| \left|\tilde{\bar{u}}\right| \left(\tilde{\rho}\Delta v - k_y \tilde{\rho}\Delta\bar{u}\right) \\ a_8 &= \left|\frac{\mathbf{grad}\ \xi}{J}\right| \left|\tilde{\bar{u}}\right| \left(\tilde{\rho}\Delta w - k_z \tilde{\rho}\Delta\bar{u}\right) \end{aligned}\right\} \tag{2.38}$$

2.4.4 AUSM scheme

The AUSM scheme, introduced by Liou and Steffen [36, 38], is based on the idea that convective flux consists of two physically distinct parts, namely the convective and the pressure part as follows:

$$F = \begin{bmatrix} \rho u \\ \rho u^2 + p \\ \rho uv \\ \rho uw \\ (\rho e + p)u \end{bmatrix} = u \begin{bmatrix} \rho \\ \rho u \\ \rho v \\ \rho w \\ (\rho e + p) \end{bmatrix} + \begin{bmatrix} 0 \\ p \\ 0 \\ 0 \\ 0 \end{bmatrix}. \tag{2.39}$$

The convective flux at the grid cell face $i + \frac{1}{2}$ is as follows:

$$F^c_{i+\frac{1}{2}} = m_{i+\frac{1}{2}} \Phi_{i+\frac{1}{2}}, \tag{2.40}$$

where

$$\Phi_{i+\frac{1}{2}} = \begin{bmatrix} \rho \\ \rho u \\ \rho v \\ \rho w \\ (\rho e + p) \end{bmatrix}_{i+\frac{1}{2}} = \begin{cases} \Phi_i & (m_{i+1/2} \geq 0) \\ \Phi_{i+1} & (m_{i+1/2} < 0) \end{cases} \tag{2.41}$$

$$m_{i+1/2} = c_{i+1/2} Ma_{i+1/2} \tag{2.42}$$

And the unified speed of sound at the cell face is written as

$$c_{i+1/2} = \begin{cases} c_i & (Ma_{i+1/2} \geq 0) \\ c_{i+1} & (Ma_{i+1/2} < 0) \end{cases}, \tag{2.43}$$

where $Ma_{i+1/2}$ is the advection Mach number at the cell face and is defined as

$$Ma_{i+1/2} = \bar{\lambda}_i^+ + \bar{\lambda}_{i+1}^-, \ \bar{\lambda}_i^\pm = \bar{\lambda}^\pm (Ma_i). \tag{2.44}$$

Similar to the Van Leer's FVS method, the advection Mach number is also split into the left and the right part

$$\bar{\lambda}^\pm (Ma) = \begin{cases} \pm \frac{1}{4}(Ma \pm 1)^2 & (|Ma| \leq 1) \\ \pm \frac{1}{2}(Ma \pm |Ma|) & (|Ma| > 1) \end{cases}. \tag{2.45}$$

The pressure at the face $i + \frac{1}{2}$ of the control volume is obtained from the splitting

$$\boldsymbol{p}_{i+1/2} = \boldsymbol{\psi}_i^+ \begin{bmatrix} 0 \\ p \\ 0 \\ 0 \\ 0 \end{bmatrix}_i + \boldsymbol{\psi}_{i+1}^- \begin{bmatrix} 0 \\ p \\ 0 \\ 0 \\ 0 \end{bmatrix}_{i+1} = \boldsymbol{\psi}_i^+ p_i + \boldsymbol{\psi}_{i+1}^- p_{i+1} \tag{2.46}$$

where $\boldsymbol{\psi}_i^\pm = \boldsymbol{\psi}^\pm (\mathrm{Ma}_i)$, and the related pressure-splitting function can be written as

$$\psi^\pm (Ma) = \begin{cases} \frac{1}{2} \frac{Ma \pm |Ma|}{Ma} & (|Ma| > 1) \\ \frac{1}{4}(Ma \pm 1)^2 (2 \mp Ma) & (|Ma| \leq 1) \end{cases}. \tag{2.47}$$

The AUSM's convective flux on the cell face $i+\frac{1}{2}$ is finally written as

$$F_{i+\frac{1}{2}} = F^{c}_{i+\frac{1}{2}} + p_{i+\frac{1}{2}}. \tag{2.48}$$

The AUSM+ scheme, an improved version of the AUSM shame, introduces the following modification in the advection Mach number and pressure splitting, that is,

$$\bar{\lambda}^{\pm}(Ma) = \begin{cases} \pm\frac{1}{4}(Ma\pm 1)^2 \pm \beta(Ma^2-1)^2 & (|Ma|\le 1) - \frac{1}{16} \le \beta \le \frac{1}{2} \\ \pm\frac{1}{2}(Ma\pm|Ma|) & (|Ma|>1) \end{cases}, \tag{2.49}$$

$$\psi^{\pm}(Ma) = \begin{cases} \frac{1}{2}\frac{Ma\pm|Ma|}{Ma} & (|Ma|>1) \\ \frac{1}{4}(Ma\pm 1)^2(2\mp Ma) \pm \alpha Ma(Ma^2-1)^2 & (|Ma|\le 1) - \frac{3}{4} \le \alpha \le \frac{3}{16} \end{cases}, \tag{2.50}$$

And the unified speed of sound at the cell face $i+\frac{1}{2}$ is written in a modified formula

$$\begin{aligned} c_{j+\frac{1}{2}} &= \frac{\left(c_j^*\right)^2}{u_j} & u_j > c_j^* \\ c_{j+\frac{1}{2}} &= \min\left(\tilde{c}_j, \tilde{c}_{j+1}\right) & u_j \le c_j^* \end{aligned}, \tag{2.51}$$

where $\tilde{c} = \frac{(c^*)^2}{\max\left(c^*, |u|\right)}$, and c^* is the critical speed of sound, defined by the total enthalpy h_t, written as

$$h_t = \frac{c^2}{\gamma-1} + \frac{1}{2}u^2 = \frac{(\gamma+1)(c^*)^2}{2(\gamma-1)}. \tag{2.52}$$

The final convective flux based on AUSM+ scheme is expressed as

$$F_{i+\frac{1}{2}} = m^{+}_{i+\frac{1}{2}} \begin{bmatrix} \rho \\ \rho u \\ \rho v \\ \rho w \\ \rho h \end{bmatrix}_l + m^{-}_{i+\frac{1}{2}} \begin{bmatrix} \rho \\ \rho u \\ \rho v \\ \rho w \\ \rho h \end{bmatrix}_R + \Psi^{+}_{i} \begin{bmatrix} 0 \\ p \\ 0 \\ 0 \\ 0 \end{bmatrix}_l + \Psi^{-}_{i+1} \begin{bmatrix} 0 \\ p \\ 0 \\ 0 \\ 0 \end{bmatrix}_R . \tag{2.53}$$

AUSM+ scheme is known to produce numerical oscillations, especially in the wall-bounded boundary layer. The scheme is found to introduce near-wall numerical oscillations, especially in the viscous boundary layer. In order to preserve the behavior in the vicinity of shocks with AUSM+ scheme, and provide sufficient dissipation for the numerical oscillations near the solid surface region, an improved version or

AUSMpw+ scheme was proposed by introducing modifications on Mach number and pressure splitting. The AUSMpw+ scheme can be written as follows:

$$F_{\frac{1}{2}} = \bar{m}_{\frac{1}{2}}^{+} c_{\frac{1}{2}} \begin{bmatrix} \rho \\ \rho u \\ \rho v \\ \rho w \\ \rho h \end{bmatrix}_l + \bar{m}_{\frac{1}{2}}^{-} c_{\frac{1}{2}} \begin{bmatrix} \rho \\ \rho u \\ \rho v \\ \rho w \\ \rho h \end{bmatrix}_R + \Psi_i^{+} \begin{bmatrix} 0 \\ p \\ 0 \\ 0 \\ 0 \end{bmatrix}_l + \Psi_{i+1}^{-} \begin{bmatrix} 0 \\ p \\ 0 \\ 0 \\ 0 \end{bmatrix}_R . \tag{2.54}$$

The two parameters, $\bar{m}_{\frac{1}{2}}^{+}$, $\bar{m}_{\frac{1}{2}}^{-}$, are based on AUSM+'s Mach number splitting method, defined as:

where $m_{\frac{1}{2}} > 0$,

$$\begin{cases} \bar{m}_{\frac{1}{2}}^{+} = m_{\frac{1}{2}}^{+} + m_{\frac{1}{2}}^{-}\left((1-\omega)(1+f_R) - f_l\right) \\ \bar{m}_{\frac{1}{2}}^{-} = m_{\frac{1}{2}}^{-} \cdot \omega(1+f_R) \end{cases} . \tag{2.55}$$

where $m_{\frac{1}{2}} < 0$,

$$\begin{cases} \bar{m}_{\frac{1}{2}}^{+} = m_{\frac{1}{2}}^{+} \cdot \omega(1+f_L) \\ \bar{m}_{\frac{1}{2}}^{-} = m_{\frac{1}{2}}^{-} + m_{\frac{1}{2}}^{+}\left((1-\omega)(1+f_l) - f_R\right) \end{cases} . \tag{2.56}$$

The parameter ω is defined as

$$\omega = 1 - \min\left(\frac{p_l}{p_R}, \frac{p_R}{p_l}\right). \tag{2.57}$$

The modified weighting function f is formulated as

$$f_{l,R} = \begin{cases} \frac{p_{l,R}}{p_s} - 1 & |M_{l,R}| < 1 \\ 0 & |M_{l,R}| \geq 1 \end{cases}, \tag{2.58}$$

where p_s is the pressure defined according to AUSM+ scheme, and the modified weighting function f is simple and easy to implement [69].

2.5 Limiter functions and monotone upstream-centered scheme for conservation laws interpolation

When upwind-type scheme is adopted in solving flow-governing equations, a polynomial interpolation approximation of the state variables on both sides of the grid cell interface is used to achieve high order accuracy. Therefore, spatial discretization accuracy is associated with the interpolation accuracy of the state variables. The linear interpolation with first-order approximation is written as

$$\begin{aligned} q^l_{i+\frac{1}{2}} &= q_{i,j,k} \\ q^R_{i+\frac{1}{2}} &= q_{i+1,j,k} \end{aligned}. \tag{2.59}$$

Van Leer proposed a second-order approximation method via the interpolation of state-variables-based monotone upstream-centered scheme for conservation laws (MUSCL) in 1979 [63, 65], shown as follows:

$$\begin{cases} q^l_{i+\frac{1}{2},j,k} = q_{i,j,k} + \frac{1}{4}[(1-k)\Delta_- + (1+k)\Delta_+]_{i,j,k} \\ q^R_{i+\frac{1}{2},j,k} = q_{i+1,j,k} - \frac{1}{4}[(1-k)\Delta_+ + (1+k)\Delta_-]_{i+,j,k} \end{cases}, \tag{2.60}$$

where Δ_- and Δ_+ are backward and forward difference operators, respectively.

With different value of $k \in [-1, 1]$, we can obtain different schemes with variant accuracy [6]: with $k = -1$, the above interpolation formula gives a fully one-side interpolation of the flow variables, resulting in a second-order accurate upwind approximation on uniformly spaced grid; with $k = 0$, a second-order accurate upwind-biased linear interpolation is formed; $k = \frac{1}{3}$ correspond to a third-order upwind-biased interpolation, and $k = 1$ corresponds to central difference. The schemes with $k = 0$ and $k = \frac{1}{3}$ are the most often used ones in practice.

The MUSCL interpolation has to be enhanced by the so-called limiter function, with the aim to suppress the nonphysical oscillation of the solution in the flow region containing strong shocks. The limiter variable can be the conservative variable or the primitive variable. Studies have shown that using primitive variables works better than using conservation variables [50]. In general, the limiter function can be expressed as follows:

$$\left. \begin{aligned} \mathbf{Q}_l &= \mathbf{Q}_i + \tfrac{1}{2}\varphi_l \Delta \mathbf{Q}_i \\ \mathbf{Q}_r &= \mathbf{Q}_{i+1} - \tfrac{1}{2}\varphi_r \Delta \mathbf{Q}_{i+1} \end{aligned} \right\}, \tag{2.61}$$

where the subscripts l and r mean the left and the right sides of the cell interface, respectively, and $\Delta \mathbf{Q}_i = \mathbf{Q}_i - \mathbf{Q}_{i-1}$. φ is the limiter function, which can be expressed for the left- and right-side parts as follows:

$$\begin{cases} \varphi_l = \frac{1}{2}\left[(1-k)\phi(r_l) + (1+k)r_l\phi\left(\frac{1}{r_l}\right)\right] \\ \varphi_r = \frac{1}{2}\left[(1-k)\phi(r_r)/r_r + (1+k)\phi\left(\frac{1}{r_r}\right)\right]. \end{cases} \tag{2.62}$$

In order to achieve second-order accuracy, the MUSCL interpolation technique has to be employed, and a sufficient condition for the scheme to have TVD properties is that the limiter should be located in the TVD region as shown in Fig. 2.2. The TVD limiter area is enclosed by a superbee and a min-mod limiter. The limiter that is closer to the

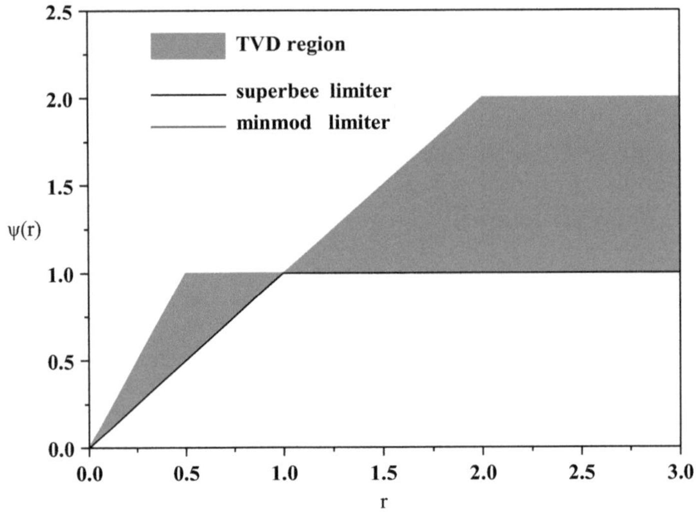

Fig. 2.2: TVD limiter region.

upper boundary of the TVD region shows lesser numerical dissipation, higher resolution, but lesser robustness. Conversely, the limiter that is closer to the lower boundary of the TVD region shows more numerical dissipation, lower resolution and more robustness. In many commercial CFD software, such as Fluent and CFDRC-Fastran, the min-mod limiter is set as a default limiter due to its desirable numerical stability.

In addition to the superbee limiter and the min-mod limiter [54], the most commonly used limiters are the Van Leer limiter and the Van Albada limiter [62], with their specific expressions as follows:

$$\left.\begin{array}{l} \text{superbee}: \quad \varphi(r) = \max[\min(2r,1), \min(r,2)] \\ min-mod: \quad \varphi(r) = \begin{cases} \min(r,1)(r>0) \\ 0(r \le 0) \end{cases} \\ \text{Van Leer}: \quad \varphi(r) = \frac{r+|r|}{1+r} \\ \text{Van Albada}: \varphi(r) = \frac{r+r^2}{1+r^2} \end{array}\right\}. \tag{2.63}$$

Figure 2.3 shows that the Van Leer limiter and the Van Albada limiter are continuous differentiable limiters, where the Van Albada limiter shows slightly more dissipation than the Van Leer limiter. It also shows that both the min-mod and the superbee limiter are nondifferentiable limiters.

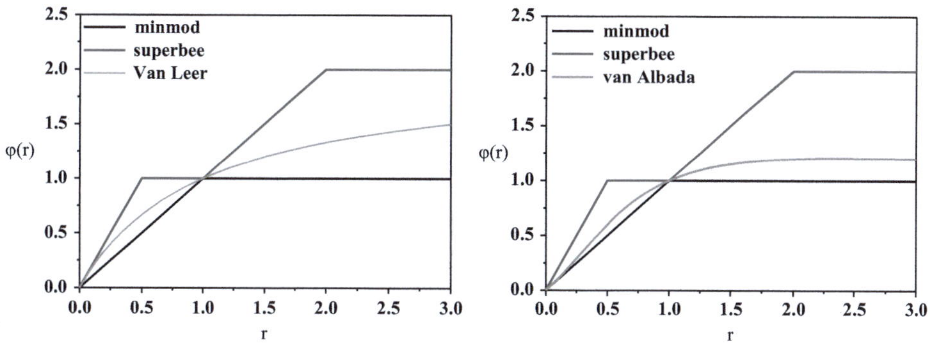

Fig. 2.3: The Van Leer and Van Albada limiter in TVD region.

2.6 Temporal discretization

The semidiscretized governing equation, eq. (2.17), can be written as the following coupled ordinary differentiable equations:

$$\left(\frac{d\hat{\mathbf{Q}}}{dt}\right)_{i,j,k} + \mathbf{RHS} = 0, \tag{2.64}$$

where RHS denotes the flow residual. To solve the equations, explicit [42] and implicit [70] time-advancing methods can be used. The advantage of the explicit method is that the implementation is straightforward, and the required memory resource, as well as the computational cost is relatively lower than its implicit method counterpart. The disadvantage of explicit method is that the time step size is strictly limited by the stability limitation resulting in low efficiency. The most commonly used time-advancing scheme is the Runge–Kutta method [15, 24, 48].

The advantage of the implicit method is that the time step size is less restrictive than explicit method, and the larger time step size can be used in computation, thereby improve the calculation efficiency. The disadvantage is that the computation cost and the required memory resource are large, because the implicit method needs to solve the linear equation for each time step advancing. It is complex to program and difficult to implement for unstructured mesh. The most commonly used implicit methods include the approximate factorization (AF) method [5], its improved version, that is, the alternating direction implicit approximation factorization method (AF-ADI) [5, 44, 46, 47], the lower–upper symmetric Gauss–Seidel (LU-SGS) method [25, 27, 70–72], LU-ADI method and various variants based on them.

2.6.1 Explicit time-advancing schemes

The Runge–Kutta four-stage time stepping method is a classical explicit time-advancing method. Equation (2.64) can be solved using the Runge–Kutta method as follows:

$$\begin{cases} \mathbf{Q}^0 = \mathbf{Q}^n \\ \mathbf{Q}^1 = \mathbf{Q}^0 - \frac{J\Delta t}{2}\mathbf{RHS}^0 \\ \mathbf{Q}^2 = \mathbf{Q}^0 - \frac{J\Delta t}{2}\mathbf{RHS}^1 \\ \mathbf{Q}^3 = \mathbf{Q}^0 - J\Delta t\mathbf{RHS}^2 \\ \mathbf{Q}^4 = \mathbf{Q}^0 - \frac{J\Delta t}{6}\left(\mathbf{RHS}^0 + 2\mathbf{RHS}^1 + 2\mathbf{RHS}^2 + \mathbf{RHS}^3\right) \\ \mathbf{Q}^{n+1} = \mathbf{Q}^4 \end{cases}. \tag{2.65}$$

The above formulation needs more memory to store the residual of each stage. To simplify the method and improve the calculation efficiency, a simplified multistage Runge–Kutta method was proposed by Jameson et al. [24] (a four-stage method shown as an example):

$$\begin{cases} \mathbf{Q}^0 = \mathbf{Q}^n \\ \mathbf{Q}^1 = \mathbf{Q}^0 - k_1 J\Delta t\mathbf{RHS}^0 \\ \mathbf{Q}^2 = \mathbf{Q}^0 - k_2 J\Delta t\mathbf{RHS}^1 \\ \mathbf{Q}^3 = \mathbf{Q}^0 - k_3 J\Delta t\mathbf{RHS}^2 \\ \mathbf{Q}^4 = \mathbf{Q}^0 - k_4 J\Delta t\mathbf{RHS}^3 \\ \mathbf{Q}^{n+1} = \mathbf{Q}^4 \end{cases}, \tag{2.66}$$

where k represents the Runge–Kutta stage coefficient. For the improved method, only the zeroth solution and the newest residual are stored to reduce the memory requirement. The coefficients can be tuned to increase the maximum time step size and improve the numerical stability for a specific spatial scheme. For an m-stage Runge–Kutta scheme, the stage coefficient is $k_m = 1$. The second-order time accuracy can be realized by setting $k_{m-1} = 0.5$; otherwise, the multistage scheme is first-order accurate as follows:

$$k_1 = 0.25,\quad k_2 = 0.3333,\quad k_3 = 0.5,\quad k_4 = 1.0000. \tag{2.67}$$

2.6.2 Implicit time-advancing schemes

The main disadvantage of explicit scheme for solving the N-S equations is that the time step size is severely restricted by the so-called CFL number of computational stability condition. Especially for the viscous effects near the viscous boundary layer, the CFD grid has to be very dense near the surface of the solid wall, thus resulting in a very small time step size defined by the CFL condition, which makes the overall computational time less desirable. Therefore, the implicit time-advancing scheme is a commonly used method for solving the N-S equations in current engineering applications.

After implicitly discretizing, eq. (2.17) can be written as the following general form (the superscript is omitted for writing convenience):

$$\Delta\mathbf{Q}^n + J\Delta t\left[\frac{\partial}{\partial\xi}(\mathbf{A}\Delta\mathbf{Q}^n) + \frac{\partial}{\partial\eta}(\mathbf{B}\Delta\mathbf{Q}^n) + \frac{\partial}{\partial\zeta}(\mathbf{C}\Delta\mathbf{Q}^n)\right] = -J\Delta t\mathbf{RHS}^n. \tag{2.68}$$

It can be furtherly derived as follows:

$$\left[\mathbf{I} + J\Delta t\left(D_\xi\mathbf{A} + D_\eta\mathbf{B} + D_\zeta\mathbf{C}\right)\right]\Delta\mathbf{Q}^n = -J\Delta t\mathbf{RHS}^n, \tag{2.69}$$

where D_ξ, D_η, D_ζ are the differentiable operators in three directions especially.

2.6.2.1 AF-ADI scheme

Based on Beam and Wage's work, Pulliam Chaussee [47] proposed an AF-ADI scheme. In the original AF scheme, each block unit contains a 5 × 5 matrix and the solution of the block tridiagonal matrix accounts for the most of the entire computational time. The ADI method is to replace each block tridiagonal matrix with a scalar diagonal matrix, which greatly reduces the computational cost.

With the approximation factorization scheme, eq. (2.68) can be split into two (in two-dimensional (2D)) or three (in three-dimensional (3D)) factors, which leads to the following formulation

$$\left(\mathbf{I} + J\Delta t D_\xi\mathbf{A}^n\right)\left(\mathbf{I} + J\Delta t D_\eta\mathbf{B}^n\right)\left(\mathbf{I} + J\Delta t D_\zeta\mathbf{C}^n\right)\Delta\mathbf{Q}^n = -J\Delta t\mathbf{RHS}^n. \tag{2.70}$$

For the sake of clarity, taking ξ direction of the grid cell (i,j,k) as an example and only the subscript i is kept for convenience, the factor in this direction can be expressed as

$$\begin{aligned}\left(\mathbf{I} + J\Delta t D_\xi\mathbf{A}^n\right)_i &\approx \mathbf{T}_{\xi,i}\left(\mathbf{I} + J\Delta t\delta_\xi\Lambda_\xi^n\right)_i\mathbf{T}_{\xi,i}^{-1} \\ &= \mathbf{T}_{\xi,i}\left(\mathbf{I} + 0.5J_{\xi,i}\Delta t_{\xi,i}\Lambda_{\xi,i+1}^n - 0.5J_{\xi,i}\Delta t_{\xi,i}\Lambda_{\xi,i-1}^n\right)\mathbf{T}_{\xi,i}^{-1}\end{aligned} \tag{2.71}$$

where the difference operator is replaced with second-order central difference, and the term $\left(\mathbf{I} + 0.5J_{\xi,i}\Delta t_{\xi,i}\Lambda_{\xi,i+1}^n - 0.5J_{\xi,i}\Delta t_{\xi,i}\Lambda_{\xi,i-1}^n\right)$ can be written for all the grid points with the following matrix form

$$\begin{bmatrix} 1 & 0.5J_2\Delta t_2\Lambda_2 & 0 & \cdots & 0 \\ -0.5J_1\Delta t_1\Lambda_1 & 1 & 0.5J_3\Delta t_3\Lambda_3 & 0 & 0 \\ \vdots & \ddots & \ddots & \ddots & \vdots \\ 0 & \ddots & \ddots & \ddots & 0.5J_{imax-1}\Delta t_{imax-1}\Lambda_{imax-1} \\ 0 & \cdots & 0 & -0.5J_{imax-2}\Delta t_{imax-2}\Lambda_{imax-2} & 1 \end{bmatrix}^{\xi,n} \tag{2.72}$$

Equation (2.72) is a tridiagonal matrix, and the most effective method to solve this kind of matrix is the so-called pursuing method based on Thomas' algorithm for solving tridiagonal system of equations. In order to increase the stability, the central difference in eq. (2.71) can be replaced by the backward difference when the eigenvalues are positive or forward difference when the eigenvalues are negative, which is analogous to the upwind scheme idea. The diagonal elements of the matrix contain the positive and negative eigenvalues simultaneously, that is,

$$\begin{aligned}(\mathbf{I}+J\Delta tD_\xi\mathbf{A}^n)_i &\approx \mathbf{T}_{\xi,i}\left(\mathbf{I}+J\Delta t\delta_\xi^-\mathbf{\Lambda}_\xi^+ + J\Delta t\delta_\xi^+\mathbf{\Lambda}_\xi^-\right)_i\mathbf{T}_{\xi,i}^{-1} \\ &= \mathbf{T}_{\xi,i}\left(\mathbf{I}+J_i\Delta t_i\mathbf{\Lambda}_{\xi,i}^+ - J_i\Delta t_i\mathbf{\Lambda}_{\xi,i-1}^+ + J_i\Delta t_i\mathbf{\Lambda}_{\xi,i+1}^- - J_i\Delta t_i\mathbf{\Lambda}_{\xi,i}^-\right)\mathbf{T}_{\xi,i}^{-1} \\ &= \mathbf{T}_{\xi,i}\left(-J_i\Delta t_i\mathbf{\Lambda}_{\xi,i-1}^+ + \mathbf{I} + J_i\Delta t_i\mathbf{\Lambda}_{\xi,i}^+ - J_i\Delta t_i\mathbf{\Lambda}_{\xi,i}^- + J_i\Delta t_i\mathbf{\Lambda}_{\xi,i+1}^-\right)\mathbf{T}_{\xi,i}^{-1}.\end{aligned}$$

The similar formulation can be seen in η and ζ directions, respectively.

2.6.2.2 LU-SGS scheme

In order to obtain a stable and reliable implicit scheme, Yoon proposed the LU-SGS method, which has been widely used in CFD field with high reputation of computational stability. Taking eq. (2.63) as an example, its Jacobian matrix can be split into

$$\begin{cases}\mathbf{A}=\mathbf{A}^+ + \mathbf{A}^- \\ \mathbf{B}=\mathbf{B}^+ + \mathbf{B}^- \\ \mathbf{C}=\mathbf{C}^+ + \mathbf{C}^-\end{cases} \tag{2.74}$$

Substituting the above formula back into eq. (2.63), we have

$$\begin{aligned}\Delta\mathbf{Q}^n + J\Delta t\bigg[&\frac{\partial}{\partial\xi}(\mathbf{A}^+\Delta\mathbf{Q}^n)+\frac{\partial}{\partial\xi}(\mathbf{A}^-\Delta\mathbf{Q}^n)+\frac{\partial}{\partial\eta}(\mathbf{B}^+\Delta\mathbf{Q}^n)+\frac{\partial}{\partial\eta}(\mathbf{B}^-\Delta\mathbf{Q}^n) \\ &+\frac{\partial}{\partial\zeta}(\mathbf{C}^+\Delta\mathbf{Q}^n)+\frac{\partial}{\partial\zeta}(\mathbf{C}^-\Delta\mathbf{Q}^n)\bigg] = -J\Delta t\mathbf{RHS}^n.\end{aligned} \tag{2.75}$$

In eq. (2.75), the positive/negative Jacobian matrix adopt backward/forward difference as the following,

$$
\begin{aligned}
&\Delta\mathbf{Q}^n + J\Delta t\Big[\big(\mathbf{A}_i^+\Delta\mathbf{Q}_i^n - \mathbf{A}_{i-1}^+\Delta\mathbf{Q}_{i-1}^n\big) + \big(\mathbf{A}_{i+1}^-\Delta\mathbf{Q}_{i+1}^n - \mathbf{A}_i^-\Delta\mathbf{Q}_i^n\big)\\
&\quad + \big(\mathbf{B}_j^+\Delta\mathbf{Q}_j^n - \mathbf{B}_{j-1}^+\Delta\mathbf{Q}_{j-1}^n\big) + \big(\mathbf{B}_{j+1}^-\Delta\mathbf{Q}_{j+1}^n - \mathbf{B}_j^-\Delta\mathbf{Q}_j^n\big)\\
&\quad + \big(\mathbf{C}_k^+\Delta\mathbf{Q}_k^n - \mathbf{C}_{k-1}^+\Delta\mathbf{Q}_{k-1}^n\big) + \big(\mathbf{C}_{k+1}^-\Delta\mathbf{Q}_{k+1}^n - \mathbf{C}_k^-\Delta\mathbf{Q}_k^n\big)\Big] = -J\Delta t\mathbf{RHS}^n .
\end{aligned} \tag{2.76}
$$

After rearrangement,

$$
\begin{aligned}
&\Big[\mathbf{I} + J\Delta t\big(\mathbf{A}_i^+ - \mathbf{A}_i^- + \mathbf{B}_j^+ - \mathbf{B}_j^- + \mathbf{C}_k^+ - \mathbf{C}_k^-\big)\Big]\Delta\mathbf{Q}^n + J\Delta t\big(\mathbf{A}_{i+1}^-\Delta\mathbf{Q}_{i+1}^n - \mathbf{A}_{i-1}^+\Delta\mathbf{Q}_{i-1}^n\\
&\quad + \mathbf{B}_{j+1}^-\Delta\mathbf{Q}_{j+1}^n - \mathbf{B}_{j-1}^+\Delta\mathbf{Q}_{j-1}^n + \mathbf{C}_{k+1}^-\Delta\mathbf{Q}_{k+1}^n - \mathbf{C}_{k-1}^+\Delta\mathbf{Q}_{k-1}^n\big) = -J\Delta t\mathbf{RHS}^n.
\end{aligned} \tag{2.77}
$$

To improve the stability of solving the equation, the Jacobian matrix in the above formula is modified to the following approximate Jacobian matrix:

$$
\begin{cases}
\mathbf{A}^\pm = \frac{1}{2}[\mathbf{A} \pm \rho(\mathbf{A})\mathbf{I}]\\
\mathbf{B}^\pm = \frac{1}{2}[\mathbf{B} \pm \rho(\mathbf{B})\mathbf{I}]\\
\mathbf{C}^\pm = \frac{1}{2}[\mathbf{C} \pm \rho(\mathbf{C})\mathbf{I}]\\
\rho(\mathbf{A}) = k\max[|\lambda(\mathbf{A})|], \quad k \ge 1
\end{cases} . \tag{2.78}
$$

The approximate Jacobian matrix defined in eq. (2.78) is a diagonal dominant matrix. And the parameter k is used to adjust the stability and convergence in computation. $\lambda(\mathbf{A})$ is the eigenvalue of the matrix A. $\mathbf{D}$, $\bar{\mathbf{L}}\Delta\mathbf{Q}^n$ and $\bar{\mathbf{U}}\Delta\mathbf{Q}^n$ are defined as follows:

$$
\begin{cases}
\mathbf{D} = \mathbf{I} + J\Delta t\big(\mathbf{A}_i^+ - \mathbf{A}_i^- + \mathbf{B}_j^+ - \mathbf{B}_j^- + \mathbf{C}_k^+ - \mathbf{C}_k^-\big)\\
\bar{\mathbf{L}}\Delta\mathbf{Q}^n = -J\Delta t\big(\mathbf{A}_{i-1}^+\Delta\mathbf{Q}_{i-1}^n + \mathbf{B}_{j-1}^+\Delta\mathbf{Q}_{j-1}^n + \mathbf{C}_{k-1}^+\Delta\mathbf{Q}_{k-1}^n\big)\\
\bar{\mathbf{U}}\Delta\mathbf{Q}^n = J\Delta t\big(\mathbf{A}_{i+1}^-\Delta\mathbf{Q}_{i+1}^n + \mathbf{B}_{j+1}^-\Delta\mathbf{Q}_{j+1}^n + \mathbf{C}_{k+1}^-\Delta\mathbf{Q}_{k+1}^n\big)
\end{cases} , \tag{2.79}
$$

where the matrixed $\mathbf{D}$, $\bar{\mathbf{L}}\Delta\mathbf{Q}^n$ and $\bar{\mathbf{U}}\Delta\mathbf{Q}^n$ are diagonal, lower triangular and upper triangular matrix, respectively. Substituting formula (2.79) into formula (2.77), we have

$$
(\mathbf{D} + \bar{\mathbf{L}} + \bar{\mathbf{U}})\Delta\mathbf{Q}^n = (\mathbf{D} + \bar{\mathbf{L}})\mathbf{D}^{-1}(\mathbf{D} + \bar{\mathbf{U}})\Delta\mathbf{Q}^n = \mathbf{L}\mathbf{D}^{-1}\mathbf{U}\Delta\mathbf{Q}^n = -J\Delta t\mathbf{RHS}^n. \tag{2.80}
$$

According to eq. (2.79), the matrix $\mathbf{D}, \mathbf{L}, \mathbf{U}$ can be simplified as

$$
\begin{cases}
\mathbf{D} = \mathbf{I} + J\Delta t(\rho(\mathbf{A})\mathbf{I} + \rho(\mathbf{B})\mathbf{I} + \rho(\mathbf{C})\mathbf{I})\\
\mathbf{L} = \mathbf{I} + J\Delta t\big(\rho(\mathbf{A})\mathbf{I} + \rho(\mathbf{B})\mathbf{I} + \rho(\mathbf{C})\mathbf{I} - \mathbf{A}_{i-1}^+ - \mathbf{B}_{j-1}^+ - \mathbf{C}_{k-1}^+\big)\\
\mathbf{U} = \mathbf{I} + J\Delta t\big(\rho(\mathbf{A})\mathbf{I} + \rho(\mathbf{B})\mathbf{I} + \rho(\mathbf{C})\mathbf{I} + \mathbf{A}_{i+1}^- + \mathbf{B}_{j+1}^- + \mathbf{C}_{k+1}^-\big)
\end{cases} . \tag{2.81}
$$

The solution of eq. (2.81) can be solved by two steps.

Forward sweeping for L block

$$\mathbf{L}\Delta\mathbf{Q}^{*} = -J\Delta t\mathbf{RHS}^{n}. \tag{2.82}$$

Backward sweeping for U block

$$\mathbf{U}\Delta\mathbf{Q} = \mathbf{D}\Delta\mathbf{Q}^{*}. \tag{2.83}$$

2.6.3 Second-order implicit time-advancing schemes

In the previous sections, we discussed the time-advancing schemes of the semidiscrete equations, but all have only the first-order accuracy in time. When the time step size is large, the accuracy of the methods is compromised. By introducing the concept of pseudo-time step, the dual time-advancing method proposed by Jameson [23] can extend the time advancement accuracy to the second-order. For the semidiscrete equation (2.19), a second-order tree-point backward difference is used, which has the following expression:

$$\frac{1}{J}\frac{3\mathbf{Q}_{i,j,k}^{n+1} - 4\mathbf{Q}_{i,j,k}^{n} + \mathbf{Q}_{i,j,k}^{n-1}}{2\Delta t} + \mathbf{Q}_{i,j,k}^{n+1}\frac{\partial J^{-1}}{\partial t} + \mathbf{RHS} = 0. \tag{2.84}$$

To improve the time-advancing accuracy from the nth to the $n+1$th time layer, a pseudo-time step τ is introduced in the above equation, which becomes

$$\frac{1}{J}\frac{d\mathbf{Q}_{i,j,k}}{d\tau} + \frac{1}{J}\frac{3\mathbf{Q}_{i,j,k}^{n+1} - 4\mathbf{Q}_{i,j,k}^{n} + \mathbf{Q}_{i,j,k}^{n-1}}{2\Delta t} + \mathbf{Q}_{i,j,k}^{n+1}\frac{\partial J^{-1}}{\partial t} + \mathbf{RHS} = 0. \tag{2.85}$$

The first-order forward difference approximation is applied to the introduced pseudo-time term, and as $m \to \infty$, the pseudo-time term approaches zero and $\mathbf{Q}_{i,j,k}^{m+1} \to \mathbf{Q}_{i,j,k}^{n+1}$. Equation (2.85) can be rewritten as follows:

$$\frac{1}{J}\frac{\mathbf{Q}_{i,j,k}^{m+1} - \mathbf{Q}_{i,j,k}^{m}}{\Delta\tau} + \frac{1}{J}\frac{3\mathbf{Q}_{i,j,k}^{m+1} - 4\mathbf{Q}_{i,j,k}^{n} + \mathbf{Q}_{i,j,k}^{n-1}}{2\Delta t} + \mathbf{Q}_{i,j,k}^{m+1}\frac{\partial J^{-1}}{\partial t} + \mathbf{RHS} = 0. \tag{2.86}$$

For explicit time-advancing schemes, eq. (2.86) can be simplified to

$$\frac{1}{J}\frac{\mathbf{Q}_{i,j,k}^{m+1} - \mathbf{Q}_{i,j,k}^{m}}{\Delta\tau} = -\mathbf{RHS} - \frac{1}{J}\frac{3\mathbf{Q}_{i,j,k}^{m} - 4\mathbf{Q}_{i,j,k}^{n} + \mathbf{Q}_{i,j,k}^{n-1}}{2\Delta t} - \mathbf{Q}_{i,j,k}^{m}\frac{\partial J^{-1}}{\partial t}. \tag{2.87}$$

The second-order time scheme can be derived by directly implementing the above formula in Section 2.6.1.

For the AF-ADI time-advancing scheme, after rearranging eq. (2.86), we have

$$\left(\frac{1}{J\Delta\tau}+\frac{3}{2J\Delta t}+\frac{\partial J^{-1}}{\partial t}\right)\Delta\mathbf{Q}^m_{i,j,k}+\frac{1}{J}\frac{3\mathbf{Q}^m_{i,j,k}-4\mathbf{Q}^n_{i,j,k}+\mathbf{Q}^{n-1}_{i,j,k}}{2\Delta t}+\mathbf{Q}^m_{i,j,k}\frac{\partial J^{-1}}{\partial t}+\mathbf{RHS}=0. \quad (2.88)$$

After the residual is linearized, the RHS term becomes

$$\mathbf{RHS}\left(\mathbf{Q}^{m+1}_{i,j,k}\right)\cong\mathbf{RHS}\left(\mathbf{Q}^m_{i,j,k}\right)+\frac{\partial\mathbf{RHS}}{\partial\mathbf{Q}}\Delta\mathbf{Q}^m_{i,j,k}. \quad (2.89)$$

Substituting eq. (2.89) into eq. (2.88), we obtain the following equation, which is similar to eq. (2.64):

$$\begin{aligned}&\left[\left(\frac{1}{J\Delta\tau}+\frac{3}{2J\Delta t}+\frac{\partial J^{-1}}{\partial t}\right)\mathbf{I}+\left(D_\xi\mathbf{A}+D_\eta\mathbf{B}+D_\zeta\mathbf{C}\right)\right]\Delta\mathbf{Q}^m\\&=-\mathbf{RHS}\left(\mathbf{Q}^m_{i,j,k}\right)-\frac{1}{J}\frac{3\mathbf{Q}^m_{i,j,k}-4\mathbf{Q}^n_{i,j,k}+\mathbf{Q}^{n-1}_{i,j,k}}{2\Delta t}-\mathbf{Q}^m_{i,j,k}\frac{\partial J^{-1}}{\partial t}.\end{aligned} \quad (2.90)$$

The method for solving eq. (2.90) has been described in Section 2.6.2.1.

For the LU-SGS scheme, by introducing the pseudo-time, we have the following expression:

$$\begin{aligned}&\left(\frac{1}{J\Delta\tau}+\frac{3}{2J\Delta t}+\frac{\partial J^{-1}}{\partial t}\right)\Delta\mathbf{Q}^m+\frac{\partial}{\partial\xi}(\mathbf{A}\Delta\mathbf{Q}^m)+\frac{\partial}{\partial\eta}(\mathbf{B}\Delta\mathbf{Q}^m)+\frac{\partial}{\partial\zeta}(\mathbf{C}\Delta\mathbf{Q}^m)\\&=-\mathbf{RHS}\left(\mathbf{Q}^m_{i,j,k}\right)-\frac{1}{J}\frac{3\mathbf{Q}^m_{i,j,k}-4\mathbf{Q}^n_{i,j,k}+\mathbf{Q}^{n-1}_{i,j,k}}{2\Delta t}-\mathbf{Q}^m_{i,j,k}\frac{\partial J^{-1}}{\partial t}.\end{aligned} \quad (2.91)$$

Similar to the first-order LU-SGS scheme, the matrixes **L**, **D** and **U** in eq. (2.79) can be written as

$$\begin{cases}\mathbf{D}=\left(\frac{1}{J\Delta\tau}+\frac{3}{2J\Delta t}+\frac{\partial J^{-1}}{\partial t}+\rho(\mathbf{A})+\rho(\mathbf{B})+\rho(\mathbf{C})\right)\mathbf{I}\\\mathbf{L}=\left(\frac{1}{J\Delta\tau}+\frac{3}{2J\Delta t}+\frac{\partial J^{-1}}{\partial t}+\rho(\mathbf{A})+\rho(\mathbf{B})+\rho(\mathbf{C})\right)\mathbf{I}-\mathbf{A}^+_{i-1}-\mathbf{B}^+_{j-1}-\mathbf{C}^+_{k-1}\\\mathbf{U}=\left(\frac{1}{J\Delta\tau}+\frac{3}{2J\Delta t}+\frac{\partial J^{-1}}{\partial t}+\rho(\mathbf{A})+\rho(\mathbf{B})+\rho(\mathbf{C})\right)\mathbf{I}+\mathbf{A}^-_{i+1}+\mathbf{B}^-_{j+1}+\mathbf{C}^-_{k+1}\end{cases}. \quad (2.92)$$

The method for solving eq. (2.92) has been described in Section 2.6.2.2 as well.

2.7 Turbulence modeling

According to the different research purpose in turbulence modeling, there are different fineness levels for the turbulence numerical simulation. In order to have a deep understanding of the physical properties of turbulence, the numerical simulation needs to be performed in the most detailed level, and the turbulent flow motion of all scales must be numerically simulated based on the fully completed flow governing equations. The direct simulation of turbulence by the time-dependent N-S equations known as the direct numerical simulation [45] is applicable only to the relatively simple flow problems at low Reynolds numbers. For engineering problems, it is common to predict the averaged turbulence variables, such as the averaged velocity and force. Therefore, the Reynolds averaged N-S equations can be solved using the turbulence models, and we consider the effects of turbulence in an approximation manner. The basic idea is to introduce the various hypotheses on Reynolds stress to close the Reynolds averaged equations to obtain the averaged flow solutions, which is referred as Reynolds averaged numerical simulation (RANS) [52]. Based on the idea of RANS, a variety of turbulence models were developed. The eddy viscosity model based on the hypothesis of Boussinesq [30] is the most commonly used method in turbulence modeling community. The turbulence model based on the eddy viscosity hypothesis can be roughly divided into the following categories:

- Algebraic model (Baldwin–Lomax model as the representation) [3, 9];
- One-equation model (Spalart–Allmaras model as the representation) [2, 55];
- Two-equation model (k–ω SST (Small-Scale Turbulence) model as the representation) [26, 68].

For certain applications, there are also other turbulence models based on nonlinear eddy viscosity formulations which is beyond the scope of this section, and the interested reader is referred to the relevant literature.

2.7.1 Baldwin–Lomax algebraic model

Baldwin–Lomax (B-L) turbulence model applies different mixing length assumptions to the inner and outer layers of the boundary layer, which is described as follows:

$$\mu_t = \begin{cases} (\mu_t)_{\text{inn}} & (y \le y_c) \\ (\mu_t)_{\text{out}} & (y > y_c) \end{cases}, \tag{2.93}$$

where y_c is the normal distance to the nearest wall when $(\mu_t)_{\text{inn}} = (\mu_t)_{\text{out}}$, and μ_t is the turbulence viscosity based on the eddy viscosity concept of Boussinesq.

$$(\mu_t)_{\text{inn}} = \rho l^2 \Omega \frac{\text{Re}}{M_\infty} (y \le y_c). \tag{2.94}$$

In the above formula, Ω is the vorticity and l is the length-scale function, which are expressed as

$$l = ky[1 - \exp(-y^+/A^+)] \tag{2.95}$$

$$\Omega = \sqrt{(w_y - v_z)^2 + (u_z - w_x)^2 + (v_x - u_y)^2},$$

where $k = 0.4\, A^+ = 26.0$, and y^+ is a dimensionless normal distance with the detailed formulation as

$$u_\tau = \sqrt{\left|\frac{\mu}{\rho}\frac{\partial u}{\partial y}\right|_w \frac{M_\infty}{\mathrm{Re}}}$$
$$y^+ = u_\tau y \rho / \mu_w (\mathrm{Re}/M_\infty) = \frac{\sqrt{\rho \tau_w}}{\mu_w} y \left(\frac{\mathrm{Re}}{M_\infty}\right)^{\frac{1}{2}}, \tag{2.96}$$

where the subscript w indicates the wall surface.

For the outer layer, where $y > y_c$, we have

$$(\mu_t)_{\mathrm{out}} = 0.0168(1.6) F_{wake} F_{kleb}(y) \frac{\mathrm{Re}}{M_\infty}, \tag{2.97}$$

where

$$F_{wake} = \min\left(y_{\max} F_{\max}, C_{wk} y_{\max} u_{dif}^2 \middle/ F_{\max}\right), \tag{2.98}$$

$y_{\max}$ is the maximum position of y when the function $F(y)$ reaches its maximum value of $F_{\max}$, and the detailed expression of $F(y)$ is as follows:

$$F(y) = y\Omega[1 - \exp(-y^+/A^+)]. \tag{2.99}$$

In eq. (2.97), the Klebanoff interval function $F_{kleb}(y)$ is expressed as

$$F_{kleb}(y) = \left[1 + 5.5\left(\frac{C_{kleb} \cdot y}{y_{\max}}\right)^6\right]^{-1}, \tag{2.100}$$

where $C_{kleb} = 0.3 C_{wk} = 1.0$.

2.7.2 Spalart–Allmaras one-equation model

The one equation Spalart–Allmaras (S-A) turbulence model [14] employs transport equation for an eddy viscosity variable $\hat{v}$ to obtain the turbulence viscosity coefficient μ_t, and the detailed expression is as follows:

$$\mu_t = \rho \hat{v} f_{v_1}, \tag{2.101}$$

where

$$f_{v_1} = \frac{\chi^3}{\chi^3 + C_{v_1}^3}, \tag{2.102}$$

$$\chi = \frac{\hat{\nu}}{\nu}. \tag{2.103}$$

And $\hat{\nu}$ is the S-A model eddy viscosity variable to be solved and ν is the laminar kinetic viscosity coefficient.

The S-A equation can be written in the following general form [30]:

$$\frac{\partial}{\partial t}(X) + u_j \frac{\partial}{\partial x_j}(X) = S_P + S_D + D, \tag{2.104}$$

where X stands for the eddy viscosity variable of the S-A equation. S_P is the production term. S_D is the near wall destruction term and D is the diffusion term. The detailed expression for each term is as follows:

$$\left.\begin{aligned} S_P &= C_{b_1}\left(1 - f_{t_2}\right)\Omega\hat{\nu} \\ S_D &= \tfrac{M_\infty}{\mathrm{Re}}\left\{C_{b_1}\left[\left(1 - f_{t_2}\right)f_{v_2} + f_{t_2}\right]\tfrac{1}{\kappa^2} - C_{w_1}f_w\right\}\left(\tfrac{\hat{\nu}}{d}\right)^2 \\ D &= -\tfrac{M_\infty}{\mathrm{Re}}\tfrac{C_{b_2}}{\sigma}\hat{\nu}\tfrac{\partial\hat{\nu}}{\partial x_j^2} + \tfrac{M_\infty}{\mathrm{Re}}\tfrac{1}{\sigma}\tfrac{\partial}{\partial x_j}\left\{\left[\nu + \left(1 + C_{b_2}\right)\hat{\nu}\right]\tfrac{\partial\hat{\nu}}{\partial x_j}\right\} \end{aligned}\right\}, \tag{2.105}$$

where d is the nearest distance from the grid point to the wall surface and the other functions are expressed as follows:

$$f_{v_2} = 1 - \frac{\chi}{1 + \chi f_{v_1}}, \tag{2.106}$$

$$\hat{S} = \Omega + \frac{\hat{\nu} f_{v_2}}{\left(\frac{\mathrm{Re}}{M_\infty}\right)\kappa^2 d^2}, \tag{2.107}$$

$$\left.\begin{aligned} f_{t_2} &= C_{t_3}\exp\left(-C_{t_4}\chi^2\right) \\ f_w &= g\left(\frac{1 + C_{w_3}^6}{g^6 + C_{w_3}^6}\right) \\ g &= r + C_{w_2}\left(r^6 - r\right) \end{aligned}\right\}, \tag{2.108}$$

$$r = \frac{\hat{\nu}}{\hat{S}\left(\frac{\mathrm{Re}}{M_\infty}\right)\kappa^2 d^2}. \tag{2.109}$$

The constants in the above formulations are

$$\begin{aligned} &C_{b_1} = 0.1355 \quad \sigma = \frac{2}{3} \quad C_{b_2} = 0.622 \quad \kappa = 0.41 \\ &C_{w_3} = 2.0 \quad C_{v_1} = 7.1 \quad C_{t_3} = 1.2 \quad C_{t_4} = 0.5 \\ &C_{w_2} = 0.3 \quad C_{w_1} = \frac{C_{b_1}}{\kappa^2} + \frac{\left(1 + C_{b_2}\right)}{\sigma} \end{aligned}.$$

2.7.3 *k–ω* SST two-equation model

In the frame of k–ω SST turbulence model, the turbulence viscosity μ_t is defined as

$$\mu_t = \min\left(\frac{\rho k}{\boldsymbol{\omega}}, \frac{a_1\rho k}{\Omega F_2}\left(\frac{\mathrm{Re}}{M_\infty}\right)\right). \tag{2.110}$$

Similar to eq. (2.104), the k–ω SST model can be written in the following general form [40]:

$$\frac{\partial}{\partial t}(X) + u_j\frac{\partial}{\partial x_j}(X) = S_P + S_D + D. \tag{2.111}$$

The various symbols in the two-equation model can be expressed as

$$\left.\begin{aligned} X &= k \\ S_{P,k} &= \tfrac{1}{\rho}\mu_t\Omega^2\left(\tfrac{M_\infty}{\mathrm{Re}}\right) \\ S_{D,k} &= -\beta' k\omega\left(\tfrac{\mathrm{Re}}{M_\infty}\right) \\ D_k &= \tfrac{1}{\rho}\tfrac{\partial}{\partial x_j}\left[\left(\mu + \tfrac{\mu_t}{\sigma_k}\right)\tfrac{\partial k}{\partial x_j}\right]\left(\tfrac{M_\infty}{\mathrm{Re}}\right) \end{aligned}\right\}, \tag{2.112}$$

$$\left.\begin{aligned} X &= \boldsymbol{\omega} \\ S_{P,\omega} &= \gamma\Omega^2\left(\tfrac{M_\infty}{\mathrm{Re}}\right) \\ S_{D,\omega} &= -\beta\omega^2\left(\tfrac{\mathrm{Re}}{M_\infty}\right) + 2(1-F_1)\sigma_{\omega 2}\tfrac{1}{\omega}\tfrac{\partial k}{\partial x_j}\tfrac{\partial \omega}{\partial x_j}\left(\tfrac{M_\infty}{\mathrm{Re}}\right) \\ D_k &= \tfrac{1}{\rho}\tfrac{\partial}{\partial x_j}\left[\left(\mu + \tfrac{\mu_t}{\sigma_\omega}\right)\tfrac{\partial \omega}{\partial x_j}\right]\left(\tfrac{M_\infty}{\mathrm{Re}}\right) \end{aligned}\right\}, \tag{2.113}$$

where

$F_1 = \tanh(\Gamma^4)$, $\Gamma = \min[\max(\Gamma_1, \Gamma_3), \Gamma_2]$,

$$\left\{\begin{aligned} \Gamma_1 &= \tfrac{500\nu}{d^2\boldsymbol{\omega}}\left(\tfrac{M_\infty}{\mathrm{Re}}\right)^2 \\ \Gamma_2 &= \frac{4\rho\sigma_{\boldsymbol{\omega}2}k}{d^2(CD_{k-\omega})} \\ \Gamma_3 &= \tfrac{\sqrt{k}}{C_\mu\boldsymbol{\omega}d}\left(\tfrac{M_\infty}{\mathrm{Re}}\right) \end{aligned}\right. . \tag{2.114}$$

In the above-mentioned formulas,

$$CD_{k-\boldsymbol{\omega}} = \max\left(\rho\frac{2\sigma_{\omega 2}}{\boldsymbol{\omega}}\frac{\partial k}{\partial x_j}\frac{\partial \omega}{\partial x_j}, 1\times 10^{-20}\right),$$

$$F_2 = \tan\, h(\Pi^2),$$

$$\Pi = \max(2\Gamma_3, \Gamma_1).$$

The parameters γ, σ_k, σ_ω and β are expressed as:

$$\phi = F_1\phi_1 + (1-\phi_1)F_2\phi = \{\gamma, \sigma_k, \sigma_\omega, \beta\}, \tag{2.115}$$

$$\sigma_{k_1} = 1/0.85,$$

$$\sigma_{k_2} = 1.0,$$

$$\sigma_{\omega_1} = 1/0.5,$$

$$\sigma_{\omega_2} = 1/0.856,$$

$$\beta_1 = 0.075,$$

$$\beta_2 = 0.0828,$$

$$\kappa = 0.41,$$

$$a_1 = 0.31,$$

where γ_1 and γ_2 are expressed as

$$\begin{aligned} \gamma_1 &= \frac{\beta_1}{C_\mu} - \frac{\kappa^2}{\sigma_{\omega_1}\sqrt{C_\mu}} \\ \gamma_2 &= \frac{\beta_2}{C_\mu} - \frac{\kappa^2}{\sigma_{\omega_2}\sqrt{C_\mu}} \end{aligned}. \tag{2.116}$$

2.7.4 Detached eddy simulation

When the aircraft is in a state of high attack angle, the continual generation and shedding of the vortex produce a strong unsteady aerodynamic force on the aircraft surface. This unsteady aerodynamic force induces strong disturbance to the aircraft structure, which causes the aeroelastic divergence occur early. The detached eddy simulation (DES) method is currently a relatively fast and ideal method for simulating the vortex-shedding phenomenon [17–19, 22].

The DES method was first proposed by Spalart [56] in 1997, which is a representation of the RANS/large eddy simulation (LES) hybrid method. The idea is to employ highly stretched grids together with a RANS turbulence model (the Spalart–Allmaras model or the Menter's k–ω SST model) to resolve the attached boundary layers, and to use the LES outside the wall region together with an isotropic grid to capture the detached 3D eddies [6]. DES tries to combine the strengths of both methods in a single framework.

SA-DES method is constructed based on S-A one-equation model. The S-A model contains a destruction term proportional to $(\frac{\tilde{\nu}}{d})^2$, where d is the normal distance to the wall. When the destruction term and the production term are balanced, the eddy viscosity would be proportional to the local deformation rate S and the distance d: $\tilde{\nu} \propto Sd^2$. In the framework of LES's subgrid model, the eddy viscosity is proportional to S and the grid spacing size Δ: $\nu_{SGS} \propto S\Delta^2$. Therefore, the distance to the wall d is replaced by the DES length scale Δ, and the S-A model is viewed as a one-equation SGS model for the LES outside the boundary layer. In the standard S-A model, the nearest normal distance to the wall, d is used as the length scale. A new length scale is defined for DES model as follows:

$$\tilde{d} = \min(d, C_{des}\Delta_{\max}). \tag{2.69}$$

where $\Delta_{\max}$ is the largest dimension of the control volume in three directions, and the constant C_{des} is 0.65 for a homogeneous turbulence and 0.1 is recommended for transonic and supersonic jet.

The definition of the length scale $\tilde{d}$ in eq. (2.69) makes sure that, within the boundary layer where $d < C_{\text{des}}\Delta_{\max}$, then $\tilde{d} = d$, the original S-A model is recovered. Outside the boundary layer, $\tilde{d} = C_{\text{des}}\Delta_{\max}$, the S-A model served as one-equation SGS model for the LES.

The DES is simple to construct with a uniform equation that does not explicitly distinguish RANS or LES region in simulation process. However, it will cause discontinuity in the length-scale gradient of the destruction term, which can be improved by modifying the min function. In the definition of length scale, the grid spacing is consistent with the characteristics of the filter scale in the LES method. In DES, the role of grid refinement is to increase the filter scale range and improve the resolution of the turbulence model. This is very different from the grid refinement in RANS. In RANS, even if the grid is infinitely refined, only the turbulence model plays a major role and there is no any filter scale information included. The inclusion of grid spacing in the length scale also highlights the importance of grids in the RANS/LES hybrid approach.

With the fact that the length scale in DES is related to the grid spacing, there may be such a region where $\Delta_{\max} < \delta$ in some special grids. In this case, the DES filter would solve the boundary layer in the LES mode, while the grid density here is not sufficient to perform LES simulation. Therefore, the problem of modeled stress depletion (MSD) occurs. As a response to MSD problem, Spalart proposed a delayed DES (DDES) method, with the modified length-scale definition as follows:

$$\tilde{d} = d - f_d \max(0, d - C_{des}\Delta_{max}) \tag{2.70}$$

where f_d is defined as

$$f_d = 1 - \tan h\left([8_{rd}]^3\right) \tag{2.71}$$

$$r_d = \frac{\nu_t + \nu}{\sqrt{U_{i,j}U_{i,j}}\kappa^2 d^2};\ \kappa = 0.41 \tag{2.72}$$

Compared with DES, DDES has been greatly improved. The length scale is defined from simply relying on the grid spacing to comprehensively accounting for both the grid spacing and the eddy viscosity field, and varies over time. The newly defined length scale narrows the gray area between the RANS and LES area. Although the influence of the low-quality grid is reduced in DDES model, this does not mean that the grid quality is not important for DDES model, especially for complex configurations. In addition, the grid independence has to be ensured, which has great influence to the simulation result of the DDES model.

2.8 Geometric conservation law

From the derivation in Section 2.6.3, the numerical solution of the governing equation for the unsteady flow field involves the calculation of the control volume changing rate $\partial J^{-1}/\partial t$ when the grid is in motion. In general, in order to ensure the accuracy of the flow field solution, the control volume changing rate is solved directly by the GCL [60], not by the finite difference method which cannot guarantee the conservation of the continuity equation. The GCL is derived from the continuity equation, and must hold also for a uniform fluid velocity and a constant density. Assuming the control volume is always closed, and the density is constant, we obtain the integral form of the GCL:

$$\frac{d}{dt}\int_V dV - \int_S \mathbf{u}_b \cdot \mathbf{ds} = 0 \tag{2.117}$$

From eq. (2.117), the changing rate of the control volume is equal to the sum of the volumes swept by the surfaces of the grid in a unit of time, thus the following formula is obtained:

$$\frac{\partial J^{-1}}{\partial t} = \sum_{i=1}^{6} (\mathbf{u}_b \cdot \mathbf{ds})_i \tag{2.118}$$

As shown, the GCL relates the change of the control volume to the motion of its cell faces. It is essential that the GCL in eq. (2.117) is temporally discretized using the same scheme as it is applied to the governing equations in order to obtain a self-consistent solution method. It also needs to note that the GCL has to be solved concurrently with the fluid equations [4, 21, 32, 59].

2.9 Examples of CFD in engineering

As a supplement to the content described in the previous chapter, several examples of CFD technique in practice are presented in the following section. All examples used here are from the industrial problems and the standard model tests.

2.9.1 Two-dimensional airfoil of RAE2822

Figure 2.4 shows the comparison of the pressure distribution on the airfoil surface between the CFD computation and experimental results. The flow condition is $M_\infty = 0.73$, $\alpha = 3.19°$, $Re = 6.5 \times 10^6$. The spatial schemes are Roe scheme, van Leer scheme and the

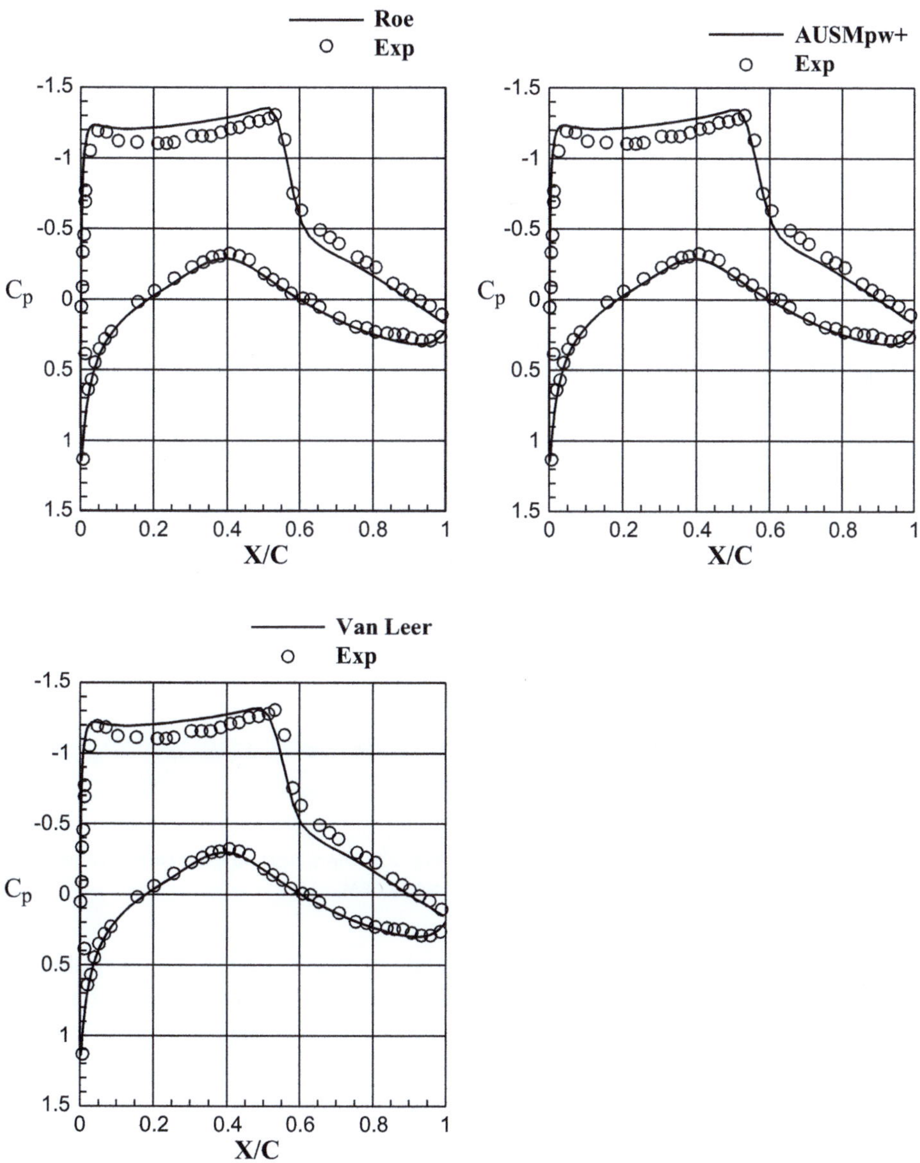

Fig. 2.4: The pressure distribution on the surface of RAE2822 airfoil.

AUSMpw+ scheme. The implicit LU-SGS scheme is used for time advancement. The turbulence model used in this example is algebraic Baldwin–Lomax model. It can be seen from Fig. 2.4 that the pressure distribution predicted by Roe, AUSMpw+ and Van Leer schemes are in good agreement with the experimental results. In a more detailed observation, the shockwave position predicted by Roe and AUSMpw + schemes is closer to the experiment. While shockwave position from van Leer scheme is earlier than the experiment. Therefore, the Roe scheme and the AUSMpw+ scheme show more shock wave resolution than the Van Leer scheme.

2.9.2 Wing/body/tail configuration of NASA CRM

NASA Customer Relationship Management (CRM) is a standard test model provided by the committee of DPW-4 (drag prediction workshop 4) [33, 66], as shown in Fig. 2.5. The computational grid was generated by the commercial software The Integrated Computer Engineering and Manufacturing code for Computational Fluid Dynamics (ICEM CFD), with a total of 54 structural grid blocks and approximately 3.8 million grid points. The Jameson's central difference scheme is adopted, and the turbulence model is one equation S-A model. The computational state is $M_\infty = 0.85$, $C_L = 0.500 \pm 0.001$ and the deflection angle of the horizontal tail is 0°, with the Reynolds number of $Re = 5.0 \times 10^6$(based on the mean aerodynamic chord length).

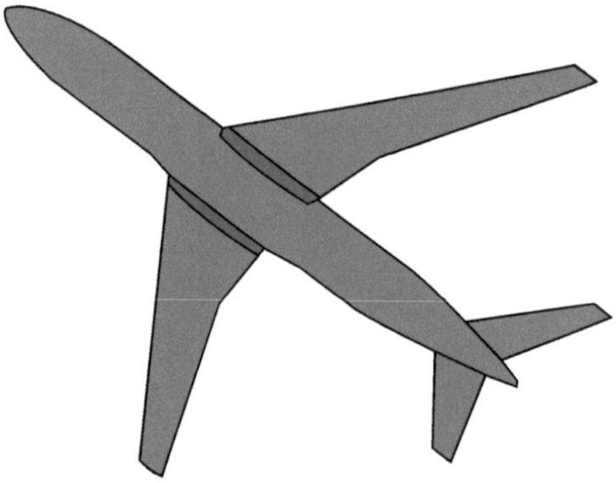

Fig. 2.5: The wing/body/tail configuration of NASA CRM.

For the current calculation state, when the angle of attack is 2.31°, the computed lift coefficient is $C_L = 0.5$. Figure 2.6 shows the pressure distribution in chord-wise direction on the six sections along the span of the wing, also the results calculated using Computational Fluids Laboratory 3-Dimensional flow solver (CFL3D) provided by B. J. Rider and E. N. Tinoco [49] based on the computational grid about 4.7 million grid points are shown in Fig. 2.6. As shown in Fig. 2.6, the pressure distribution calculated using the current CFD method is very close to CFL3D's results.

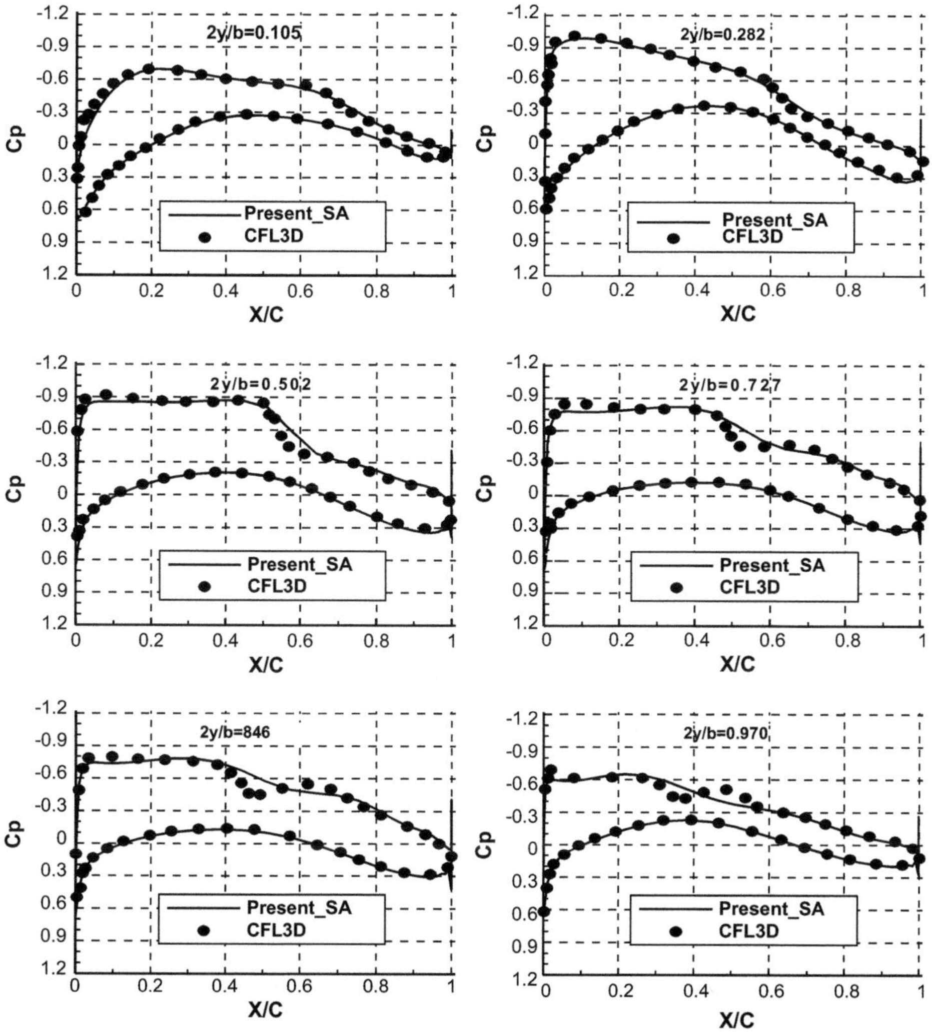

Fig. 2.6: The pressure distribution on the wing surface of NASA CRM.

2.9.3 Wing/body/pylon/nacelle configuration of DLR-F6

The wing/body/pylon/nacelle configuration of Deutsches Zentrum für Luft- und Raumfahrt (DLR-F6) is shown in Fig. 2.7, and the computational grid is a coarse grid from the website of DPW-2 generated by ICEM CFD with about 5.15 million grid points. Three CFD tools, CFL3D, TRIP2.0 and ATTF, developed by the current authors are selected to perform the numerical simulation. Figure 2.8 presents the schematic illustrations of the geometric configuration and characteristic section of the nacelle surface. The computational results from Rumsey's CFL3D [53], Yuntao Wang's TRIP2.0 [10] and

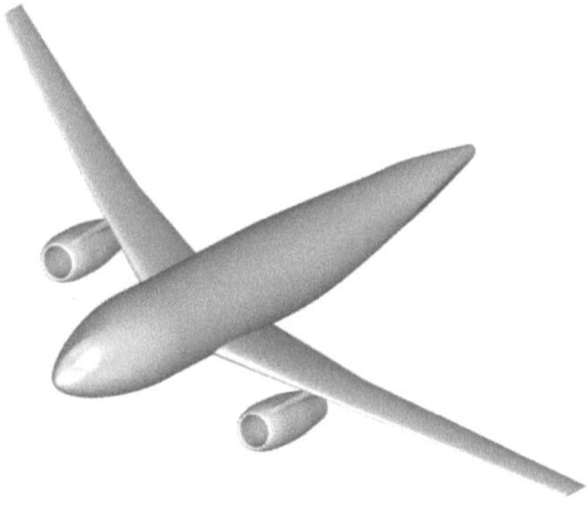

Fig. 2.7: The geometric configuration of DLR-F6.

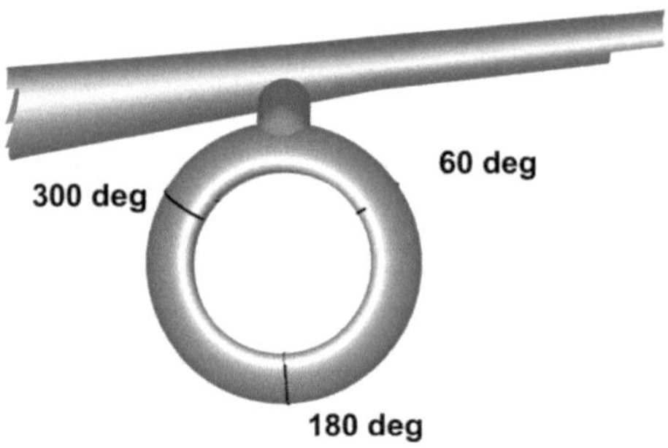

Fig. 2.8: The geometric configuration of nacelle and the illustrations of characteristic sections.

ATTF are compared with the experimental results from the ONERA S2MA wind tunnel. Table 2.1 presents the statistics of the calculation results for the computational state 1 as follows:

State 1: $M_\infty = 0.75$, $C_L = 0.500 \pm 0.001$;

State 2: $M_\infty = 0.75$, $\alpha = -3° - 2°$, $-1.5°$, $-1°$, $0°$, $1°$, $1.5°$.

Reynolds number: $\mathrm{Re} = 3.0 \times 10^6$ (based on MAC), and the turbulence model is one equation S-A turbulence model.

Table 2.1 shows that the drag coefficient predicted using CFL3D is higher about 11 counts (one count is 1.0E-4) than the experimental result, and the result from TRIP2.0 is higher about 55 counts than experiment, and the present result from ATTF is higher about nine counts than experiment. It proves that the computational capacity of ATTF developed by authors is acceptable, and its prediction accuracy is comparable to that of the similar CFD tools or software.

Figure 2.9 shows the comparison of predicted lift coefficients from experiment, CFL3D, TRIP2.0 and the currently used ATTF. Figure 2.10 shows the comparison of predicted polar curves of three CFD calculations and the wind tunnel results. It can be seen that the results from ATTF are very close to the experimental results and CFL3D's results.

Tab. 2.1: The computational results for the computational state 1.

	EXP	Present (ATTF)	CFL3D	TRIP2.0
α	1.0	0.712	0.577	1.036
C_L	0.500	0.500	0.500	0.499
$C_{D\ Total}$	0.0338	0.0347	0.0349	0.0393
$C_{D\ Pressure}$	–	0.0194	0.0187	0.0212
$C_{D\ Viscous}$	–	0.0153	0.0161	0.0181
C_M	−0.130	−0.1453	−0.1420	−0.1270

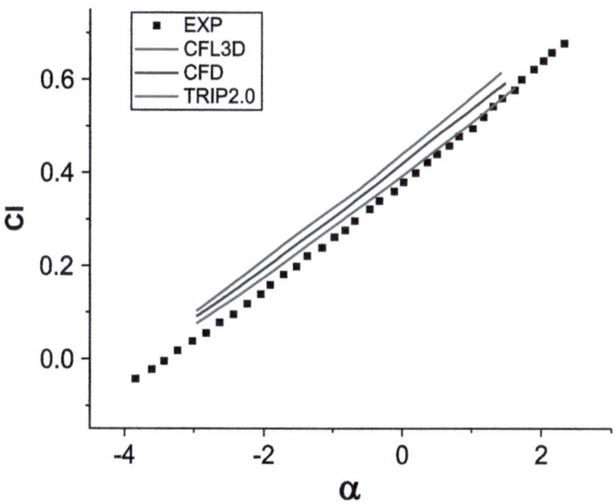

Fig. 2.9: Comparison of lift coefficients.

Two computation states are chosen to verify the pressure distribution from the CFD computation and experiment (ONERA S2MA transonic wind tunnel). The computation states are as follows:

Case 1: $\alpha = 0.712°$, $C_L = 0.500$
Case 2: $\alpha = 1.0°$, $C_L = 0.534$

A coarse computation grid is used for these two test cases. For the experiment, the angle of attack in wind tunnel test is $\alpha = 1.003°$ and the lift coefficient is $C_L = 0.4981$.

Figure 2.11 shows the comparison of the predicted pressure distribution with that from experiment at four sections along the span direction. It can be seen that the pressure distribution calculated under case 2 condition is obviously better than

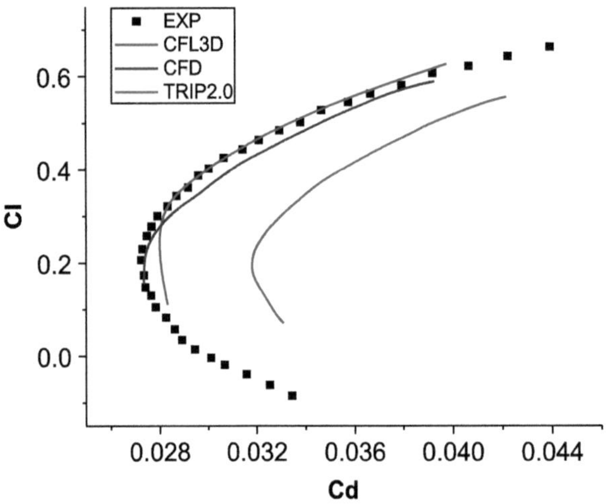

Fig. 2.10: Polar curves of Cl–Cd.

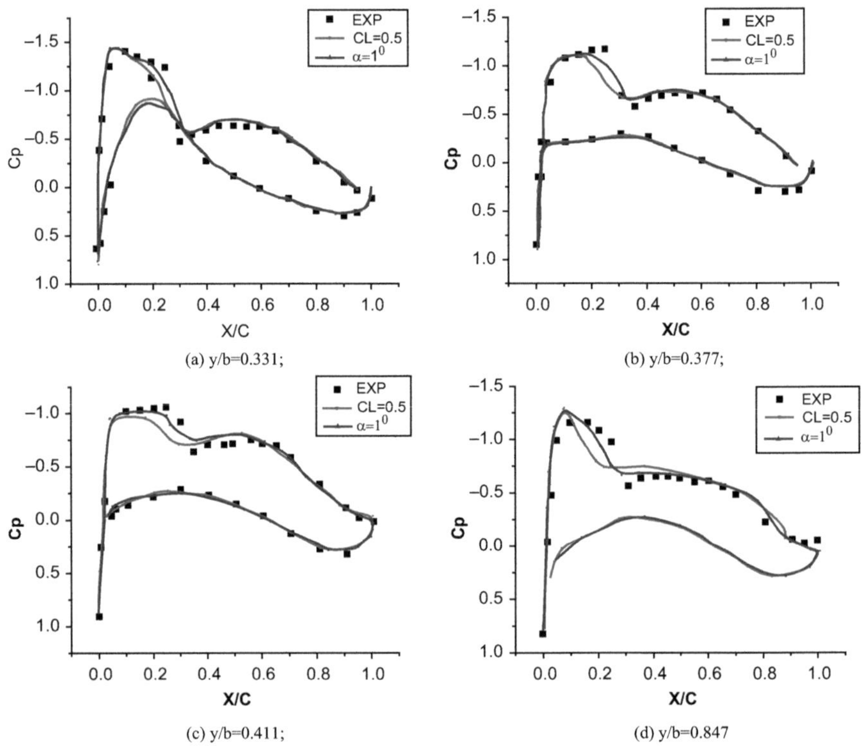

Fig. 2.11: Comparison of pressure distribution at different sections of the wing.

that of case 1 condition, which is more consistent with the experimental results. Figure 2.12 shows the pressure distribution of the three typical sections around the nacelle compared with the experiment results. It is similar to that of Fig. 2.11, that is, the results of case 2 is obviously more consistent with the experiment results than that of case 1.

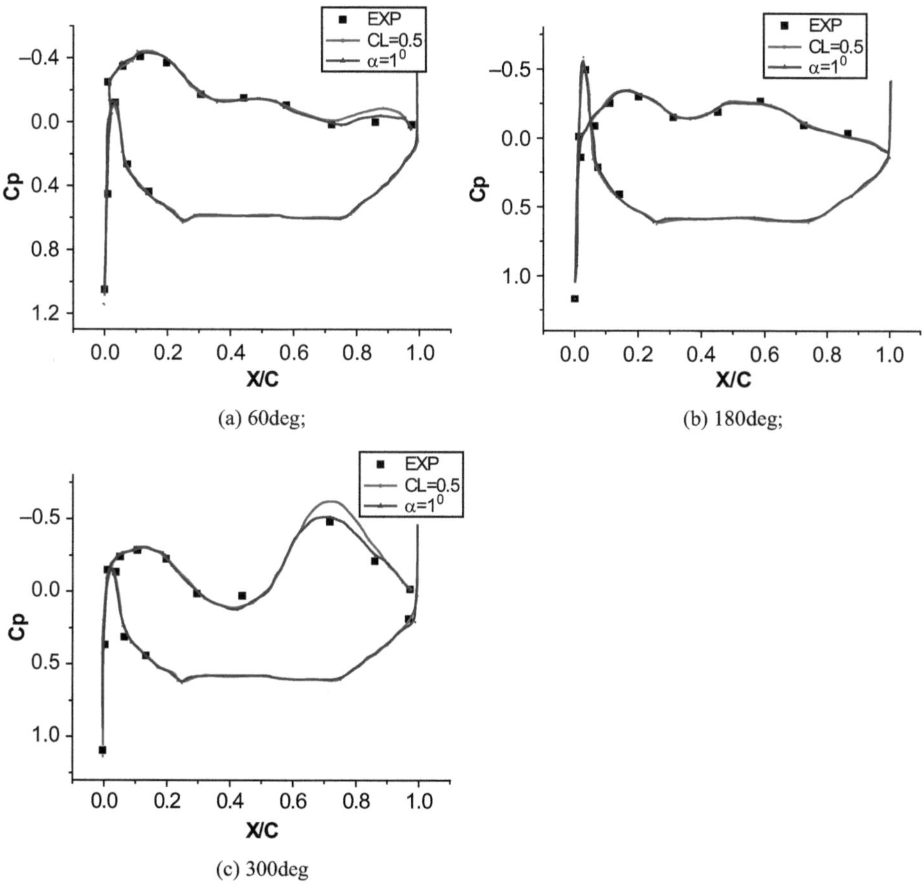

Fig. 2.12: Comparison of the pressure distribution around the nacelle.

2.9.4 Unsteady computation of AGARD CT-5

The unsteady solution technique is a key technology for aeroelastic simulation in time domain, which directly affects the accuracy of the aeroelastic numerical simulation.

AGARD CT-5 is a commonly used unsteady test case in the field of unsteady CFD technology. In this section, AGARD CT-5 is used to verify the CFD program developed by authors. Although it is a 2D case, the final effects of the verification and the assessment are the same and can be extended to 3D cases. Considering that the NACA 0012 airfoil performs pitching oscillation around the point at 1/4 chord length and the angle of attack is in motion:

$$\alpha = \alpha_0 + \Delta\alpha \times \mathbf{sin}(2kt^*),$$

where α_0 is the angle of attack at balance position, $\Delta\alpha$ is the amplitude, k is the reducing frequency and t^* is the nondimensional time. The detailed values for each parameter are defined as

$$\alpha_0 = 0.016^0, \quad \Delta\alpha = 2.51^0, \quad k = 0.814$$

The AUSMpw+ scheme is used to discrete the convective flux, the implicit LU-SGS scheme for the time-advancing method and the transfinite interpolation method for the grid deformation. Figure 2.13 shows the C-type grid used for the current AGARD CT-5 test case with the dimension of 199 × 40. Figure 2.14 shows the predicted curve of lift and moment coefficients over time, and it can be seen that the calculated results agree well with the experimental results.

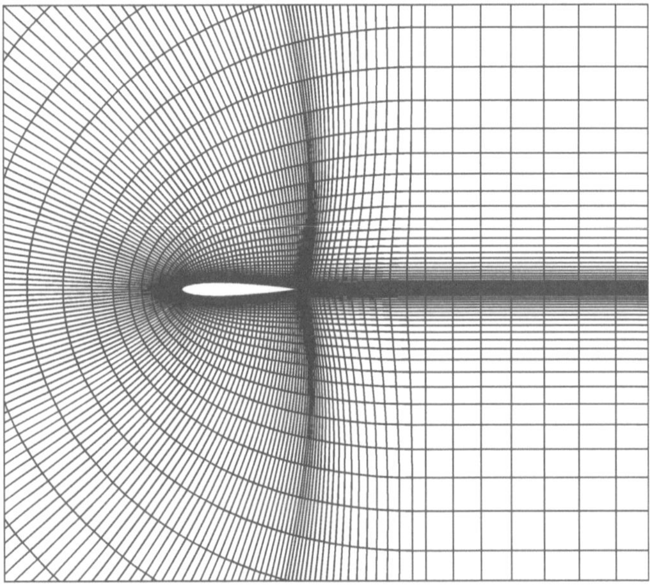

Fig. 2.13: Computational grid of AGARD CT-5.

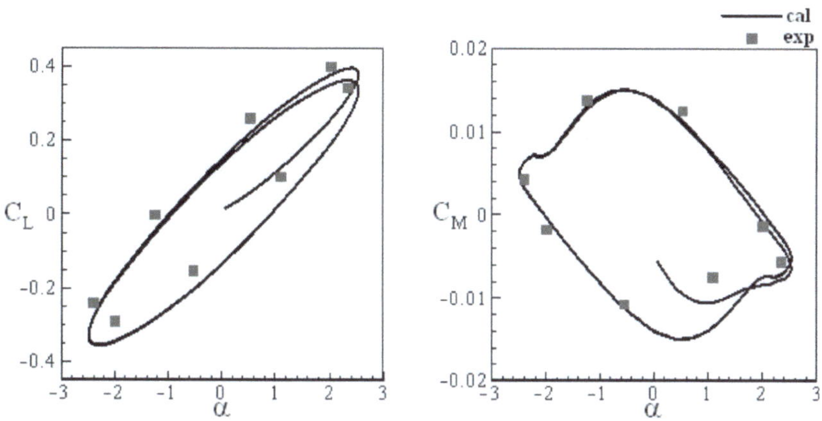

Fig. 2.14: Comparison of lift and moment coefficients of AGARD CT-5.

Figures 2.15–2.19 show the pressure coefficient distribution on the airfoil surface and the related pressure contour of flow field at different angles of attack during the oscillating of the airfoil. It can be seen that the CFD calculation results can reflect the unsteady motion of the shockwave in the flow field, and the pressure distribution on the airfoil surface at different angles of attack (i.e., different time) agrees well with the experimental results, which is similar with that of the pressure contour, and can reflect the unsteady motion of the shockwave position during oscillation.

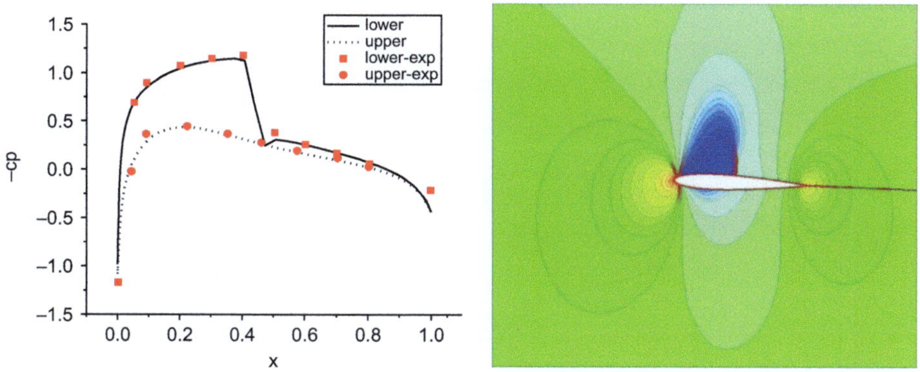

Fig. 2.15: Pressure coefficient on airfoil surface and pressure contours at AOA = 2.34°.

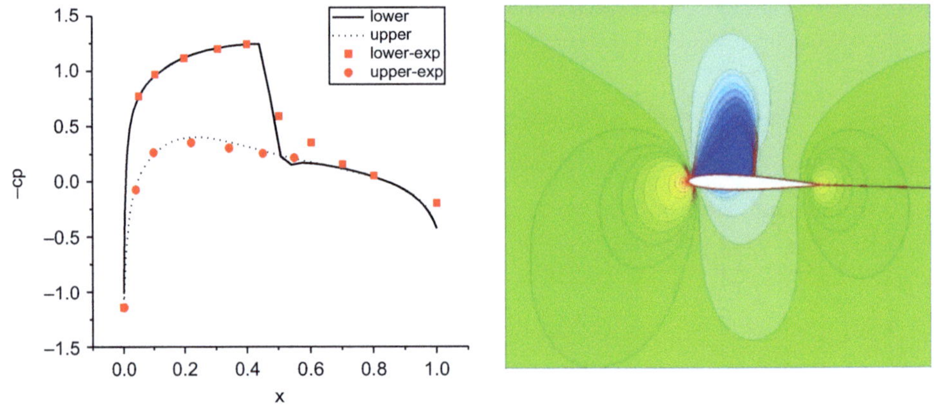

Fig. 2.16: Pressure coefficient on airfoil surface and pressure contours at AOA = 2.01°.

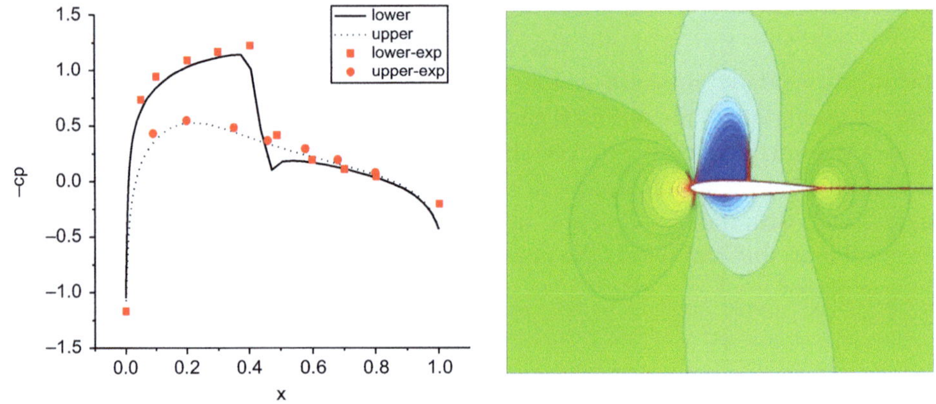

Fig. 2.17: Pressure coefficient on airfoil surface and pressure contours at AOA = 0.52°.

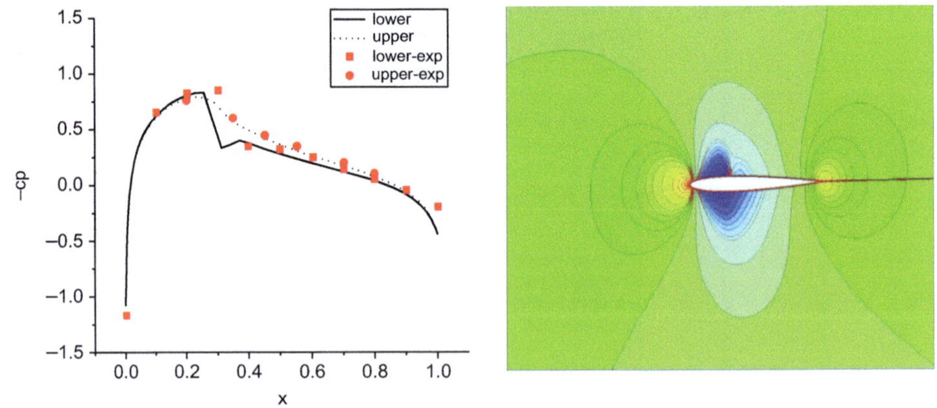

Fig. 2.18: Pressure coefficient on airfoil surface and pressure contours at AOA = –1.25°.

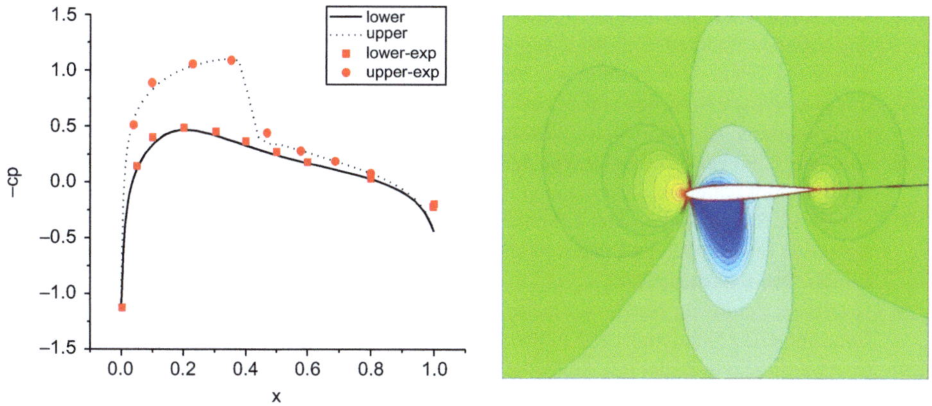

Fig. 2.19: Pressure coefficient on airfoil surface and pressure contours at AOA = −2.41°.

Bibliography

[1] Anderson JD. Computational Fluid Dynamics: The Basics with Applications. International edition: McGraw-Hill; 1995.

[2] Baldwin B, Barth T, editors. A one-equation turbulence transport model for high Reynolds number wall-bounded flows. 29th Aerospace Sciences Meeting; 1991.

[3] Baldwin B, Lomax H, editors. Thin-layer approximation and algebraic model for separated turbulent flows. 16th aerospace sciences meeting; 1978.

[4] Batina JT. Unsteady Euler airfoil solutions using unstructured dynamic meshes. AIAA Journal. 1990, 28(8), 1381–1388.

[5] Beam RM, Warming RF. An implicit finite-difference algorithm for hyperbolic systems in conservation-law form. Journal of Computational Physics. 1976, 22(1), 87–110.

[6] Blazek J. Computational fluid dynamics: principles and applications: Butterworth-Heinemann; 2015.

[7] Boussinesq J. Essai sur la théorie des eaux courantes: Impr. nationale; 1877.

[8] Boussinesq J. Théorie de l'écoulement tourbillonnant et tumultueux des liquides dans les lits rectilignes a grande section: Gauthier-Villars; 1897.

[9] Cebeci T. Analysis of Turbulent Boundary Layers. Elsevier; 2012.

[10] CHAN W, STEGER J, editors. A generalized scheme for three-dimensional hyperbolic grid generation. 10th Computational Fluid Dynamics Conference; 1991.

[11] Chang C-H, Liou M-S. A robust and accurate approach to computing compressible multiphase flow: Stratified flow model and AUSM+-up scheme. Journal of Computational Physics. 2007, 225(1), 840–873.

[12] Chen B, Cai G, editors. A new parabolized navier-stokes algorithm and its applications in some hypersonic propulsion aerodynamic problems. 42nd AIAA/ASME/SAE/ASEE Joint Propulsion Conference & Exhibit; 2006.

[13] Chi X, Li Y, Addy H, Addy G, Choo Y, Shih T, et al. editors. A Comparative Study Using CFD to Predict Iced Airfoil Aerodynamics. 43rd AIAA Aerospace Sciences Meeting and Exhibit; 2005.

[14] Deck S, Duveau P, d'Espiney P, Guillen P. Development and application of Spalart–Allmaras one equation turbulence model to three-dimensional supersonic complex configurations. Aerospace Science and Technology. 2002, 6(3), 171–183.
[15] Deese J, Agarwal R. Navier-Stokes calculations of transonic viscous flow about wing/body configurations. Journal of Aircraft. 1988, 25(12), 1106–1112.
[16] Fornasier L, Rieger H, Tremel U, Van der Weide E, editors. Time-dependent aeroelastic simulation of rapid manoeuvring aircraft. 40th AIAA Aerospace Sciences Meeting & Exhibit.
[17] Forsythe J, Squires K, Wurtzler K, Spalart P, editors. Detached-Eddy Simulation of Fighter Aircraft at High Alpha. AIAA Paper 02-0591. 40th AIAA Aerospace Sciences Meeting and Exhibit, Reno, Nevada; 2002.
[18] Forsythe J, Squires K, Wurzler K, Spalart P, editors. Prescribed Spin of the F-15E Using Detached-Eddy Simulation. 41st Aerospace Sciences Meeting and Exhibit; 2003.
[19] Georgiadis NJ, Alexander JI D, Reshotko E. Hybrid Reynolds-averaged Navier-Stokes/ large-eddy simulations of supersonic turbulent mixing. AIAA Journal. 2003, 41(2), 218–229.
[20] Godunov SK. A difference method for numerical calculation of discontinuous solutions of the equations of hydrodynamics. Matematicheskii Sbornik. 1959, 89(3), 271–306.
[21] Guillard H, Farhat C, editors. On the significance of the GCL for flow computations on moving meshes. 37th Aerospace Sciences Meeting and Exhibit; 1998.
[22] Hamba F. A hybrid RANS/LES simulation of turbulent channel flow. Theoretical and Computational Fluid Dynamics. 2003, 16(5), 387–403.
[23] Jameson A, editor. Time dependent calculations using multigrid, with applications to unsteady flows past airfoils and wings. 10th Computational Fluid Dynamics Conference; 1991.
[24] Jameson A, Schmidt W, Turkel E, editors. Numerical SOLUTIONS of the Euler Equations by finite volume methods using Runge-Kutta Time-Stepping Schemes. AIAA Paper 81-1259(1981). 14th AIAA Fluid and Plasma Dynamic Conference, Palo Alto, California.
[25] Jameson A, Yoon S. Lower-upper implicit schemes with multiple grids for the Euler equations. AIAA Journal. 1987, 25(7), 929–935.
[26] Jones W, Launder BE. The prediction of laminarization with a two-equation model of turbulence. International Journal of Heat and Mass Transfer. 1972, 15(2), 301–314.
[27] Kandula M, Buning P, editors. Implementation of LU-SGS algorithm and Roe upwinding scheme in overflow thin-layer Navier-Stokes code. Fluid Dynamics Conference; 1994.
[28] Kermani M, Plett E, editors. Modified entropy correction formula for the Roe scheme. 39th Aerospace Sciences Meeting and Exhibit; 2001.
[29] Kim KH, Kim C, Rho O-H. Methods for the accurate computations of hypersonic flows: I. AUSM PW+ scheme. Journal of Computational Physics. 2001, 174(1), 38–80.
[30] Krist SL, Biedron RT, Rumsey CL. CFL3D user's manual (version 5.0). 1998.
[31] Küchemann D. Vorlesungen uber theoretische Gasdynamik. By J. Z IEREP. Karlsruhe: G. Braun, 1963. pp. DM. 42. Journal of Fluid Mechanics. 1963, 16(1), 160.
[32] Lesoinne M, Farhat C. Geometric Conservation Laws for flow problems with moving boundaries and deformable meshes, and their impact on aeroelastic computations. Computer Methods in Applied Mechanics and Engineering. 1996, 134(1–2), 71–90.
[33] Li G, Li F, Zhou Z, Sang W, editors. Validation of a Multigrid-Based Navier-Stokes Solver for Transonic Flows. 28th AIAA Applied Aerodynamics Conference; 2010.
[34] Li J, Huang S, Jiang S, Li F, editors. Unsteady Viscous Flow Simulations by a Fully Implicit Method with Deforming Grid. 43rd AIAA Aerospace Sciences Meeting and Exhibit; 2005.
[35] Liepmann HW, Roshko A. Elements of Gasdynamics: Courier Corporation. 2001.
[36] Liou M-S, editor. On a new class of flux splittings. Thirteenth International Conference on Numerical Methods in Fluid Dynamics; 1993: Springer.

[37] Liou M-S, editor. Progress towards an improved CFD method-AUSM+. 12th Computational Fluid Dynamics Conference; 1995.
[38] Liou M-S, Steffen JC. A new flux splitting scheme. Journal of Computational Physics. 1993, 107 (1), 23–39.
[39] Liou M-S. A sequel to ausm: Ausm+. Journal of Computational Physics. 1996, 129(2), 364–382.
[40] McDonald P, editor. The computation of transonic flow through two-dimensional gas turbine cascades. ASME 1971 International Gas Turbine Conference and Products Show; 1971: American Society of Mechanical Engineers Digital Collection.
[41] Menter FR. Two-equation eddy-viscosity turbulence models for engineering applications. AIAA Journal. 1994, 32(8), 1598–1605.
[42] Müller B, editor. Simple improvements of an upwind TVD scheme for hypersonic flow. 9th Computational Fluid Dynamics Conference; 1989.
[43] Osher S, Solomon F. Upwind difference schemes for hyperbolic systems of conservation laws. Mathematics of Computation. 1982, 38(158), 339–374.
[44] Peyret R. Handbook of Computational Fluid Mechanics. Elsevier; 1996.
[45] Piller M, Nobile E, Hanratty TJ. DNS study of turbulent transport at low Prandtl numbers in a channel flow. Journal of Fluid Mechanics. 2002, 458, 419–441.
[46] Pulliam T, Steger J, editors. Recent improvements in efficiency, accuracy, and convergence for implicit approximate factorization algorithms. 23rd Aerospace Sciences Meeting; 1985.
[47] Pulliam TH, Chaussee D. A diagonal form of an implicit approximate-factorization algorithm. Journal of Computational Physics. 1981, 39(2), 347–363.
[48] Radespiel R A cell-vertex multigrid method for the Navier-Stokes equations. 1989.
[49] Rider B, Tinoco E. CFL3D Analysis of the NASA Common Research Model for the 4th Drag Prediction Workshop. June; 2009.
[50] Rider WJ. Methods for extending high-resolution schemes to non-linear systems of hyperbolic conservation laws. International Journal for Numerical Methods in Fluids. 1993, 17(10), 861–885.
[51] Roe PL. Approximate Riemann solvers, parameter vectors, and difference schemes. Journal of Computational Physics. 1981, 43(2), 357–372.
[52] Rollet-Miet P, Laurence D, Ferziger J. LES and RANS of turbulent flow in tube bundles. International Journal of Heat and Fluid Flow. 1999, 20(3), 241–254.
[53] Rumsey CL, Rivers SM, Morrison JH. Study of CFD variation on transport configurations from the second drag-prediction workshop. Computers & Fluids. 2005, 34(7), 785–816.
[54] Scott J, Niu -Y-Y, editors. Comparison of limiters in flux-split algorithms for Euler equations. 31st Aerospace Sciences Meeting; 1993.
[55] Spalart P, Allmaras S, editors. A one-equation turbulence model for aerodynamic flows. 30th aerospace sciences meeting and exhibit; 1992.
[56] Spalart PR. Strategies for turbulence modelling and simulations. International Journal of Heat and Fluid Flow. 2000, 21(3), 252–263.
[57] Steger JL, Warming R. Flux vector splitting of the inviscid gasdynamic equations with application to finite-difference methods. Journal of Computational Physics. 1981, 40(2), 263–293.
[58] Swanson R, Turkel E, editors. Artificial dissipation and central difference schemes for the Euler and Navier-Stokes equations. 8th Computational Fluid Dynamics Conference; 1987.
[59] Tamura Y, Fujii K, editors. Conservation law for moving and transformed grids. 11th Computational Fluid Dynamics Conference; 1993.
[60] Thomas PD, Lombard CK. Geometric Conservation Law and its application to flow computations on moving grids. AIAA Journal. 1979, 17(10), 1030–1037.
[61] Turkel E, Swanson R, Vatsa V, White J, editors. Multigrid for hypersonic viscous two-and three-dimensional flows. 10th Computational Fluid Dynamics Conference; 1991.

[62] Van Albada GD, Van Leer B, Roberts W. A comparative study of computational methods in cosmic gas dynamics. Upwind and High-Resolution Schemes: Springer; 1997. 95–103.
[63] Van Leer B, editor. Towards the ultimate conservative difference scheme I. The quest of monotonicity. Proceedings of the Third International Conference on Numerical Methods in Fluid Mechanics; 1973: Springer.
[64] Van Leer B. Flux-vector splitting for the Euler equation. Upwind and High-Resolution Schemes: Springer; 1997. 80–89.
[65] Van Leer B. Towards the ultimate conservative difference scheme. V. A second-order sequel to Godunov's method. Journal of Computational Physics. 1979, 32(1), 101–136.
[66] Vassberg J, Dehaan M, Rivers M, Wahls R, editors. Development of a common research model for applied CFD validation studies. 26th AIAA Applied Aerodynamics Conference; 2008.
[67] Wada Y, Liou M-S, editors. A flux splitting scheme with high-resolution and robustness for discontinuities. 32nd Aerospace Sciences Meeting and Exhibit; 1994.
[68] Wilcox DC. Reassessment of the scale-determining equation for advanced turbulence models. AIAA Journal. 1988, 26(11), 1299–1310.
[69] Yao W, Xu M, editors. Modified AUSMPW+ scheme and its application. 2008 Asia Simulation Conference-7th International Conference on System Simulation and Scientific Computing; 2008: IEEE.
[70] Yoon S, Jameson A. Lower-upper symmetric-Gauss-Seidel method for the Euler and Navier-Stokes equations. AIAA Journal. 1988, 26(9), 1025–1026.
[71] Yoon S, Kwak D. Three-dimensional incompressible Navier-Stokes solver using lower-upper symmetric-Gauss-Seidel algorithm. AIAA Journal. 1991, 29(6), 874–875.
[72] Yoon S, Kwakt D. Implicit Navier-Stokes solver for three-dimensional compressible flows. AIAA Journal. 1992, 30(11), 2653–2659.

Chapter 3
Numerical method for nonlinear structural response computation with geometric nonlinearity

3.1 Introduction

Structural mechanics mainly studies the changes of stress, strain and displacement of engineering structures under external loads. Generally, it is divided into structural statics, structural dynamics, structural stability theory, structural fracture and fatigue theory according to its research properties and objects. Structural statics mainly studies the elastoplastic deformation and stress state of engineering structures under static loading. Structural dynamics is the study of the dynamic characteristics of structural systems and their dynamic response analysis under dynamic loads. Under the dynamic load, the stress, strain and displacement of the structure are functions of time. The dynamic response is to calculate a series of solutions at all time points, and because the position of the structure changes rapidly with time, the inertial force is generated, which has an important influence on the structural response. The basic linear theory of structural analysis has matured currently, and nonlinear solving techniques are constantly developed. With these theories, many large-scale commercial softwares, such as NASTRAN and ANSYS, have been developed, which have been widely applied in many engineering fields.

In nonlinear aeroelastic research, nonlinearity in structural analysis is often one of the causes for nonlinear aeroelastic research. The structural nonlinearity mainly has three aspects: (1) because of large elastic deformation of the structure, shape change of the structural element has a nonnegligible influence on the internal force balance and the external load condition, that is, geometric nonlinearity. When the structural response exceeds the linear relationship between stress and strain, it enters the elastoplastic range, that is, material nonlinearity; (2) the discontinuity of the structure, such as the installation gap of the control surface hinge of the aircraft mechanical control system and the fittings, aging and loosening of structural components, usually manifests as the free play nonlinearity with bilinear features; (3) when there are both free play and solid friction in the system, it is characterized by hysteresis nonlinearity. The emergence of these nonlinear phenomena makes the study of the problem extremely complicated. Especially for the characteristics of large, lightweight and variable parameters of modern aircrafts, the flexible characteristic is noticeably obvious, and hence, the geometric nonlinear effect cannot be ignored in structural analysis, which has an important impact on the static and dynamic stabilities.

https://doi.org/10.1515/9783110576689-003

Geometric nonlinear analysis involves two types: one is the small strain–large displacement relationship, where the stress and strain remain in linear relationship, while the strain and displacement are nonlinearly related; the other one is large strain–large displacement relationship, where there is not only the nonlinear relationship between strain and displacement but also the nonlinear relationship between stress and strain. The large deformation of the traditional aircraft often belongs to the first type of geometric nonlinearity. Due to the nonlinear relationship between strain and displacement, the structural stiffness matrix no longer maintains a constant value, but changes with the structural displacement. The change, thus, forms a nonlinear structural equilibrium equation, and for solving the equilibrium equation, it is necessary to obtain the tangent stiffness matrix and the element internal force at each equilibrium moment.

This chapter mainly describes the nonlinear structure finite element method. First, a coordinate system describing the basic variables of elastic mechanics and nonlinear stresses and strains are defined. Then, the elastic mechanics' variational principle as the basis of the traditional finite method is introduced. By analyzing the shortcomings of the traditional variational principle method, the improved variational principle method–multivariate variational principle is introduced. The multivariate finite element method based on the multivariate variational principle treats the structural stress, strain and displacement fields as independent variables, which overcomes the problems that the stress and strain are both dependent on displacement in the traditional displacement-based finite element method. Shape functions are simultaneously assumed for stress, strain and displacement to improve the solution accuracy of each variable. This advantage makes the multivariate finite element method particularly suitable for solving nonlinear problems.

Then, for the geometric nonlinear problem of the rod-type structure and the thin-walled structure, the geometric nonlinear stiffness matrix and internal force formula are derived and developed based on the corotational (CR) theory for the two-dimensional beam element and three-dimensional shell element. Comparing with the traditional global incremental method, the two nonlinear incremental methods used for structural static analysis are discussed. Based on the energy conservation in dynamic processes, by introducing a predictor–corrector step, a dynamic solution technique with approximate energy conservation for structural geometric nonlinearity is developed. Compared with the traditional Newmark dynamic solution technology, stability and computational accuracy of the developed algorithm are significantly improved, which hence can be conveniently used for the analysis of nonlinear aeroelasticity.

3.2 Deformation and movement

3.2.1 Definition of variables and coordinate system

Consider the initial state of an object at the time of $t = 0$, as shown in Fig. 3.1. The object is represented in the domain of Ω_0 at initial moment which is called the initial configuration. In describing the motion and deformation of the object, a configuration as a reference for various equations is also required, referred to as a reference configuration. The initial configuration is generally used as a reference configuration, the meaning of which lies in that the actual motion is defined with reference to this configuration.

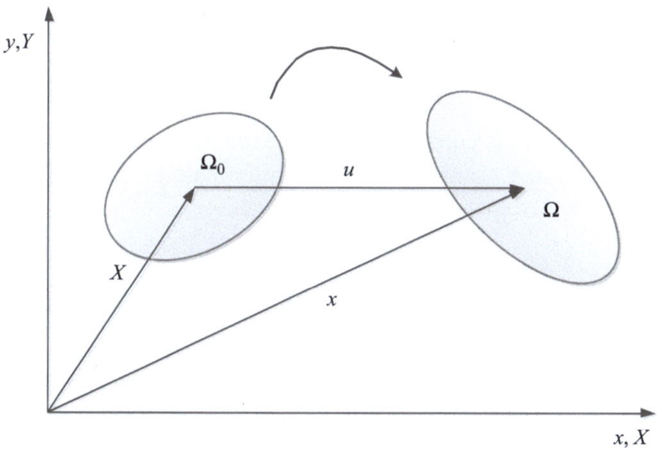

Fig. 3.1: Undeformed (initial) configuration and deformed (current) configuration of an object.

In addition, an undeformed configuration in the entire domain of Ω_0 needs to be specified. The current configuration is represented by Ω, often referred to as a deformed configuration, which can be one-, two- or three-dimensional, and correspondingly, Ω represents a line, an area or a volume. The boundary of the field Ω is represented by Γ, corresponding to the two endpoints of a line segment in the one-dimensional case, a curve in the two-dimensional case and a surface in the three-dimensional case.

The position vector of the material point in the reference configuration is represented by X

$$X = X_i e_i \tag{3.1}$$

where X_i is the component of the position vector in the reference configuration; e_i is the unit basis vector of the Cartesian coordinate system. Note that the Einstein summation convention is used here.

For a given material point, the vector variable X does not change with time, which is called material coordinate or Lagrangian coordinate for providing material identification point. Therefore, in order to track a function $f(X,t)$ at the given material point, we can simply track the function with X as a constant. The point position in the current configuration is

$$x = x_i e_i \tag{3.2}$$

where x_i is the component of position vector in the current configuration.

The motion of the body is described as

$$x = \phi(X,t), \tag{3.3}$$

where x is the location of the material point X at time t, and the component x_i gives the spatial position, which is called spatial coordinate or Eulerian coordinate. The function $\phi(X,t)$ maps the reference configuration to the current configuration at time t, which is named mapping, or a mapping from the initial configuration to the current configuration.

When the reference configuration is consistent with the initial configuration, at time $t=0$ the position vector x at any point coincides with its material coordinate, that is,

$$X = x(X,0) = \phi(X,0). \tag{3.4}$$

In this way, the mapping $\phi(X,t)$ becomes a consistent mapping.

There are two ways to describe the deformation and response of a continuum. In the first approach, the independent variables are material coordinates and time, which are described in terms of material descriptions or Lagrangian descriptions. In the second way, independent variables are spatial coordinates and time, called spatial descriptions or Eulerian descriptions. In solid mechanics, stress generally depends on deformation and its history, so one of them must be specified as a deformed configuration. Because most solids have the historical dependence, Lagrangian descriptions are commonly used in solid mechanics.

Mathematically, for the same field variable, different symbols are often used when represented by different independent variables, that is, described by Eulerian or Lagrangian. According to this convention, the function described as $f(x,t)$ in the Eulerian description will be presented by $F(X,t)$ in the Lagrangian description. The relationship between these two functions is

$$F(X,t) = f(\phi(X,t),t). \tag{3.5}$$

In the finite element method, to define functions, it is required to specify three or more independent variables to associate a symbol with a field. In this way, $f(x,t)$ is the function that describes the field f in terms of independent variables x and t, and $f(X,t)$ is a different function that describes the same field in the material coordinates.

The displacement is given by the difference between the current position and the initial position of the material point (Fig. 3.1)

$$\begin{aligned} u(X,t) &= \phi(X,t) - \phi(X,0) = \phi(X,t) - X, \\ u_i &= \phi_i(X_j,t) - X_i, \end{aligned} \tag{3.6}$$

where $u(X,t) = u_i e_i$.

Velocity refers to the change rate of the position vector at a material point, such as the derivative of X with time when X remains constant. The time derivative of a constant X is called the material time derivative or the material derivative. The material time derivative is also called the full derivative, and the velocity can be written in various forms

$$v(X,t) = \frac{\partial \phi(X,t)}{\partial t} = \frac{\partial u(X,t)}{\partial t} \equiv \dot{u}\,. \tag{3.7}$$

In eq. (3.7), because X is independent of time, the motion in the third term is replaced by the displacement u. It is also used as a normal time derivative when the variable is only a function of time.

Acceleration is the change rate of material point velocity, which is the material time derivative of the velocity and can be written as

$$a(X,t) = \frac{\partial v(X,t)}{\partial t} = \frac{\partial^2 u(X,t)}{\partial t^2} \equiv \dot{v}\,. \tag{3.8}$$

It is referred to as the material form of acceleration.

When the velocity is expressed in the form of spatial coordinates and time, such as in the Eulerian description as $v(x,t)$, the material time derivative is obtained by the chain derivation law

$$\begin{aligned} \frac{Dv_i(x,t)}{Dt} &= \frac{\partial v_i(x,t)}{\partial t} + \frac{\partial v_i(x,t)}{\partial x_j}\frac{\partial \phi_j(X,t)}{\partial t} \\ &= \frac{\partial v_i}{\partial t} + \frac{\partial v_i}{\partial x_j} v_j, \end{aligned} \tag{3.9}$$

where the second term on the right is a convection term, also called a migration term; $\frac{\partial v_i}{\partial t}$ is a spatial time derivative. Formula (3.9) can be expressed in tensor form as

$$\frac{Dv(x,t)}{Dt} = \frac{\partial v(x,t)}{\partial t} + v \cdot \nabla v = \frac{\partial v}{\partial t} + v \cdot \mathrm{grad}\, v, \tag{3.10}$$

where ∇v and *grad* v are the left gradients of the vector field, the matrix of which is

$$[\nabla v] = [\mathrm{grad}\, v] = \begin{bmatrix} v_{x,x} & v_{y,x} \\ v_{x,y} & v_{y,y} \end{bmatrix}. \tag{3.11}$$

The material time derivative of any function in terms of the spatial variable and time can be similarly obtained by the chain derivation rule. Hence, for scalar function $f(x,t)$ and tensor function $\sigma_{ij}(x,t)$, the material time derivatives are written as

$$\begin{aligned}\frac{Df}{Dt} &= \frac{\partial f}{\partial t} + v_i \frac{\partial f}{\partial x_i} \\ &= \frac{\partial f}{\partial t} + v \cdot \nabla f \\ &= \frac{\partial f}{\partial t} + v \cdot \operatorname{grad} f,\end{aligned} \tag{3.12}$$

$$\begin{aligned}\frac{D\sigma_{ij}}{Dt} &= \frac{\partial \sigma_{ij}}{\partial t} + v_k \frac{\partial \sigma_{ij}}{\partial x_k} \\ &= \frac{\partial \sigma}{\partial t} + v \cdot \nabla \sigma \\ &= \frac{\partial \sigma}{\partial t} + v \cdot \operatorname{grad} \sigma.\end{aligned} \tag{3.13}$$

In the Eulerian description, establishing a material time derivative does not require a complete description of the motion. Each instantaneous motion can also be described when the reference configuration coincides with the configuration at a fixed time t. Therefore, the configuration of the fixed time $t=\tau$ is taken as the reference configuration, and the position vector of the material point is represented by reference coordinate X^τ, which is

$$X^\tau = \phi(X, \tau). \tag{3.14}$$

The position vector at time t is represented with X^τ, and the upper reference τ is used to distinguish these reference coordinate systems from the initial reference coordinate system. Motion can be described by these reference coordinate systems as

$$x = \phi^\tau(X^\tau, t), t \geq \tau. \tag{3.15}$$

Considering the current configuration as a reference configuration, the acceleration can be obtained in another expression as

$$\begin{aligned}\frac{Dv_i}{Dt} &= \frac{\partial v_i(x,t)}{\partial t} + \frac{\partial v_i(x,t)}{\partial x_j}\frac{\partial \phi_j^\tau}{\partial t} \\ &= \frac{\partial v_i}{\partial t} + \frac{\partial v_i}{\partial x_j} v_j.\end{aligned} \tag{3.16}$$

The description of deformation and strain metric are the foundations of nonlinear finite element, and deformation gradient is an important variable of deformation characteristics. Deformation gradient tensor is then defined as

$$F_{ij} = \frac{\partial \phi_i}{\partial X_j} = \frac{\partial x_i}{\partial X_j}. \tag{3.17}$$

Mathematically, the deformation gradient F is the Jacobian matrix of motion $\phi(X,t)$, with the first indicator of F_{ij} in eq. (3.17) representing motion and the second indicator representing partial derivative.

Considering an infinitesimal line segment dX in the reference configuration, the corresponding line segment dx in the current configuration can be expressed as

$$\begin{aligned} dx &= F \cdot dX, \\ dx_i &= F_{ij} dX_j. \end{aligned} \tag{3.18}$$

The deformation gradient in a two-dimensional Cartesian coordinate system is

$$F = \begin{bmatrix} \frac{\partial x_1}{\partial X_1} & \frac{\partial x_1}{\partial X_2} \\ \frac{\partial x_2}{\partial X_1} & \frac{\partial x_2}{\partial X_2} \end{bmatrix} = \begin{bmatrix} \frac{\partial x}{\partial X} & \frac{\partial x}{\partial Y} \\ \frac{\partial y}{\partial X} & \frac{\partial y}{\partial Y} \end{bmatrix}. \tag{3.19}$$

Assume that the mapping $\phi(X,t)$ describing motion and object deformation satisfies the following conditions:
(1) The function $\phi(X,t)$ is continuously differentiable.
(2) The function $\phi(X,t)$ is one-to-one correspondence.
(3) The Jacobian determinant satisfies the condition $J > 0$.

These conditions guarantee $\phi(X,t)$ to be sufficiently smooth to satisfy the coordination that there are no gaps or overlaps in the deformed body. The second condition requires that each point on the reference configuration Ω_0 has only one corresponding point in Ω, and vice versa. This is a necessary condition for the F rule, that is, F is reversible. The third condition is to follow the mass conservation.

Rigid body rotation plays a vital role in nonlinear finite element. The motion of a rigid body includes translation $x_T(t)$ and rotation, which can be written as

$$\begin{aligned} x(X,t) &= R(t) \cdot X + x_T(t), \\ x_i(X,t) &= R_{ij}(t) X_j + x_{Ti}(t), \end{aligned} \tag{3.20}$$

where $R(t)$ is the rotation tensor, also known as the rotation matrix which is an orthogonal matrix.

3.2.2 Strain measurement

Comparing with linear elasticity, there are many different measurements of strain and strain rate in nonlinear finite elements. Only two of them are considered here:
(1) Green (Green–Lagrangian) strain E
(2) Deformation rate tensor D

Green strain tensor is defined as

$$ds^2 - dS^2 = 2dX \cdot E \cdot dX \tag{3.21}$$

or

$$dx_i dx_i - dX_i dX_i = 2dX_i E_{ij} dX_j. \tag{3.22}$$

It gives the square change the material vector length dX. Green strain measures the squared difference of a tiny line segment in the current (deformed) configuration and the reference (undeformed) configuration using the deformation gradient tensor of eq. (3.22), which can be rewritten as

$$\begin{aligned} dx \cdot dx &= (F \cdot dX) \cdot (F \cdot dX) \\ &= (FdX)^T (FdX) \\ &= dX^T F^T FdX \\ &= dX \cdot \left(F^T \cdot F\right) \cdot dX. \end{aligned} \tag{3.23}$$

Write eq. (3.23) as the indicator mark form

$$\begin{aligned} dx \cdot dx &= dx_i dx_i \\ &= F_{ij} dX_j F_{ik} dX_k \\ &= dX_j F_{ji}^T F_{ik} dX_k. \end{aligned} \tag{3.24}$$

Combining eq. (3.24) with eq. (3.19), we can obtain

$$dX \cdot F^T \cdot F \cdot dX - dX \cdot I \cdot dX - dX \cdot 2E \cdot dX = 0. \tag{3.25}$$

Then, it can be simplified to

$$dX \cdot (F^T \cdot F - I - 2E) \cdot dX = 0. \tag{3.26}$$

Since eq. (3.26) is applicable for any dX, there is

$$E = \frac{1}{2}(F^T \cdot F - I) \tag{3.27}$$

$$E_{ij}=\frac{1}{2}(F_{ik}^{T}\cdot F_{kj}-\delta_{ij}). \tag{3.28}$$

Green strain tensor can also be expressed as a form of displacement gradient

$$E=\frac{1}{2}((\nabla_0 u)^T+\nabla_0 u+\nabla_0 u(\nabla_0 u)^T). \tag{3.29}$$

The deformation rate D also becomes the velocity strain, which is a measure of the deformation rate. The velocity gradient is first defined to establish the expression of the deformation rate

$$\begin{aligned} L &= \frac{\partial V}{\partial x}=(\nabla v)^T=(\mathrm{grad}\, v)^T, \\ dv &= L\cdot dx. \end{aligned} \tag{3.30}$$

The velocity gradient tensor can be decomposed into a symmetrical part and a symmetrical part

$$L=\frac{1}{2}(L+L^T)+\frac{1}{2}(L-L^T). \tag{3.31}$$

This is a standard decomposition of a second-order tensor or square matrix. The deformation rate tensor D is defined as the symmetrical part of L, and the rotational tensor W is defined as the symmetrical part. According to this definition, we can obtain

$$\begin{aligned} L &= (\nabla v)^T = D+W, \\ D &= \frac{1}{2}(L+L^T), \\ W &= \frac{1}{2}(L-L^T). \end{aligned} \tag{3.32}$$

The deformation rate measures the change rate of the length square of a small material segment:

$$\frac{\partial}{\partial t}(ds^2)=\frac{\partial}{\partial t}(dx(X,t)\cdot dx(X,t))=2dx\cdot D\cdot dx. \tag{3.33}$$

The deformation rate can be associated with the Green strain tensor. First, the material gradient of the velocity field is obtained by

$$L=\frac{\partial v}{\partial x}=\frac{\partial v}{\partial X}\cdot\frac{\partial X}{\partial x}. \tag{3.34}$$

Then, take the time derivative of the deformation gradient

$$\dot{F}=\frac{\partial}{\partial t}\left(\frac{\partial \phi(X,t)}{\partial X}\right)=\frac{\partial v}{\partial X}. \tag{3.35}$$

According to the above formula, the following relationship can be obtained:

$$L = \dot{F} \cdot F^{-1}, \tag{3.36}$$

$$D = \frac{1}{2}(\dot{F} \cdot F^{-1} + F^{-T} \cdot \dot{F}^{T}), \tag{3.37}$$

$$\dot{E} = \frac{1}{2}\frac{D}{Dt}(F^{T} \cdot F - I) = \frac{1}{2}(F^{T} \cdot \dot{F} + \dot{F}^{T} \cdot F). \tag{3.38}$$

3.2.3 Stress measurement

In nonlinear problems, various stress measurements can be defined, but only three stress metrics are considered here:

1) Cauchy stress σ
2) Nominal stress tensor P
3) The second type of Piola–Kirchhoff (PK2) stress tensor S

Stress is defined by Cauchy's theorem (Fig. 3.2):

$$n \cdot \sigma d\Gamma = t d\Gamma. \tag{3.39}$$

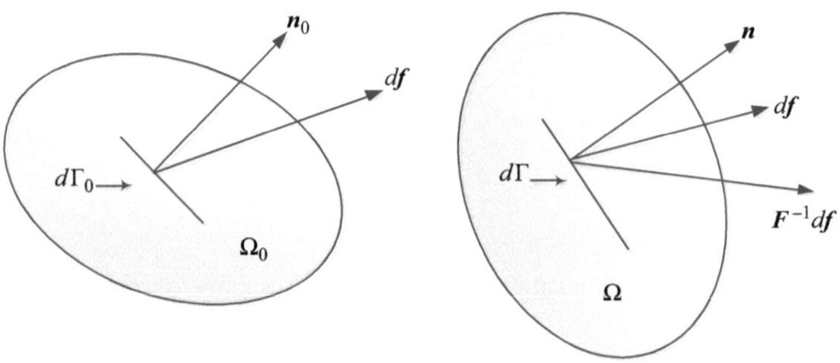

Fig. 3.2: Definition of stress metric: (a) Reference configuration and (b) current configuration.

Nominal stress is defined as

$$n_0 \cdot P d\Gamma_0 = df = t_0 d\Gamma_0. \tag{3.40}$$

The second type of Piola–Kirchhoff stress is defined as

$$\begin{aligned} n_0 \cdot S d\Gamma_0 &= F^{-1} \cdot df = F^{-1} \cdot t_0 d\Gamma_0, \\ df &= t d\Gamma = t_0 d\Gamma_0. \end{aligned} \tag{3.41}$$

Different stress tensors are interrelated, and each stress tensor has the following conversion relationships:

$$P = JF^{-1} \cdot \sigma, \tag{3.42}$$

$$P = S \cdot F^T, \tag{3.43}$$

$$\sigma = J^{-1} F \cdot S \cdot F^T, \tag{3.44}$$

$$S = JF^{-1} \cdot \sigma \cdot F^{-T}. \tag{3.45}$$

3.2.4 Lagrangian conservation equation

Using the Lagrangian measurements of stress and strain, the conservation equations are directly established in the reference configuration. These equations are called Lagrangian descriptions or Lagrangian formats. In the complete Lagrangian format, the independent variables are Lagrangian (material) coordinates X and time t, and the main related variables are initial density $\rho_0(X,t)$, displacement $u(X,t)$ and Lagrangian measurements for stress and strain. The applied loads are defined in the reference configuration, such as the surface force defined as t_0, and the volume force expressed as b.

1 Conservation of momentum

In the Lagrangian description, the linear momentum of an object is given as the integral form over the entire reference configuration

$$p(t) = \int_{\Omega_0} \rho_0 v(X,t) d\Omega_0. \tag{3.46}$$

The total force on the object is obtained by integrating the volume force over the entire reference domain and by integrating the surface force over the entire reference boundary

$$f(t) = \int_{\Omega_0} \rho_0 b(X,t) d\Omega_0 + \int_{\Gamma_0} t_0(X,t) d\Gamma_0. \tag{3.47}$$

Newton's second law gives

$$\frac{dp}{dt} = f. \tag{3.48}$$

Substituting eqs. (3.46) and (3.47) into the above equation, we obtain

$$\frac{d}{dt} \int_{\Omega_0} \rho_0 v d\Omega_0 = \int_{\Omega_0} \rho_0 b d\Omega_0 + \int_{\Gamma_0} t_0 d\Gamma_0. \tag{3.49}$$

Applying Cauchy's law and Gauss' theorem, we obtain

$$\int_{\Gamma_0} t_0 d\Gamma_0 = \int_{\Gamma_0} n \cdot p d\Gamma_0 = \int_{\Omega_0} \nabla_0 \cdot p d\Omega_0. \tag{3.50}$$

Substituting eq. (3.50) into eq. (3.49) results in

$$\int_{\Omega_0} \left(\rho_0 \frac{\partial v(X,t)}{\partial t} - \rho_0 b - \nabla_0 \cdot p \right) d\Omega_0 = 0. \tag{3.51}$$

Due to the arbitrariness of Ω_0, there will be

$$\rho_0 \frac{\partial v(X,t)}{\partial t} = \rho_0 b + \nabla_0 \cdot p. \tag{3.52}$$

The above equation is called the Lagrangian form of the momentum equation.

2 Conservation of energy

The change rate of total energy in the motion of an object is

$$\begin{aligned} p^{\text{tot}} &= p^{\text{int}} + p^{\text{kin}}, \\ p^{\text{int}} &= \frac{D}{Dt} \int_{\Omega} \rho w^{\text{int}} d\Omega, \\ p^{\text{kin}} &= \frac{D}{Dt} \int_{\Omega} \frac{1}{2} \rho v \cdot v d\Omega, \end{aligned} \tag{3.53}$$

where p^{int} is the change rate of internal energy, p^{kin} represents the change rate of kinetic energy, w^{int} is the internal energy per unit mass, and the work done by the volume force in the domain and by the surface force on the surface is

$$p^{\text{ext}} = \int_{\Omega} v \cdot \rho b d\Omega + \int_{\Gamma} v \cdot t d\Gamma. \tag{3.54}$$

The power provided by the heat source s and heat flux q is

$$p^{\text{heat}} = \int_{\Omega} \rho s d\Omega - \int_{\Gamma} n \cdot q d\Gamma \tag{3.55}$$

The expression of energy conservation is

$$p^{\text{tot}} = p^{\text{ext}} + p^{\text{heat}} \tag{3.56}$$

Substituting eqs. (3.53)–(3.55) into eq. (3.56), we obtain

$$\begin{aligned}&\frac{d}{dt}\int_{\Omega_0}\left(\rho w^{\text{int}}+\frac{1}{2}\rho_0 v\cdot v\right)d\Omega_0\\&=\int_{\Omega_0} v\cdot\rho_0 b d\Omega_0+\int_{\Gamma_0} v\cdot t_0 d\Gamma_0+\int_{\Omega_0}\rho_0 s d\Omega_0-\int_{\Gamma_0} n_0\cdot\tilde{\boldsymbol{q}}d\Gamma_0\end{aligned}\tag{3.57}$$

where $\tilde{\boldsymbol{q}}$ is defined as heat flux per unit area in the reference configuration.

Applying the Gauss theorem to transform the above equation into the following form:

$$\int_{\Omega_0}\left[\rho_0\frac{\partial w^{\text{int}}}{\partial t}-\frac{\partial F^T}{\partial t}:P+\nabla_0\cdot\tilde{\boldsymbol{q}}-\rho_0 s+\left(\rho_0\frac{\partial v(X,t)}{\partial t}-\rho_0 b-\nabla_0\cdot p\right)\cdot v\right]d\Omega_0=0.\tag{3.58}$$

Using the momentum conservation for eq. (3.58), there is

$$\rho_0\dot{w}^{\text{int}}=\dot{F}^T:P-\nabla_0\cdot\tilde{\boldsymbol{q}}+\rho_0 s.\tag{3.59}$$

Equation (3.59) shows that the material time derivatives of the nominal stress and the deformation gradient are coupled in power. Further derivation shows that the rate of Green strain tensor is also coupled with PK2 in terms of power.

3.3 Variational principle

3.3.1 Classical variational principle of elasticity theory

For a given differential equation boundary value problem, the variational solving method can not only provide an equivalent and convenient method, but the more important point is that when the exact solution for the given problem is unavailable, the variational method can provide an approximate solution, which is important for some engineering problems. The problem of transforming the boundary value problem of the differential equation of elasticity theory into the extreme value of the functional is called the variational principle of elastic theory. The variational principle is not only a powerful tool for finding solutions to elastic mechanics problems, but also a mathematical basis for the establishment of various types of finite element methods.

Three types of variables are required to solve the elastic theory problem: stress, strain and displacement. However, in practical problems, it is often necessary to solve one part of the variables first, and then the rest. In this way, one or two of the three variables are first selected as independent variables. Based on the first choice of variables, the classical variational principle can be divided into the minimum potential energy principle and the minimum complementary energy principle.

1 Principle of minimum potential energy

By the principle of virtual displacement, when the elastic body is in equilibrium, any allowable displacement δu_i and strain $\delta\varepsilon_{ij}$ that satisfies the continuous condition should satisfy the following formula:

$$\int_V \sigma_{ij}\delta\varepsilon_{ij}dV = \int_V \bar{F}_i\delta u_i dV + \int_{S\sigma} \bar{T}_i\delta u_i dS. \tag{3.60}$$

At the same time, based on the relationship between stress and strain, we can know

$$\sigma_{ij}\delta\varepsilon_{ij} = \frac{\partial A}{\partial\varepsilon_{ij}}\delta\varepsilon_{ij} = \delta A(\varepsilon), \tag{3.61}$$

where $A(\varepsilon)$ represents the strain energy density, so the principle of virtual work can be written as

$$\delta\int_V A(\varepsilon)dV - \int_V \bar{F}_i\delta u_i dV - \int_{S\sigma} \bar{T}_i\delta u_i dS = 0. \tag{3.62}$$

When the volume force and surface force can be both derived by potential functions $\phi(u)$ and $\psi(u)$ as

$$\begin{aligned} -\delta\phi &= \bar{F}_i\delta u_i, \\ -\delta\psi &= \bar{T}_i\delta u_i. \end{aligned} \tag{3.63}$$

The principle of virtual displacement becomes

$$\begin{aligned} &\delta\pi = 0, \\ &\pi = \int_V (A + \phi)dV + \int_{S\sigma} \psi dS. \end{aligned} \tag{3.64}$$

For the elastic theory problem with small displacement deformation, it is assumed that the volume force $\bar{F}$, surface force $\bar{T}$ and surface displacement $\bar{\boldsymbol{u}}$ all keep constant for variation, and for such external force, the potential function is

$$\begin{aligned} -\phi &= \bar{F}_i u_i, \\ -\psi &= \bar{T}_i u_i. \end{aligned} \tag{3.65}$$

Then, the principle of virtual displacement is equivalent to

$$\begin{aligned} &\delta\pi = 0, \\ &\pi = \int_V (A - \bar{F}_i u_i)dV - \int_{S\sigma} \bar{T}_i u_i dS. \end{aligned} \tag{3.66}$$

Equation (3.66) is the principle of minimum potential energy.

In all allowable strains ε_{ij} and allowable displacements u_i that are sufficiently smooth and satisfy the strain–displacement equation and the known boundary conditions of the displacement, the actual ε_{ij} and u_i must make the total potential energy of the elastic body extremely small. The variational function of the principle of minimum potential energy is

$$\pi_p = \int_V \left[A(\varepsilon) - \bar{F}_i u_i\right] dV - \int_{S_\sigma} \bar{T}_i u_i dS = \min. \tag{3.67}$$

The variational constraints are

$$\varepsilon_{ij} = \frac{1}{2}(u_{i,j} + u_{j,i}) \quad (in\ the\ V),$$

$$u_i = \bar{u}_i \quad (on\ the\ S_\sigma).$$

For the variational principle, the difference between the true displacement and the possible displacement of the deformation lies in that the true displacement satisfies the equilibrium condition, so the principle of the minimum potential energy can be regarded as the variational expression of the equilibrium condition. In addition, in this variational principle, the stress does not participate in the variational operation, and the stress–strain relationship is used only when the stress needs to be calculated finally. The principle of minimum potential energy is the basis of the Rayleigh–Ritz approximation. The widely used hypothetical displacement finite element is based on the discretized principle of minimum potential energy.

2 Principle of minimum complementary energy

It can be known from the principle of virtual stress that for the elastic body in equilibrium, all allowable stresses $\delta\sigma_{ij}$ and allowable surface forces δT_i should satisfy

$$\int_V \varepsilon_{ij}\delta\sigma_{ij} dV - \int_{S_u} \bar{u}_i \delta T_i dS = 0. \tag{3.68}$$

It can be obtained from the stress–strain relationship

$$\varepsilon_{ij}\delta\sigma_{ij} = \frac{\partial B}{\partial \sigma_{ij}}\delta\sigma_{ij} = \delta B(\sigma), \tag{3.69}$$

where $B(\sigma)$ is strain residual energy, and thus the principle of virtual stress can be written as

$$\begin{aligned} &\delta\pi = 0, \\ &\pi = \int_V B(\sigma) dV - \int_{S_u} \bar{u}_i \delta T_i dS. \end{aligned} \tag{3.70}$$

In all allowable stresses σ that are sufficiently smooth and satisfy the equilibrium equation as well as the boundary conditions of known external force, the true stress σ must minimize the total complementary energy of the elastic system. The functional of the principle of minimum complementary energy is

$$\pi_C = \int_V B(\sigma)dV - \int_{S_u} \bar{u}_i \sigma_{ij} \upsilon_j dS = \min. \tag{3.71}$$

The variational constraints are

$$\sigma_{ij,j} + \bar{F}_i = 0 \quad (in\ the\ V),$$

$$\sigma_{ij}\upsilon_j - \bar{T}_i = 0 \quad (on\ the\ S_\sigma).$$

This variational principle shows that the difference between the true stress and the possible static stress is that the true stress satisfies the continuous deformation condition. Therefore, the principle of minimum complementary energy can be considered as a variational expression of continuous conditions, which is a necessary and sufficient condition for solving the displacement from the strain–displacement equation and the boundary conditions of known displacement. The principle of complementary energy is only related to stress σ. The Euler equations obtained by variation and the natural boundary conditions also involve stress and displacement only. Strain ε does not participate in variation, so the stress–strain relationship is not required if strain is not needed. The stress–strain relationship just as an additional condition for solving strain does not participate in this principle, so the stress–strain relationship is the nonvariational constraint of this variational principle. Applying this variational principle and starting to select the allowable stress σ, it is sometimes difficult to satisfy both the known boundary conditions of the external force and the equilibrium equation. For this reason, the stress function expression of the stress component can be introduced to automatically satisfy the equilibrium equation. The principle of complementary energy provides another effective way for the approximate analysis of elastic theory. When the solution of stress σ is obtained by using this principle, there is no difficulty to utilize the stress–strain relationship to solve strain ε if it is required. However, it will be difficult and in fact it is unfeasible to solve the displacement u. Because the σ is an approximate value and the resulting ε is approximate too, which generally cannot make the strain coordination equation strictly satisfied. Therefore, to obtain the displacement u, some other methods such as unit virtual load are needed.

For the static elastic problem with small displacement deformation, the principles of minimum potential energy and minimum complementary energy derived above are collectively referred to as the classical variational principle of elastic theory. The principle of minimum potential energy is to find the true displacement field from all the allowed displacement fields, and the principle of minimum complementary energy is to find the true stress field from all the allowed stress fields. They both transform the problem of solving small displacement deformation elasticity into solving the minimum value of the function. In these two variational principles, the variables participating in the variation are nine, and the variables not participating in the variation are six, and the stress–strain relationship is their nonvariational constraint. The nonvariational constraint equation is a supplement to the variational principle, when solving the variables not included in the function, the nonvariational constraint or other methods can be used. If the two variational principles are used to approximate the solution, the constraints before the variation must be strictly satisfied, thus the Euler equations after the variation and the natural boundary conditions are automatically approximated. The functions of the two variational principles both do not contain all three kinds of field variables for elasticity problems. Therefore, they both have shortcomings, that is, the comparison object has certain restrictions: π_p only compares the possible u and ε allowed by deformations while π_C only compares the possible stress σ allowed by static force. The advantages of the elastic theory variational principle are: (1) the variational functional usually has a clear physical meaning and remains invariant in the coordinate transformation; (2) the variational principle converts a given problem into a much easier equivalent problem; (3) the variational principle can sometimes give the upper or lower boundary of the exact solution of the problem; and (4) the variational principle often provides an approximate solution when the elastic theory problem cannot find an exact solution.

3.3.2 Generalized variational principle of elasticity theory

For the classical variational principle of the elastic theory mentioned above, the self-variable function needs to satisfy different constraints in the variation, hence, the variational principle with new functions will be further discussed below. Applying the classical variational principle discussed in Section 3.3.1, it is often easy to find the approximate solution for some boundary value problems of differential equations. However, they all find the minimum value of total potential energy or total complementary energy under certain constraints. In some cases, such as complex boundary conditions, it is difficult to satisfy the constraints of the two variational principles in advance. Therefore, it is natural to think whether some processing methods can be used to remove the constraints of the functional variational principle. The functional variational principle of this unconstrained constraint is called the generalized variational principle in China, and the modified variational principle or extended variational principle in the United States, which by Zienkiewiez are called mixed variational principles. These are the brand-new words used in the theory of elasticity in the twentieth century.

Generally speaking, the so-called generalized variational principle is to use the Lagrangian multiplier method or other methods to remove the constraints of the original basic variational principle, thereby establishing a variational principle with no constraint or less constraint. The generalized variational principle is the variational principle obtained by removing some or all of the constraints of the basic variational principle. Generalization is relative to the basic variational principle and the degree of generalization has its relativity. If a generalized variational principle completely removes the constraint of the corresponding basic variational principle, it is a complete generalized variational principle or an unconstrained variational principle. If only some of the constraints in the basic variational principle are removed, it is called the incomplete generalized variational principle. As early as 1759, Lagrange proposed a method which currently is called the Lagrangian multiplier method to solve the extreme value problem of the function with constraints. Courant and Hilert mathematically explained the use of the Lagrange multiplier method to remove the constraints in the variational method. However, the introduction of these methods into the mechanics and removing the various constraints of the original classical variational principle in the elastic theory step by step, so as to systematically establish the incomplete generalized variational principle and the complete variational principle, is just the achievements in the past 50 years. But so far, people still have an in-depth discussion on the generalized variational principle, and there are still some incomprehensible points for the Lagrange multiplier method, which will promote the further development of the generalized variational principle.

1 Hellinger–Reissner variational principle

The Hellinger–Reissner variational principle can be established by using the Lagrangian multiplier method to remove the variational constraint condition of the principle of minimum complementary energy, that is, the equilibrium equation and the boundary condition of the known external force.

The variational constraints of the principle of minimum complementary energy are as follows:

Equilibrium equation

$$\sigma_{ij,j} + \bar{F}_i = 0 \ (in\ the\ V).$$

Boundary condition of known external force

$$\sigma_{ij} v_j - \bar{T}_i = 0 \ \ (on\ the\ S_\sigma).$$

The function is

$$\pi_C(\sigma_{ij}) = \int_V B(\sigma) dV - \int_{S_u} \sigma_{ij} v_j \bar{u}_i dS = \min.$$

The Euler equation is given after the variation: the deformation coordination condition expressed by the stress

$$\frac{\partial B}{\partial \sigma_{ij}} = \frac{1}{2}(u_{i,j} + u_{j,i}) \text{ (in the } V).$$

Natural boundary conditions: boundary condition of known displacement known

$$u_i = \bar{u}_i \text{ (on the } S_\sigma).$$

Stress–strain relationship is the nonvariational constraint condition of this variational principle

$$\frac{\partial B}{\partial \sigma_{ij}} = \varepsilon_{ij}. \tag{3.72}$$

Substituting the stress–strain relationship into the deformation coordination equation, the strain–displacement equation is obtained as

$$\varepsilon_{ij} = \frac{1}{2}(u_{i,j} + u_{j,i}). \tag{3.73}$$

Now, introduce two types of Lagrangian multipliers $\lambda_i(x_i)$ and $\eta_i(x_i)$ $(i = 1, 2, 3)$ to remove the two types of constraints of π_C, thus forming a new function π^*

$$\begin{aligned}\pi^*(\sigma_{ij}, \lambda_i, \eta_i) = & \int_V [B(\sigma) + \lambda_i(\sigma_{ij,j} + \bar{F}_i)] dV \\ & - \int_{S_u} \sigma_{ij} v_j \bar{u}_i dS \int + \int_{S_\sigma} \eta_i(\sigma_{ij} v_j - \bar{T}_i) dS.\end{aligned} \tag{3.74}$$

Note that $\lambda_i(x_i)$ and $\eta_i(x_i)$ are also the independent variables of the new function π^*, so π^* has three types of independent variables: σ_{ij}, $\lambda_i(x_i)$ and $\eta_i(x_i)$.

Let the variation of eq. (3.74) be zero, and we get

$$\begin{aligned}\delta\pi^* = & \int_V \left[\frac{\partial B(\sigma)}{\partial \sigma_{ij}} \delta\sigma_{ij} + \lambda_i \delta\sigma_{ij,j} + (\sigma_{ij,j} + \bar{F}_i)\delta\lambda_i\right] dV \\ & - \int_{S_u} \bar{u}_i \delta(\sigma_{ij} v_j) dS + \int_{S_\sigma} [\eta_i \delta(\sigma_{ij} v_j) + (\sigma_{ij} v_j - \bar{T}_i)\delta\eta_i] dS \\ = & 0,\end{aligned} \tag{3.75}$$

where

$$\int_V \lambda_i \delta\sigma_{ij,j} dV = \int_{S_u + S_\sigma} \lambda_i \delta(\sigma_{ij} v_j) dS - \int_V \lambda_{i,j} \delta\sigma_{ij} dV. \tag{3.76}$$

Substituting eq. (3.76) into eq. (3.75) and using dummy variable replacement, we obtain

$$\begin{aligned}\delta\pi^{*} = &\int_V \left\{\left[\frac{\partial B}{\partial\sigma_{ij}} - \frac{1}{2}(\lambda_{i,j}+\lambda_{j,i})\right]\delta\sigma_{ij} + (\sigma_{ij,j}+\bar{F}_i)\delta\lambda_i\right\}dV \\ &+ \int_{S_u}(\lambda_i - \bar{u}_i)\delta(\sigma_{ij}v_j)dS \\ &+ \int_{S_\sigma}\left[(\eta_i+\lambda_i)\delta(\sigma_{ij}v_j) + (\sigma_{ij}v_j - \bar{T}_i)\delta\eta_i\right]dS \\ = &\, 0.\end{aligned} \tag{3.77}$$

Since $\delta\sigma_{ij}$ and $\delta\lambda_i$ in the V, $\delta(\sigma_{ij}v_j)$ and $\delta\eta_i$ on the S_σ and $\delta(\sigma_{ij}v_j)$ on the S_u are all independent variation, so as to make the variation zero, there must be:

Euler equation

$$\begin{cases}\frac{\partial B}{\partial\sigma_{ij}} - \frac{1}{2}(\lambda_{i,j}+\lambda_{j,i}) = 0 \\ \sigma_{ij,j} + \bar{F}_i = 0\end{cases} \quad \text{(in the } V\text{)}.$$

Natural boundary condition

$$\begin{cases}\lambda_i - \bar{u}_i = 0 & \text{(on the } S_u\text{)} \\ \eta_i + \lambda_i = 0 & \text{(on the } S_\sigma\text{)}. \\ \sigma_{ij}v_j - \bar{T}_i = 0 & \end{cases}$$

By identifying the Lagrange multiplier, we can get

$$\begin{cases}\lambda_i = u_i \text{ (in the } V\text{)} \\ \eta_i = -u_i \text{ (on the } S_\sigma\text{)}.\end{cases}$$

Substituting the identified Lagrangian multiplier into the new functional π^* gives the Hellinger–Reissner variational principle

$$\begin{aligned}\pi_{HR}(\sigma, u) = &\int_V \left[B(\sigma) + (\sigma_{ij,j}+\bar{F}_i)u_i\right]dV \\ &- \int_{S_u}\sigma_{ij}v_j\bar{u}_i dS - \int_{S_\sigma}(\sigma_{ij}v_j - \bar{T}_i)u_i dS \\ = &\text{ stagnation value.}\end{aligned} \tag{3.78}$$

With regard to this generalized variational principle, it should be noted that since the Lagrange multiplier is introduced to remove the constraint of the equilibrium equation, the six stress components are all independent functions. The reason why π_{HR} is more popular than π_C lies in that it is not necessary for u to satisfy the displacement boundary condition and the external force known boundary condition in advance, and the displacement is no longer limited to the possible displacement. Also, it is not necessary for σ to satisfy the equilibrium equation and the external force known boundary condition in advance, that is, the stress is no longer limited to the stress allowed by possible static force; σ and u do not need satisfy any relationship in advance, and they can be generalized functions, with some discontinuities. It should be noted that the Hellinger–Reissner variational principle derived by the principle of minimum complementary energy has not introduced the Lagrangian multiplier to eliminate the stress–strain relationship. Therefore, the stress–strain relationship is still the nonvariational constraint of the Hellinger–Reissner variational principle.

2 Hu–Washizu variational principle

The generalized variational principle function was first proposed by Hu Haichang in 1954, who obtained the function using the trial algorithm, and later, Ichiro used the Lagrange multiplier method to obtain the same function. First, two Lagrangian multipliers λ_{ij} and β_i are introduced for eliminating two sets of constraints of the minimum potential energy: the strain–displacement equation in the domain and the known displacement conditions on the boundary, and then the following new functional π^* can be obtained:

$$\begin{aligned}\pi^*(u,\varepsilon,\lambda,\beta) = &\int_V \left[A(\varepsilon) + \lambda_{ij}(\varepsilon_{ij} - \frac{1}{2}u_{i,j} - \frac{1}{2}u_{j,i}) - \bar{F}_i u_i\right] dV \\ &+ \int_{S_u} \beta_i(u - \bar{u}_i) dS - \int_{S_\sigma} \bar{T}_i u_i dS.\end{aligned} \tag{3.79}$$

By the variational stagnation condition of π^*, there is

$$\begin{aligned}\delta\pi^* = &\int_V \left[\left(\frac{\partial A}{\partial \varepsilon_{ij}} + \lambda_{ij}\right)\delta\varepsilon_{ij} + \left(\varepsilon_{ij} - \frac{1}{2}u_{i,j} - \frac{1}{2}u_{j,i}\right)\delta\lambda_{ij} + (\lambda_{ij,j} - \bar{F}_i)\delta u_i\right] dV \\ &+ \int_{S_u} \left[(u - \bar{u}_i)\delta\beta_i + (\beta_i - \lambda_{ij}v_j)\delta u_i\right] dS \\ &- \int_{S_\sigma} (\lambda_{ij}v_j + \bar{T}_i)\delta u_i dS \\ = &\,0.\end{aligned} \tag{3.80}$$

Since $\delta\varepsilon_{ij}$, $\delta\lambda_{ij}$ and δu_i in the V, δu_i and $\delta\beta_i$ on the S_u and δu_i on the S_σ are all independent variables, the resulting Euler equation is

$$\begin{cases} \frac{\partial A}{\partial \varepsilon_{ij}} + \lambda_{ij} = 0 \\ \varepsilon_{ij} = \frac{1}{2}(u_{i,j} + u_{j,i}) \quad (\text{in the } V). \\ \lambda_{ij,j} - \bar{F} = 0 \end{cases}$$

Natural boundary condition

$$\begin{cases} u - \bar{u}_i = 0 \quad (\text{on the } S_u) \\ \beta_i - \lambda_{ij} v_j = 0 \\ \lambda_{ij} v_j + \bar{T}_i = 0 \quad (\text{on the } S_\sigma). \end{cases}$$

By identifying the Lagrange multiplier, we can get

$$\begin{cases} \lambda_{ij} = -\sigma_{ij} \\ \beta_i = -\lambda_{ij} v_j. \end{cases}$$

Substituting the identified Lagrangian multiplier into the new functional π^* gives the Hu–Washizu generalized variational principle

$$\begin{aligned} \pi_{\mathrm{HW}}(\sigma, \varepsilon, u) &= \int_V \left[A(\varepsilon) - \sigma_{ij}(\varepsilon_{ij} - \frac{1}{2} u_{i,j} - \frac{1}{2} u_{j,i}) - \bar{F}_i u_i \right] dV \\ &\quad - \int_{S_u} \sigma_{ij} v_j (u - \bar{u}_i) dS - \int_{S_\sigma} \bar{T}_i u_i dS \\ &= \text{stagnation value}. \end{aligned} \tag{3.81}$$

The Hu–Washizu generalized variational principle is a generalized variational principle with three independent functions u, ε and σ. In general, the Hu–Washizu generalized variational principle is not equal to the Hellinger–Reissner variational principle. Only when the strain ε depends on the stress σ following the stress–strain relationship and thus loses its independence, the relation $\pi_{\mathrm{HW}} = \pi_{\mathrm{HR}}$ is true. In other words, the Hu–Washizu generalized variational principle degenerates into the Hellinger–Reissner principle only under the condition of obeying the stress–strain relationship. There are three kinds of independent variables in the function π_{HW}. The stress–strain relationship is the Euler equation after the variation when π_{HW} takes the stagnation value. The Hu–Washizu variational principle is the most general variational principle in elastic mechanics, and other variational principles are all its special case. From the application point of view, if the Rayleigh–Ritz method is applied to find the approximate solution and the approximate stress–strain relationship needs to be satisfied, only the Hu–Washizu generalized variational principle can be used. Such as the beam, plate and shell theory in engineering, for considering the transverse shear stress but neglecting the transverse shear strain, so they are approximate elastic theories that do not satisfy the stress–strain relationship. This kind of approximation theory has been excluded from the energy method for a long time.

Only after establishing the Hu–Washizu generalized variational principle, the Rayleigh–Ritz method can be used to derive the classical beam, plate and shell theory from the elastic theory space problem, and the exact elastic theory and the approximate classical theory can be unified completely.

3.4 Multivariate solid-shell element

3.4.1 Three fields Fraeijs de Veubeke–Hu–Washizu variational principle

The functional expression of the variational principle is [11]

$$\begin{aligned}\prod(u,E,S) = &\int_{B_0} W_s(E)dV + \int_{B_0} S{:}[E^c(u) - E]dV \\ &- \int_{B_0} u\cdot(b^* - \ddot{u})\rho dV + \int_{S_u}(u^* - u)\cdot t dS - \int_{S_\sigma} u\cdot t^* dS\end{aligned} \tag{3.82}$$

Two special treatments are adopted in the finite element formula of the solid-shell element so that the element can be unaffected by various locking phenomena and tested by the fragmentation experiment. The two methods are enhanced assumed strain and assumed natural strain. The enhanced assumed strain method is expressed as follows:

$$E = E^c + \tilde{\boldsymbol{E}} \tag{3.83}$$

The assumed natural strain method is

$$\begin{Bmatrix} E_{13}^{ANS} \\ E_{23}^{ANS} \\ E_{33}^{ANS} \end{Bmatrix} = \begin{Bmatrix} (1-\xi^2)E_{13}^c(\xi_A) + (1+\xi^2)E_{13}^c(\xi_C) \\ (1-\xi^1)E_{23}^c(\xi_D) + (1+\xi^1)E_{23}^c(\xi_B) \\ \sum_{i=1}^{4} N_i(\xi^1,\xi^2)E_{33}^c(\xi_i) \end{Bmatrix} \tag{3.84}$$

Substitutting eqs. (3.83) and (3.84) into the function of the Fraeijs de Veubeke–Hu–Washizu variational principle we obtain

$$\begin{aligned}\prod(u,\tilde{E},S) = &\int_{B_0} W_s(E^c + \tilde{E})dV - \int_{B_0} S{:}\tilde{E}\,dV \\ &- \int_{B_0} u\cdot(b^* - \ddot{u})\rho dV + \int_{S_u}(u^* - u)\cdot t(S)dS - \int_{S_\sigma} u\cdot t^* dS\end{aligned} \tag{3.85}$$

3.4.2 Nonlinear finite element discretization

In the solution of the dynamic nonlinear equations of the solid-shell element, the classical Newton method is adopted. First, conduct the variation to the two sides of the functional expression (3.85) and obtain

$$\delta\pi(u,\tilde{\boldsymbol{E}}) = \delta\pi_{\text{mass}}(u) + \delta\pi_{\text{stiff}}(u,\tilde{\boldsymbol{E}}) + \delta\pi_{\text{ext}}(u) = 0 \tag{3.86}$$

where

$$\delta\pi_{\text{mass}} = \int_{B_0} \delta u \cdot \ddot{\boldsymbol{u}}\rho dV,$$

$$\delta\pi_{\text{stiff}} = \int_{B_0} (\delta\tilde{\boldsymbol{E}}^C + \delta\tilde{\boldsymbol{E}}):\frac{\partial}{\partial E} W_S(E^c(u) + \tilde{\boldsymbol{E}})dV,$$

$$\delta\pi_{\text{ext}}(u) = -\int_{B_0} \delta u \cdot b^* \rho dV - \int_{S_\sigma} \delta u \cdot t^* dS.$$

Then, linearize the expression after the variation

$$D\delta\prod_{\text{stiff}}^{(e)} \cdot (\Delta d^{(e)}, \Delta\alpha^{(e)}) + D\delta\prod_{\text{mass}}^{(e)} \cdot \Delta d^{(e)} + \delta\prod_{\text{stiff}}^{(e)} + \delta\prod_{\text{mass}}^{(e)} + \delta\prod_{\text{ext}}^{(e)} = 0 \tag{3.87}$$

After simplification, there is

$$\begin{bmatrix} k_{uu}^{(e)} + \frac{\alpha_m}{\beta\Delta t^2} m^{(e)} & k_{u\alpha}^{(e)} \\ k_{\alpha u}^{(e)} & k_{\alpha\alpha}^{(e)} \end{bmatrix} \begin{Bmatrix} \Delta d^{(e)} \\ \Delta\alpha^{(e)} \end{Bmatrix} = \begin{Bmatrix} f_{\text{ext}}^{(e)} - f_{\text{stiff}}^{(e)} - f_{\text{mass}}^{(e)} \\ -f_{\text{EAS}}^{(e)} \end{Bmatrix}. \tag{3.88}$$

Equation (3.88) is the equation to be solved.

3.4.3 Procedure of solving algorithm

The nonlinear finite element method of solid-shell element is an incremental solution method. In the solution, the displacement field, stress field and strain field of the element are obtained at the same time. The specific solution procedure is presented in Fig. 3.3.

The advantage of the solid-shell element is that the element is based on the multivariate variational principle, specifically an element of three field variables, that is, the displacement field, stress field and strain field are independent variables. It is possible to simultaneously assume shape functions for the three variables, which can greatly improve the accuracy of each variable, especially the accuracy of stress and strain.

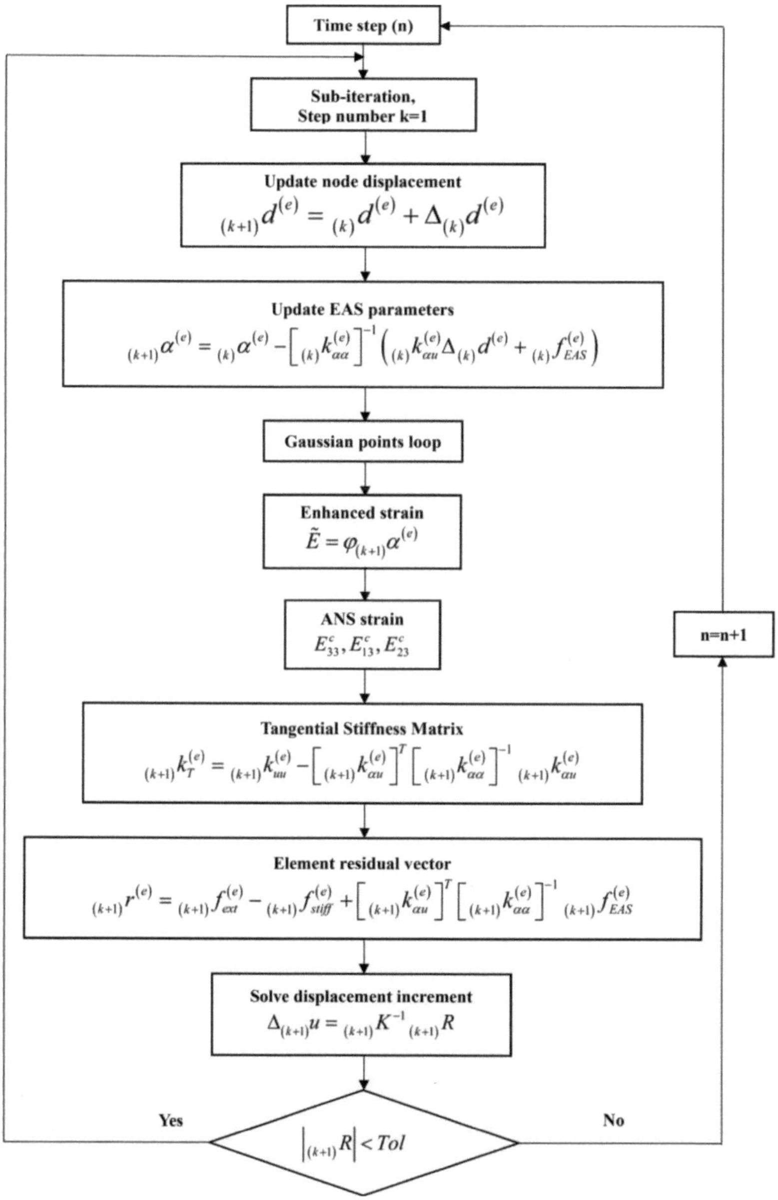

Fig. 3.3: Solid-shell nonlinear finite element solving procedure.

In the traditional displacement-based finite element method, only the displacement field is an independent variable, and other physical quantities, such as stress and strain, are derived based on the displacement field. Since the solutions of the strain and stress fields require the displacement to be derived, but the displacement field is a numerical solution, so the accuracies of strain and stress obtained are thus reducing. Another advantage of the solid-shell element is that when the variational principle of the three fields' variables is used, special corrections or improvements can be made to the physical quantities of stress and strain to better meet the accuracy requirements. In general, the large geometric deformation of the elastic body brings various locking phenomena, such as volume locking, membrane locking and shear locking. These locking phenomena make certain terms in the finite element structural stiffness matrix becomes very large, and the structure exhibits a large rigidity, so that the deformation under the external load is very small, and it seems like the structure is locked. Certainly, this will not happen in practice, but is caused by the unreasonable solution method used in the finite element formula. To deal with these problems, the solid-shell element uses the enhanced assumed strain method and the assumed natural strain method to eliminate the various locking phenomena, so that the solution in the geometric nonlinear problem can be very close to the real result. In addition, the solid-shell element has the ability to simultaneously model the solid, plate, shell and beam elements. Taking into account the above advantages, using the solid-shell element as the structural modeling element in the aeroelastic model is suggested.

3.3.4 Results and analysis

Here, a cantilever plate is taken as an example to verify the ability of the solid-shell element to model the plate and shell structure, and to verify the ability to solve the geometric nonlinear problems using the multivariate finite element method. The finite element model of the cantilever plate is shown in Fig. 3.4, for which the elastic modulus is 1.0E + 007 MPa and the Poisson's ratio is 0.4.

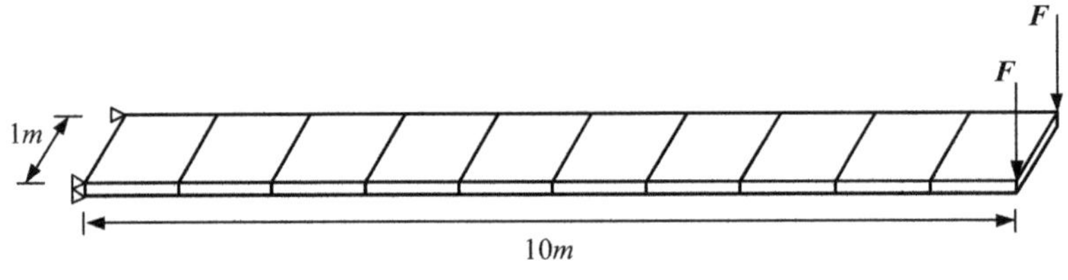

Fig. 3.4: Finite element model of cantilever plate.

The solid-shell element method, NASTRAN's linear solver and nonlinear solver are all used for comparison. With the gradually increasing value of the load F, the displacement of the plate end and the in-plane Von Mises stress are investigated. Figures. 3.5 and 3.6 show the comparison of stress of the cantilever plate between the nonlinear solver and solid element method for $F = 1\,\mathrm{N}$ and $F = 10\,\mathrm{N}$, respectively. Figure 3.7 compares the three different solvers in terms of the displacement of the plate end.

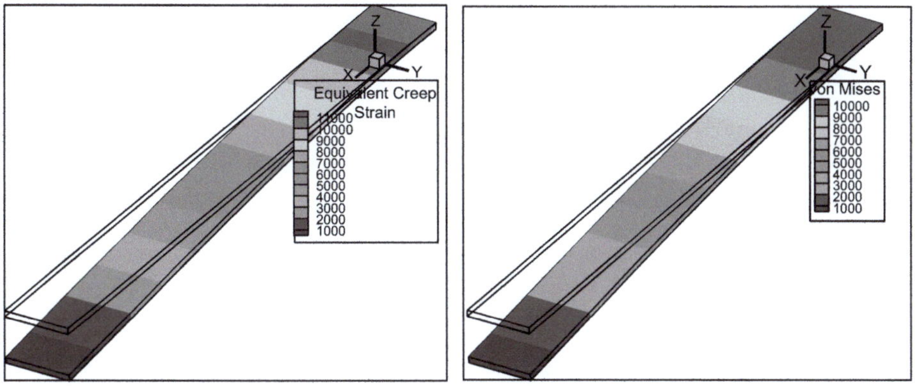

(a) Nastran; (b) Solid shell element

Fig. 3.5: $F = 1$ N Von Mises stress (Pa): (a) NASTRAN and (b) solid-shell element.

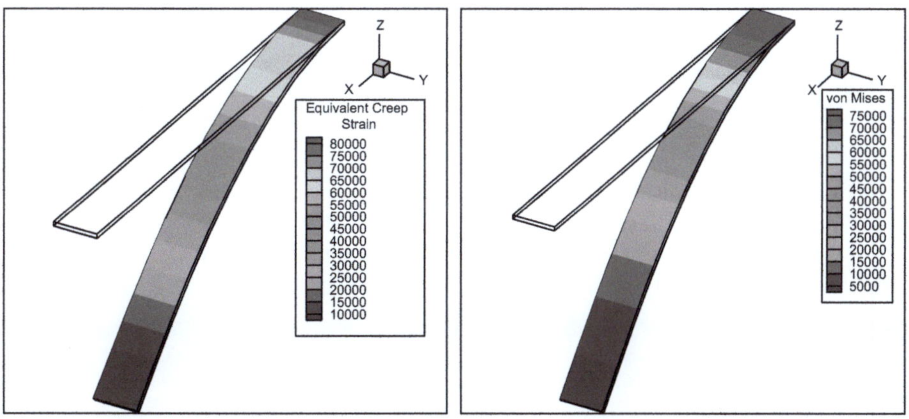

(a) Nastran; (b) Solid shell element

Fig. 3.6: $F = 10$ N Von Mises stress (Pa): (a) NASTRAN and (b) solid-shell element.

Figures 3.5 and 3.6 show that under various load conditions the Von Mises equivalent stresses calculated by NASTRAN and the solid-shell element are very close to each other, which tells that the calculation result of the solid-shell element is credible. Figure 3.7(a) presents the displacement at the plate end in the x-direction. The result of the linear finite element method shows that the displacement in the x-direction is small, which indicates the length and area of the plate are gradually increasing with the increasing load, which is not true in the actual situation. When

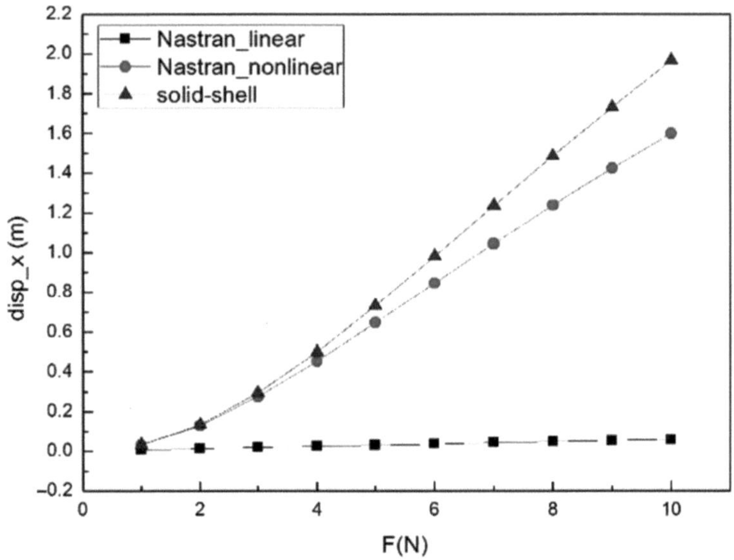

(a) Absolute value of displacement in the x-direction;

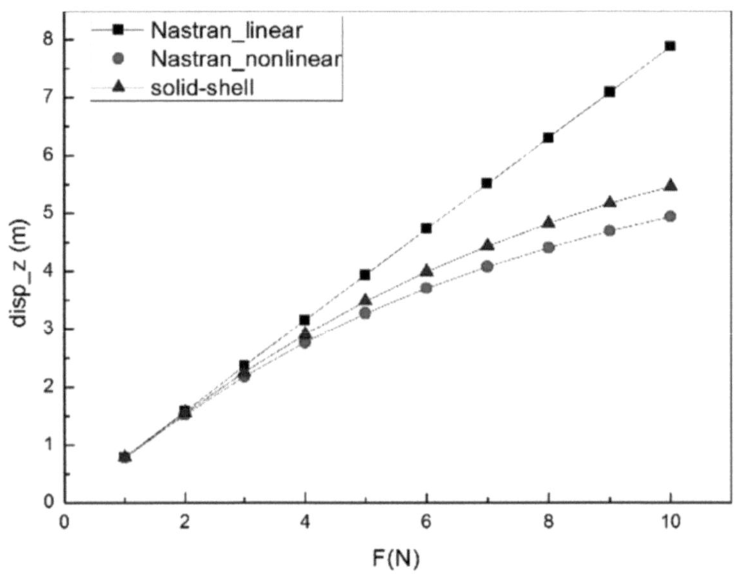

(b) Absolute value of displacement in the z-direction

Fig. 3.7: The comparison of the displacements at the plate end between the three solvers: (a) absolute value of displacement in the x-direction and (b) absolute value of displacement in the z-direction.

the nonlinear finite element method is used, the plate end has a large displacement in the x-direction, indicating that the plate is deformed in the longitudinal direction and the projection is constantly decreasing with the gradually increasing load. The results in Fig. 3.7(b) show that as the load increases, the result of linear finite element increases linearly, while the result of nonlinear finite element shows a nonlinear growth trend, and the growth rate decreases with the load increases. This shows that the geometric nonlinearity increases the structure stiffness.

3.5 Finite element model of geometric nonlinearity of beam-shell element based on CR theory

At present, the description of geometric nonlinear problems can be roughly divided into three categories according to the different reference configurations: (1) total Lagrangian (TL) method; (2) updated Lagrangian method and (3) CR method. The CR method was first applied to the finite element structural analysis by Belytschko in the 1970s. Horrigmoe and Bergan applied the CR system to establish a finite element model of a single element. Based on the continuous CR formula, Crisfield developed a unified CR framework for solids, shells and beams, which accelerated the application of CR theory in finite element modeling. Battini and Pacoste derived the finite tangent stiffness matrix of different elements for analysis of geometric nonlinearity based on consistent CR formulation. The main idea of the CR method is that the motion of the structure is decomposed into rigid body motion and pure elastic deformation, and the elastic equation is established based on local coordinate system. The structural motion can be split into two step, first is the rotation and translation of the rigid body, and then is the deformation based on the local coordinate system. If the geometrical characteristics of the element are properly selected, the pure part of deformation still belongs to the small strain–small displacement relationship. Therefore, the linear elastic relationship can be used to establish the stiffness matrix and the internal force formula, and then assemble into the global stiffness and internal force, and thus greatly simplifies the complexity of the problem.

3.5.1 Tangent stiffness matrix derivation of two-dimensional beam element

1 Transformation between the global coordinate system and the local coordinate system

The space bar structure can be described by beam elements, taking the two-dimensional planar beam [2, 4, 6] shown in Fig. 3.8 as an example. The coordinates of nodes 1 and 2 in the global coordinate system are (x_1, z_1) and (x_2, z_2), and the displacement vector of the elements in the global and local coordinate systems are defined as

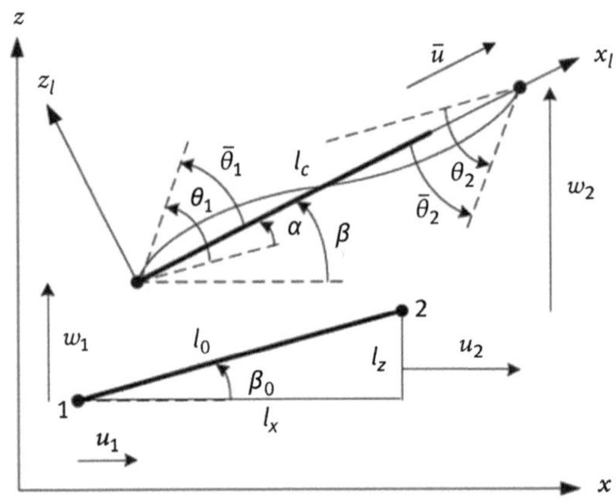

Fig. 3.8: Deformation of a two-dimensional beam.

$$d = [u_1 \quad w_1 \quad \theta_1 \quad u_2 \quad w_2 \quad \theta_2]^T, \ d_l = [\bar{u} \quad \bar{\theta}_1 \quad \bar{\theta}_2]^T. \tag{3.89}$$

Applying the geometric relationship in the figure, there is

$$\begin{cases} \bar{u} = l_c - l_0 \\ \bar{\theta}_1 = \theta_1 - \alpha, \\ \bar{\theta}_2 = \theta_2 - \alpha \end{cases} \tag{3.90}$$

where l_0 and l_c are the element lengths before and after the deformation, respectively, and α is the deformation angle, they are defined as follows:

$$l_0 = \sqrt{l_x^2 + l_z^2}, \ l_c = \sqrt{(l_x + u_2 - u_1)^2 + (l_z + w_2 - w_1)^2}, \ \alpha = \beta - \beta_0 \tag{3.91}$$

and

$$\tan \beta_0 = l_z / l_x, \ \tan \beta = (l_z + w_2 - w_1)/(l_x + u_2 - u_1) \tag{3.92}$$

Let $c = \cos\beta$ and $s = \sin\beta$. Calculating the differential of eq. (3.90) and using eq. (3.91), we obtain

$$\begin{cases} \delta\bar{u} = c(\delta u_2 - \delta\delta u_1) + s(\delta w_2 - \delta w_1) = [-c \quad -s \quad 0 \quad c \quad s \quad 0]\delta d \\ \delta\bar{\theta}_1 = \delta\theta_1 - \delta\alpha = \delta\theta_1 - \delta\beta \\ \delta\bar{\theta}_2 = \delta\theta_2 - \delta\alpha = \delta\theta_2 - \delta\beta. \end{cases} \tag{3.93}$$

Calculating the differential of eq. (3.92) and applying eqs. (3.91) and (3.92) with derivation and simplification produce

$$\delta\beta = 1/l_c[s \quad -c \quad 0 \quad -s \quad c \quad 0]\delta d. \tag{3.94}$$

Substituting eq. (3.94) into eq. (3.93), we get

$$\delta d_l = T\delta d, \tag{3.95}$$

where

$$T = \begin{bmatrix} -c & -s & 0 & c & s & 0 \\ -s/l_c & c/l_c & 1 & s/l_c & -c/l_c & 0 \\ -s/l_c & c/l_c & 0 & s/l_c & -c/l_c & 1 \end{bmatrix} \tag{3.96}$$

2 Element internal force derivation

The internal force in the local coordinate system and the global coordinate system is defined as

$$f_l = \left[\bar{N} \quad \bar{M}_1 \quad \bar{M}_2\right]^T, \; F_i = [X_1 \quad Z_1 \quad M_1 \quad X_2 \quad Z_2 \quad M_2]^T. \tag{3.97}$$

If the material is still linear, the internal forces in the local coordinate system can be described by the linear relationship of the local displacement:

$$\begin{cases} \bar{N} = EA/l_c \cdot \bar{u} \\ \bar{M}_1 = EI_z/l_c \cdot \left(4\bar{\theta}_1 + 2\bar{\theta}_2\right) \\ \bar{M}_2 = EI_z/l_c \cdot \left(2\bar{\theta}_1 + 4\bar{\theta}_2\right). \end{cases} \tag{3.98}$$

The relationship between the two can be obtained through the virtual work principle:

$$F_i = T^T f_l. \tag{3.99}$$

3 Tangent stiffness matrix derivation

The tangent stiffness matrix in the global coordinate system is defined as

$$\delta F_i = K_t \delta d. \tag{3.100}$$

Calculating the differential of eq. (3.99) derives

$$\delta F_i = T^T \delta f_l + f_l^T \delta T = T^T \delta f_l + \bar{N}\delta T_1 + \bar{M}_1 \delta T_2 + \bar{M}_2 \delta T_3 \tag{3.101}$$

T_1, T_2 and T_3 are three row vectors of T in the above equation. Introducing

$$\begin{cases} r=[-c \quad -s \quad 0 \quad c \quad s \quad 0]^T \\ p=[s \quad -c \quad 0 \quad -s \quad c \quad 0]^T \end{cases} \tag{3.102}$$

and then T can be described as

$$\begin{cases} T_1=r \\ T_2=-[0 \quad 0 \quad 1 \quad 0 \quad 0 \quad 0]^T p/l_c \\ T_3=-[0 \quad 0 \quad 0 \quad 0 \quad 0 \quad 1]^T p/l_c \end{cases} \tag{3.103}$$

Calculating the differential of eq. (3.102), we get

$$\delta r=p\delta\beta, \quad \delta p=-r\delta\beta. \tag{3.104}$$

Using eqs. (3.103) and (3.104) results in

$$\begin{cases} \delta T_1=\delta r=(pp^T)/l_c\cdot\delta d \\ \delta T_2=\delta T_3=1/l_c^2\cdot(rp^T+pr^T)\delta d \end{cases}. \tag{3.105}$$

Substituting eqs. (3.102)–(3.104) into the second right term of eq. (3.101) results in

$$f_l^T\delta T=\left[(pp^T)/l_c\cdot\bar{N}+1/l_c^2\cdot(rp^T+pr^T)(\bar{M}_1+\bar{M}_2)\right]\delta d. \tag{3.106}$$

In the local coordinate system, we can use the linear elastic structure relationship for the small strain–small displacement relationship:

$$\delta f_l=K_l\delta d_l=K_lT\delta d \tag{3.107}$$

where K_l is the linear stiffness matrix of element in the local coordinate system. Substituting eq. (3.106) into eq. (3.101), and using eqs. (3.97) and (3.107), the tangent stiffness matrix of eq. (3.100) in the global coordinate system can be obtained as

$$K_t=K_{tl}+K_{t\sigma}=T^TK_lT+\left[\bar{N}/l_c\cdot pp^T+(\bar{M}_1+\bar{M}_2)/l_c^2\cdot(rp^T+pr^T)\right] \tag{3.108}$$

where K_{tl} and $K_{t\sigma}$ are linear elastic stiffness matrix and geometric stiffness matrix, respectively. It can be seen that after introducing the local dynamic coordinate system, it is not necessary to solve the large displacement matrix in the traditional TL method, and the global tangent stiffness matrix remains as a symmetric matrix, which can be conveniently stored and solved. Moreover, K_l is the stiffness matrix of the traditional linear two-dimensional beam element, with the corresponding freedoms of the stiffness matrix are $\bar{u}$, $\bar{\theta}_1$, $\bar{\theta}_2$

$$K_l=\begin{bmatrix} EA/l_c & 0 & 0 \\ 0 & 4EI_z/l_c & 1EI_z/l_c \\ 0 & 2EI_z/l_c & 4EI_z/l_c \end{bmatrix}. \tag{3.109}$$

Substituting eq. (3.109) into eq. (3.108) and defining $\bar{A}=EA/l_c$, $\bar{B}=12EI_z/l_c^3$, $\bar{C}=6EI_z/l_c^2$, $\bar{D}=4EI_z/l_c^2$, we obtain

$$K_{tl}=\begin{bmatrix} \bar{A}c^2+\bar{B}s^2 & & & & & \\ \bar{A}sc-\bar{B}sc & \bar{A}s^2+\bar{B}c^2 & & & & \\ -\bar{C}s & \bar{C}c & \bar{D} & & & \\ -\bar{A}c^2-\bar{B}s^2 & -\bar{A}sc+\bar{B}sc & \bar{C}s & \bar{A}c^2+\bar{B}s^2 & & \\ -\bar{A}sc+\bar{B}sc & -\bar{A}s^2-\bar{B}c^2 & -\bar{C}c & \bar{A}sc-\bar{B}sc & \bar{A}s^2+\bar{B}c^2 & \\ -\bar{C}s & \bar{C}c & 1/2\bar{D} & \bar{C}s & -\bar{C}c & \bar{D} \end{bmatrix}. \tag{3.110}$$

Defining $Q_1=\bar{N}/l_c, Q_2=\left(\bar{M}_1+\bar{M}_2\right)/l_c^2$, the geometric stiffness matrix is expressed as

$$K_{t\sigma}=\begin{bmatrix} Q_1s^2-2Q_2sc & & & & & \\ -Q_1sc-Q_2(s^2-c^2) & Q_1c^2+2Q_2sc & & & & \\ 0 & 0 & 0 & 0 & & \\ -Q_1s^2+2Q_2sc & Q_1sc+Q_2(s^2-c^2) & 0 & Q_1s^2-2Q_2sc & & \\ Q_1sc+Q_2(s^2-c^2) & -Q_1c^2-2Q_2sc & 0 & -Q_1sc-Q_2(s^2-c^2) & Q_1c^2+2Q_2sc & \\ 0 & 0 & 0 & 0 & 0 & 0 \end{bmatrix}. \tag{3.111}$$

3.5.2 Derivation of tangent stiffness matrix for a three-dimensional shell element based on CR method

Most of the thin-walled structures of the aircraft can be described by shell elements. Based on the CR theory, this section deduces and develops the tangent stiffness matrix of the triangular shell element with geometric nonlinearity [1, 4, 7].

1 Establishment of CR local coordinate system of triangular shell element

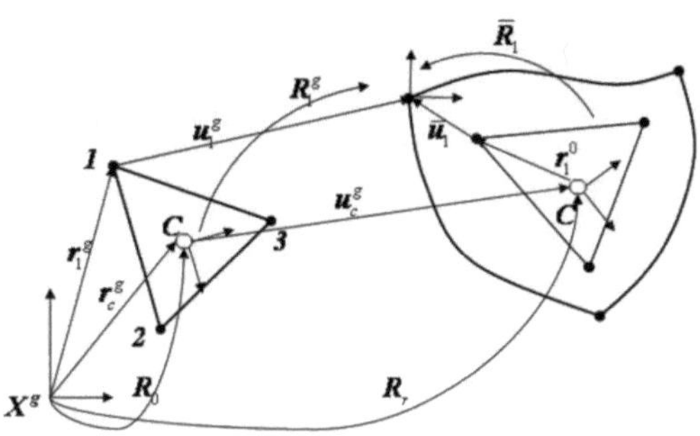

Fig. 3.9: Deformation of a three-dimensional shell element.

The local coordinate system defined in the CR method continuously changes with the rotation and translation of the element. As shown in Fig. 3.9, the initial origin position of the local coordinate system is taken at the geometric center point C of the triangular element, and its local coordinate axes vectors are defined as

$$\mathbf{e}_1 = \mathbf{r}_{12}^{\mathbf{g}} / \left\| \mathbf{r}_{12}^{\mathbf{g}} \right\|, \ \mathbf{e}_3 = \mathbf{r}_{12}^{\mathbf{g}} \times \mathbf{r}_{13}^{\mathbf{g}} / \left\| \mathbf{r}_{12}^{\mathbf{g}} \times \mathbf{r}_{13}^{\mathbf{g}} \right\|, \ \mathbf{e}_2 = \mathbf{e}_3 \times \mathbf{e}_1 \tag{3.112}$$

where $\mathbf{r}_{12}^{\mathbf{g}} = \mathbf{r}_2^{\mathbf{g}} - \mathbf{r}_1^{\mathbf{g}}$, $\mathbf{r}_1^{\mathbf{g}}$ is the coordinate of the first vertex of the triangle element in the global coordinate system, defined as

$$\mathbf{R}_0 = [\, \mathbf{e}_1 \quad \mathbf{e}_2 \quad \mathbf{e}_3 \,]. \tag{3.113}$$

The movement of the element from the initial state to the current state requires two steps: first, the translation and rotation of the rigid body. The rigid body translation is described by the displacement $\mathbf{u}_c^g$ of node C in the global coordinates, and the rigid body rotation is described by the orthogonal matrix $\mathbf{R}_r$, which is defined by the deformed local coordinate system. The expression of the rotation is similar to eq. (3.112), except that the vertex coordinates are replaced by the position of the deformed vertex in the global coordinates. Since $\mathbf{R}_r$ is defined by the deformed local coordinate system, the local coordinate system can be linked to the global coordinate system by $\mathbf{R}_r$, such that vectors and matrices in the local coordinates can be represented by the ones in the global coordinates as follows:

$$\begin{cases} \mathbf{V}^e = \mathbf{R}_r^T \mathbf{V}^g \\ \tilde{\mathbf{V}}^e = \mathbf{R}_r^T \tilde{\mathbf{V}}^g \mathbf{R}_r. \end{cases} \tag{3.114}$$

Superscripts e and g represent the local and global coordinate systems, respectively. The second step is the local deformation in the local coordinate system. The local pure deformational values of the node is described by the description $\bar{\mathbf{u}}_i$, $i = 1, 2, 3$, as shown in Fig. 3.9. The local rotation of the node is described by an orthogonal matrix $\bar{\mathbf{R}}_i$, $i = 1, 2, 3$. By the above two steps, the whole deformation process is divided into a rigid body deformation and a local deformation.

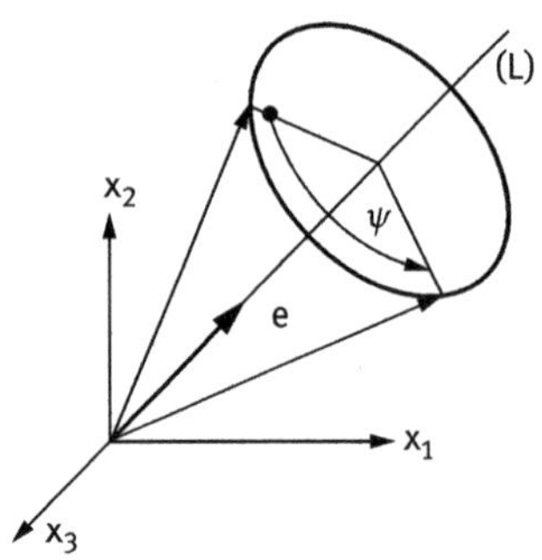

Fig. 3.10: Parameterization of the spatial rotation variable.

The traditional rotation variable is described by the variation of the spatial angle, which limits its application in matrix operations. The parametric method proposed by Pacoste [7] expresses the rotation variable with the axis vector. Any limited rotation can be described by a unique angle ψ along the axis L(defined by the unit vector **e**) as shown in Fig. 3.10.

Its expression is

$$\boldsymbol{\psi} = \mathbf{e}\psi, \quad \psi = \sqrt{q_1^2 + q_2^2 + q_3^2} \tag{3.115}$$

where q_i, $i = 1, 2, 3$ are the three components of $\boldsymbol{\psi}$, expressed as

$$\tilde{\boldsymbol{\psi}} = \begin{bmatrix} 0 & -q_3 & q_2 \\ q_3 & 0 & -q_1 \\ -q_2 & q_1 & 0 \end{bmatrix}. \tag{3.116}$$

The rotation variable is parameterized by an orthogonal matrix as follows:

$$R_i^g = \exp\left(\tilde{\psi}_i\right) = I + \frac{\sin(\psi)}{\psi}\tilde{\psi}_i + \frac{1}{2}\left[\frac{\sin(\psi/2)}{\psi/2}\right]^2 \tilde{\psi}_i \tag{3.117}$$

After derivation, we get,

$$R_i^g = 2\begin{bmatrix} q_0^2 + q_1^2 - 1/2 & q_1q_2 - q_3q_0 & q_1q_3 + q_2q_0 \\ q_2q_1 + q_3q_0 & q_0^2 + q_2^2 - 1/2 & q_2q_3 - q_1q_0 \\ q_3q_1 - q_2q_0 & q_3q_2 + q_1q_0 & q_0^2 + q_3^2 - 1/2 \end{bmatrix} \tag{3.118}$$

where $q_0^2 + q_1^2 + q_2^2 + q_3^2 = 1$, and thus the local deformation can be derived as

$$\bar{\mathbf{u}}_i = \mathbf{R}_r^T\left(\mathbf{r}_i^g + \mathbf{u}_i^g - \mathbf{r}_c^g - \mathbf{u}_c^g\right) - \mathbf{r}_i^0, \quad i = 1, 2, 3 \tag{3.119}$$

where $\mathbf{u}_i^g$, $i = 1, 2, 3$ are the node displacements in global coordinates, $\mathbf{r}_c^g$ is the initial coordinate of node C and $\mathbf{r}_i^0$, $i = 1, 2, 3$ is the vector from the triangle vertex to the geometric center C at the initial time, which is defined as

$$\mathbf{r}_i^0 = \mathbf{R}_0^T\left(\mathbf{r}_i^g - \mathbf{r}_c^g\right), \quad i = 1, 2, 3. \tag{3.120}$$

The following relationship is shown in Fig. 3.9:

$$\begin{aligned} &\mathbf{R}_r\bar{\mathbf{R}}_i = \mathbf{R}_i^g\mathbf{R}_0 \\ &\bar{\mathbf{R}}_i = \mathbf{R}_r^T\mathbf{R}_i^g\mathbf{R}_0, \quad i = 1, 2, 3. \end{aligned} \tag{3.121}$$

The following differential expressions have been derived by Pacoste:

$$\begin{cases} \delta \mathbf{R}_r = \delta \tilde{\theta}_r^g \mathbf{R}_r \\ \delta \mathbf{R}_i^g = \delta \tilde{\theta}_i^g \mathbf{R}_i^g , \\ \delta \bar{\mathbf{R}}_i = \delta \tilde{\bar{\theta}}_i \bar{\mathbf{R}}_i \end{cases} \tag{3.122}$$

where $\delta\tilde{\theta}_r^g$, $\delta\tilde{\theta}_i^g$ and $\delta\tilde{\bar{\theta}}_i$ are the antisymmetric matrices consisting of the spatial angle increment $\delta\theta_r^g$ of the rigid body rotation, the rotation spatial angle increment $\delta\theta_i^g$ of the node deformation and the spatial angle increment $\delta\bar{\theta}_i$ of the local node deformation rotation, respectively, which are defined as eq. (3.116). Thus, the differential of local deformation in eq. (3.119) is obtained:

$$\delta \bar{\mathbf{u}}_i = \delta \mathbf{R}_r^T \left(\mathbf{r}_i^g + \mathbf{u}_i^g - \mathbf{r}_c^g - \mathbf{u}_c^g \right) + \mathbf{R}_r^T \left(\delta \mathbf{u}_i^g - \delta \mathbf{u}_c^g \right). \tag{3.123}$$

In local coordinate system, the deformation increment $\mathbf{R}_r^T \delta \mathbf{u}_c^g \approx 0$ can be ignored. Using eq. (3.122), eq. (3.123) becomes

$$\begin{aligned} \delta \bar{\mathbf{u}}_i &= -\mathbf{R}_r^T \delta\tilde{\theta}_r^g \left(\mathbf{r}_i^g + \mathbf{u}_i^g - \mathbf{r}_c^g - \mathbf{u}_c^g \right) + \mathbf{R}_r^T \delta \mathbf{u}_i^g \\ &= -\mathbf{R}_r^T \delta\tilde{\theta}_r^g \mathbf{R}_r \mathbf{R}_r^T \left(\mathbf{r}_i^g + \mathbf{u}_i^g - \mathbf{r}_c^g - \mathbf{u}_c^g \right) + \mathbf{R}_r^T \delta \mathbf{u}_i^g . \end{aligned} \tag{3.124}$$

Considering eqs. (3.114) and (3.119), using definition $\mathbf{a}_i = \bar{\mathbf{u}}_i + \mathbf{r}_i^0$ and relation $-\delta\tilde{\theta}_r^e \mathbf{a}_i = \tilde{\mathbf{a}}_i \delta\theta_r^e$, then eq. (3.124) can be written as

$$\delta \bar{\mathbf{u}}_i = \tilde{\mathbf{a}}_i \delta\theta_r^e + \delta \mathbf{u}_i^e . \tag{3.125}$$

The differential of the first equation in eq. (3.121) is

$$\delta \bar{\mathbf{R}}_i = \delta \mathbf{R}_r^T \mathbf{R}_i^g \mathbf{R}_0 + \mathbf{R}_r^T \delta \mathbf{R}_i^g \mathbf{R}_0 . \tag{3.126}$$

Substitutting eq. (3.122) into eq. (3.126) results in

$$\begin{aligned} \delta \bar{\mathbf{R}}_i &= \delta\tilde{\theta}_r^g \mathbf{R}_r^T \mathbf{R}_i^g \mathbf{R}_0 + \mathbf{R}_r^T \delta\tilde{\theta}_i^g \mathbf{R}_i^g \mathbf{R}_0 = -\mathbf{R}_r^T \delta\tilde{\theta}_r^g \mathbf{R}_r \mathbf{R}_r^T \mathbf{R}_i^g \mathbf{R}_0 + \mathbf{R}_r^T \delta\tilde{\theta}_i^g \mathbf{R}_r \mathbf{R}_r^T \mathbf{R}_i^g \mathbf{R}_0 \\ &= -\delta\tilde{\theta}_r^e \mathbf{R}_r^T \mathbf{R}_i^g \mathbf{R}_0 + \delta\tilde{\theta}_i^e \mathbf{R}_r^T \mathbf{R}_i^g \mathbf{R}_0 = \left(\delta\tilde{\theta}_i^e - \delta\tilde{\theta}_r^e \right) \bar{\mathbf{R}}_i . \end{aligned} \tag{3.127}$$

Introducing the rotational deformation $\bar{\theta}_i$ in local coordinates from eq. (3.122), and then

$$\delta\bar{\theta}_i = \delta\theta_i^e - \delta\theta_r^e . \tag{3.128}$$

The displacement increment in local coordinates can be obtained from eqs. (3.125) and (3.128).

2. Stiffness matrix and internal force solution in local coordinate system
For a triangular shell element, the local displacement vector $\bar{\mathbf{u}}_i$ is taken as follows:

$$\bar{\mathbf{u}}_i = \begin{bmatrix} \bar{u}_i & \bar{v}_i & \bar{w}_i \end{bmatrix}, \quad i = 1, 2, 3. \tag{3.129}$$

According to the definition of the orthogonal matrix $\bar{\mathbf{R}}_i$, that is, $\bar{\mathbf{R}}_i = \exp\left(\tilde{\bar{\vartheta}}_i\right)$, the local rotation is

$$\tilde{\bar{\vartheta}}_i = \log_e\left(\bar{\mathbf{R}}_i\right), \quad i = 1, 2, 3. \tag{3.130}$$

Based on eq. (3.130), ignoring the high-order terms, a more concise expression can be obtained after derivation as

$$\bar{\vartheta}_i = \frac{1}{2}\begin{bmatrix} \bar{R}_{32} - \bar{R}_{23} \\ \bar{R}_{13} - \bar{R}_{31} \\ \bar{R}_{21} - \bar{R}_{12} \end{bmatrix}, \quad i = 1, 2, 3. \tag{3.131}$$

All translational and rotational displacements in local coordinates can be obtained from eqs. (3.129) and (3.131)

$$\mathbf{d}_l = \begin{bmatrix} \bar{\mathbf{u}}_1^T & \bar{\vartheta}_1^T & \bar{\mathbf{u}}_2^T & \bar{\vartheta}_2^T & \bar{\mathbf{u}}_3^T & \bar{\vartheta}_3^T \end{bmatrix}^T. \tag{3.132}$$

The internal force in the local coordinate system is defined as

$$\mathbf{f}_l = \begin{bmatrix} \bar{\mathbf{F}}_1^T & \bar{M}_1^T & \bar{\mathbf{F}}_2^T & \bar{\mathbf{M}}_2^T & \bar{\mathbf{F}}_3^T & \bar{\mathbf{M}}_3^T \end{bmatrix}^T \tag{3.133}$$

where

$$\begin{cases} \bar{\mathbf{F}}_i = \begin{bmatrix} \bar{f}_x & \bar{f}_y & \bar{f}_z \end{bmatrix} \\ \bar{\mathbf{M}}_i = \begin{bmatrix} \bar{m}_x & \bar{m}_y & \bar{m}_z \end{bmatrix} \end{cases}, \quad i = 1, 2, 3. \tag{3.134}$$

In local coordinates, without considering rigid body displacement and rotation, there will be a linear relationship between the displacement and strain as

$$\mathbf{f}_l = K_l \mathbf{d}_l \tag{3.135}$$

where K_l is the linear elastic rigid matrix of triangular shell element, which is composed of plane stress element and plate-bending element. The specific expression can be found in many finite element literatures.

3.6 Solution of nonlinear finite element equations

Whether it is a material nonlinear problem or a geometric nonlinear problem, it ultimately comes down to solving a set of nonlinear equations with N variables and N equations after discretization by finite element method, that is,

$$\begin{cases} \varphi_1(a_1, \ldots, a_N) = 0 \\ \vdots \\ \varphi_N(a_1, \ldots, a_N) = 0 \end{cases}$$

where $a_1, \ldots, a_N$ are unknown variables and $\varphi_1, \ldots, \varphi_N$ are real-valued functions defined in the open domain D of the N-dimensional Euclid space. Generally they are nonlinear functions of $a_1, \ldots, a_N$, using vector notation

$$a = [a_1, \ldots, a_N]^T,$$

$$\varphi = [\varphi_1, \ldots, \varphi_N]^T.$$

The above equations can be represented by a vector equation

$$\varphi(a) = \mathbf{0}.$$

In order to highlight its mechanical meaning, this equation is rewritten as follows:

$$\varphi(a) = P(a) - R = \mathbf{0}. \tag{3.136}$$

Here, $P(a)$ is a vector function of an unknown vector, but R is a known vector. In the finite element representation of all the variables problem, a is the unknown node displacement vector, $P(a)$ is the equivalent nodal force vector of the internal force, R is the equivalent nodal force vector of the load, and the vector equation $\varphi(a) = \mathbf{0}$ represents the node equilibrium equation, where each component equation corresponds to the equilibrium of a degree of freedom. For incremental problems, a represents the displacement increment vector, and R is the load increment vector.

In the univariate case, the relationship of $\varphi(a)$ and $P(a)$ is shown in Fig. 3.11. It can be easily seen that the coordinate a^* of the intersection of the curve $z = \varphi(a)$ and the horizontal axis $z = 0$ is the solution to the problem. Similarly, the horizontal coordinate a^* of intersection of the curve $z = P(a)$ and the horizontal line $z = R$ is also the solution to the problem.

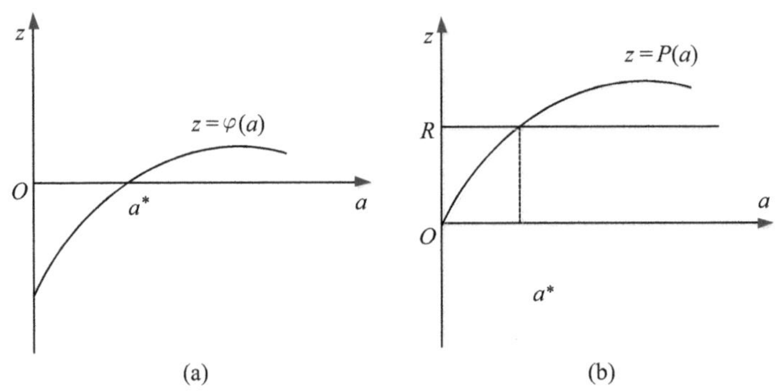

Fig. 3.11: Nonlinear vector equation.

The problem of solving nonlinear equations is not as mature or effective as the linear equations in theory or in solving algorithm. In the solving algorithm, except for the extremely special nonlinear equations, the direct solving method is almost impossible. Generally, various numerical methodologies are applied and a series of solutions of linear equations are used to approximate the solution of the nonlinear equations.

3.6.1 Nonlinear solution technique for structural static analysis

The tangent stiffness matrix of the structural element is obtained in the above derivation, and the incremental stiffness equation in the structural static analysis is obtained as follows:

$$\mathrm{K}_t \Delta u = \Delta F. \tag{3.137}$$

Currently, Newton–Raphson or the modified Newton–Raphson method is employed to solve this nonlinear equation, and the incremental form can be selected as load increment or displacement increment.

1 Load increment method [5, 8]

For the accuracy of the solution, the external force F_e is loaded with multiple steps, and the Newton–Raphson method is used in each loading step. The incremental equation of the ith iteration in the nth loading step is

$$K_{\mathrm{tn}}^{i-1} \Delta u_n^i = \Delta \lambda_n^i F_e + R_n^{i-1} \tag{3.138}$$

where K_{tn}^{i-1} and Δu_n^i represent the global tangent stiffness matrix and displacement increment, respectively; $\Delta\lambda_n^i$ is the force loading factor, which is the control quantity and remains constant $\Delta\lambda_n^i = \Delta\lambda_n$ during each loading step; $R_n^{i-1} = \lambda_0 F_e - F_{\mathrm{i}}$ is the unbalanced load, $\lambda_0 = \sum \Delta\lambda_{n-1}$ is the sum of the load increment coefficients in the previous $n-1$ loading steps and F_i is the internal force vector of the structural element. After the initial value of the displacement Δu_n^0 is determined in each loading step (the initial value of the initial displacement can be selected as the displacement value of the linear solution), the tangent stiffness matrix K_{tn} and the internal force F_i in the global coordinate can be calculated according to the formulation derived in Section 3.5. When the load increment coefficients are given, the entire displacement can be obtained by the convergent solution of eq. (3.138) within each loading step:

$$u = u + \sum \Delta u_n^i. \tag{3.139}$$

The general nonlinear static analysis of the structure can obtain satisfactory results by the load increment method, but when solving the static instability characteristics, the load increment is often difficult to determine the instability point.

2 Displacement increment method [5, 8]

Compared to the load increment method, the incremental equation is the same as eq. (3.138), except that the control quantity changes to the displacement increment and it remains constant but the force loading coefficient is no longer constant in each loading step. At the initial moment in the nth loading step, $R_n^0 = 0$, and $\Delta\lambda_n^i$ can be calculated from the following formula:

$$\Delta\lambda_n^1 = (-1)^n \Delta\lambda_1^1 |\mathrm{GSP}|^{1/2} \tag{3.140}$$

where GSP is stiffness parameter, n is a notation of the number of GSP changes and $\Delta\lambda_1^1$ is the coefficient specified by the initial loading step, which can be tried in advance. *GSP* is calculated as follows:

$$\mathrm{GSP} = \frac{\left(u_1^1\right)^T u_1^1}{\left(u_{n-1}^1\right)^T u_n^1} \tag{3.141}$$

where u_1^1, u_{n-1}^1 and u_n^1 can be calculated from $K_0^1 u_1^1 = F_e$, $K_{n-1}^0 u_{n-1}^1 = F_e$ and $K_n^0 u_n^1 = F_e$, and the initial incremental displacement of the nth loading step is

$$K_n^0 \Delta u_n^1 = \Delta\lambda_n^1 F_e. \tag{3.142}$$

The incremental equation of the ith iteration in the nth loading step can be written in convenient form,

$$\begin{cases} K_n^{i-1} u_n^{i1} = F_e \\ K_n^{i-1} u_n^{i2} = R_n^{i-1} \end{cases}. \tag{3.143}$$

The incremental displacement is

$$\Delta u_n^i = \Delta\lambda_n^i u_n^{i1} + u_n^{i2}, \tag{3.144}$$

where the load increment coefficient is

$$\Delta\lambda_n^i = -\frac{\left(u_{n-1}^{11}\right)^T u_n^{i2}}{\left(u_{n-1}^{11}\right)^T u_n^{i1}}. \tag{3.145}$$

Repeat the steps above until the convergence criteria are satisfied. The displacement of each loading step is

$$u = u + \sum \Delta u_n^i. \tag{3.146}$$

In practical application, assuming that the kth displacement of the node variables is the control displacement, and the displacement increment is ΔU, and then the initial load increment coefficient in the nth loading step and the load increment coefficient of the ith iteration can be calculated as follows:

$$\Delta\lambda_n^1 = \frac{\Delta U}{u_1^{11}}, \quad \Delta\lambda_n^i = -\frac{u_n^{i2}}{u_n^{i1}}. \tag{3.147}$$

In each loading step, ΔU and k remain constant. Incremental equations are as eqs. (3.138), (3.143) and (3.144). The displacement increment method can obtain a more accurate critical point, which however, needs to calculate loading coefficient in each time step, thus resulting a larger computational cost.

3.6.2 Nonlinear solution technology for structural dynamic analysis

The structural dynamic equation of motion can be expressed as follows:

$$M\ddot{u} + C\dot{u} + Ku = F_e \tag{3.148}$$

where M is mass matrix, C is damping matrix, K is stiffness matrix, F_e is the total external forces of the structure and $\ddot{u}, \dot{u}, u$ are the acceleration, velocity and displacement, respectively. Equation (3.148) can be solved by finite element discretization of space variables and time domain solution of time variables to obtain the dynamic response.

Three general categories of time domain solving methods are commonly used for structural dynamic analysis: (1) numerical integration methods, the basic idea for which is that assuming the displacement, velocity and acceleration conform to a simple relationship in each time interval Δt. Since it is only assumed that the structural constitutive relation is linear in a small-time step, this method can thus deal with nonlinear structural dynamic problems. (2) Modal superposition method introduces modal generalized coordinates and transforms the finite element coordinate basis to the eigenvector basis. This method is applicable to the case where the structure is relatively simple, the load is periodic and the vibration mode is less, and especially has practical significance for analyzing the frequency and mode of the structure and the effect of each mode in the response, and performing dynamic repair and optimization. Currently, this method is most used for analysis of aeroelastic dynamic stability. Since the modal generalized coordinates in the nonlinear dynamic response problem are difficult to establish, the method is difficult to deal with nonlinear cases. In addition, when there are more high-frequency components in the response, the method requires more high-order modes and the calculation cost is large. (3) State space method, which converts n second-order ordinary differential equations that have been discretized spatially into $2n$ first-order ordinary differential equations. The state variable can be the real displacement and velocity of the finite element node in the structure or the generalized displacement and its derivative (also a kind of modal superposition method). For a given structure, the load form, structural modeling, accuracy requirements and degree of nonlinear influence should be considered comprehensively in performing dynamic analysis, and then an appropriate time domain method is selected. A good numerical analysis method must be convergent, accurate, stable and efficient.

1. Solution of nonlinear dynamics – Newmark algorithm

Writing eq. (3.148) into the nonlinear dynamic equilibrium equation of structure

$$M\ddot{u} + C\dot{u} + F_i = F_e, \tag{3.149}$$

where $F_i = \int \bar{B}^T \sigma dV$ is element internal force. Because of the nonlinearity of F_i, eq. (3.149) needs to be linearized

$$F_{i,n+1} = F_{i,n} + K_{T,n}\Delta u \tag{3.150}$$

where $\Delta u = u_{n+1} - u_n$, and thus the linearized dynamic equilibrium equation is obtained as

$$M\ddot{u}_{n+1} + C\dot{u}_{n+1} + K_{T,n}\Delta u = F_{e,n+1} - F_{i,n}. \tag{3.151}$$

Because the linearization of eq. (3.150) is not accurate enough, the direct solution of eq. (3.151) may produce errors. Thus, the Newmark method with a subiteration is considered, and then the basic equation is

$$M\ddot{u}^i_{n+1} + C\dot{u}^i_{n+1} + K_T\Delta u^i = F_{e,n+1} - F^i_{i,n}. \tag{3.152}$$

The main steps are as follows:
1) Calculate the initial values of u_0, $\dot{u}_0$ and $\ddot{u}_0$.
2) Calculate the tangent stiffness matrix K_{Tl}, mass matrix and element internal force f_i in the local coordinate system.
3) Calculate the transformation matrix T, assemble the global tangent stiffness matrix K_T, mass matrix and node internal forces F_i.
4) Perform the Newmark method with a subiteration.
5) Obtain the node displacement, velocity and acceleration at the next moment.
6) Determine whether the convergence criterion is met, otherwise return to step (2) to continue the calculation.

The Newmark method subiteration:

① Form the equivalent stiffness matrix by the global tangent stiffness matrix, mass matrix and damping matrix at the nth moment

$$\hat{K}_n = a_0 M_n + a_1 C_n + K_{T,n}. \tag{3.153}$$

② Form equivalent load array

$$\hat{R}_{n+1} = F_{e,n+1} + M_n(a_2 \dot{u}_n + a_3 \ddot{u}_n) + C_n(a_4 \dot{u}_n + a_5 \ddot{u}_n) - F_{i,n}. \tag{3.154}$$

③ Solve the initial displacement value

$$\hat{K}_n \Delta u = \hat{R}_{n+1}. \tag{3.155}$$

④ Calculate the displacement, velocity and acceleration for next iteration based on Δu

$$\begin{cases} u_{n+1}^i = u_{n+1}^{i-1} + \Delta u^i \\ \dot{u}_{n+1}^i = a_1 \Delta u^i - a_4 \dot{u}_n^i - a_5 \ddot{u}_n^i \\ \ddot{u}_{n+1}^i = a_0 \Delta u^i - a_2 \dot{u}_n^i - a_3 \ddot{u}_n^i \end{cases}. \tag{3.156}$$

⑤ Solve the internal forces of the structural nodes and calculate the unbalanced forces

$$\psi_{n+1}^i = F_{e,n+1} - \left(M \ddot{u}_{n+1}^i + C \dot{u}_{n+1}^i + F_{i,n+1}^{\,i}\right). \tag{3.157}$$

⑥ Solve the correction of the displacement increment for the ith iteration to get the displacement increment for the $i+1$th iteration

$$\hat{K}_n \Delta u'^i = \psi_{n+1}^i, \tag{3.158}$$

$$\Delta u^{i+1} = \Delta u^i + \Delta u'^i. \tag{3.159}$$

⑦ Determine whether the displacement increment satisfies the convergence criterion, otherwise return to step ④.

The coefficients of the Newmark method are $a_0 = 1/(\beta \Delta t^2)$; $a_1 = \gamma/(\beta \Delta t)$; $a_2 = 1/(\beta \Delta t)$; $a_3 = 1/(2\beta) - 1$; $a_4 = \gamma/\beta - 1$ and $a_5 = \Delta t/2 \cdot (\gamma/\beta - 2)$.

For the damping matrix C, Rayleigh damping is applied, and it is assumed that the damping matrix is a combination of mass and stiffness matrices as

$$C = \alpha_0 M + \alpha_1 K_T \tag{3.160}$$

where a_0 and a_1 are two scale factors, with dimensions of s^{-1} and s, respectively, which are given by

$$\begin{Bmatrix} \alpha_0 \\ \alpha_1 \end{Bmatrix} = \frac{2\omega_i \omega_j}{\omega_j^2 - \omega_i^2} \begin{pmatrix} \omega_j & -\omega_i \\ -\frac{1}{\omega_j} & \frac{1}{\omega_i} \end{pmatrix} \begin{Bmatrix} \zeta_i \\ \zeta_j \end{Bmatrix}, \tag{3.161}$$

where ω_i and ω_j are the ith and jth natural frequencies of the structure, ζ_i and ζ_j are the ith and jth damping ratio for the corresponding modes. When $\zeta_i = \zeta_j = \zeta$, the above equation is simplified as

$$\begin{Bmatrix} \alpha_0 \\ \alpha_1 \end{Bmatrix} = \frac{2\zeta}{\omega_i + \omega_j} \begin{pmatrix} \omega_i \omega_j \\ 1 \end{pmatrix}. \tag{3.162}$$

It should be noted that frequencies of ω_i and ω_j need to cover the interested frequency bands in structure analysis. In the nonlinear analysis, the average frequency can be considered to calculate the damping matrix. The Newmark method is widely used in solving structural dynamics. Although the method is unconditionally stable for linear problems, the stability of this method is limited for nonlinear problems and requires a step. Moreover, for high-frequency responses, the results obtained by this method may appear deadlock or unreal amplification.

2 Solution of nonlinear dynamics – approximate energy conservation algorithm

In the process of structural dynamic response, the energy change is introduced by neglecting damp [1, 9, 10]:

$$\Delta E = \Delta E_\varepsilon + \Delta E_v + \Delta E_p, \tag{3.163}$$

where ΔE_ε, ΔE_v and ΔE_p are the increments of structural strain energy, kinetic energy and potential energy, respectively. The above equation should equal to zero to satisfy the energy conservation. For nonlinear structural dynamic equation, it is assumed that the displacement, velocity and acceleration confirm to the following relationships:

$$\begin{cases} (\dot{u}_{n+1} + \dot{u}_n)/2 = (u_{n+1} - u_n)/\Delta t \\ (\ddot{u}_{n+1} + \ddot{u}_n)/2 = (\dot{u}_{n+1} - \dot{u}_n)/\Delta t \end{cases}. \tag{3.164}$$

The nonlinear equilibrium equation can be written as

$$g_{n+1} = F_{i,n+1} + M\ddot{u}_{n+1} - F_{e,n+1} = 0. \tag{3.165}$$

Considering eq. (3.164), there is

$$g_{n+1} = T_{n+1}^T f_{il} + \frac{2}{\Delta t} M(\dot{u}_{n+1} - \dot{u}_n) - M\ddot{u}_n - F_{e,n+1} = 0. \tag{3.166}$$

Taking a two-dimensional beam as an example, considering the configuration shown in Fig. 3.12, the intermediate values before and after the deformation are introduced as

$$f_{il,n+1/2} = \frac{1}{2}(f_{il,n} + f_{il,n+1}) = \begin{bmatrix} \bar{N}_{n+1/2} & \bar{M}_{1,n+1/2} & \bar{M}_{2,n+1/2} \end{bmatrix}^T, \tag{3.167}$$

$$F_{i,n+1/2} = T_{n+1/2}^T f_{il,n+1/2}, \quad T_{n+1/2} = 1/2(T_n + T_{n+1}), \tag{3.168}$$

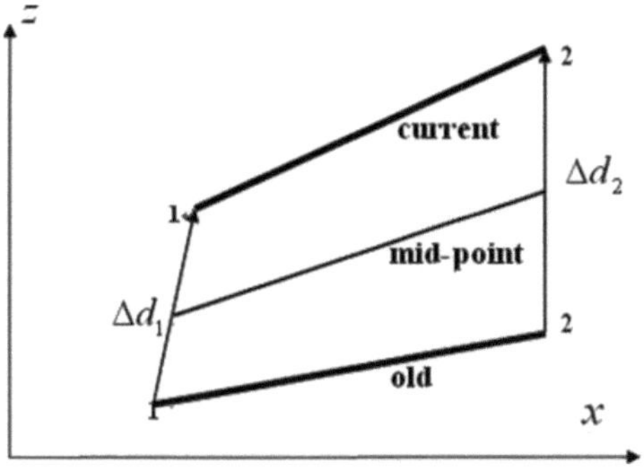

Fig. 3.12: Motion of a two-dimensional beam.

$$\ddot{u}_{n+1/2} = (\dot{u}_{n+1} - \dot{u}_n)/\Delta t. \tag{3.169}$$

Substituting eqs. (3.167)–(3.169) into eq. (3.166) produces

$$g_{n+1/2} = F_{i,n+1/2} + \frac{1}{\Delta t} M(\dot{u}_{n+1} - \dot{u}_n) - F_{e,n+1/2} = 0. \tag{3.170}$$

Then, the energy change is

$$\begin{cases} \Delta E_\varepsilon = F^T_{i,n+1/2} \Delta u_{n+1/2} \\ \Delta E_v = \frac{1}{2} M(\dot{u}_{n+1} - \dot{u}_n)^2 = (\dot{u}_{n+1} - \dot{u}_n)^T M\left(\frac{\dot{u}_{n+1}+\dot{u}_n}{2}\right) = \left(\frac{\dot{u}_{n+1}-\dot{u}_n}{\Delta t}\right)^T M \Delta u_{n+1/2} \\ \Delta E_p = -F^T_{e,n+1/2} \Delta u_{n+1/2} \end{cases}. \tag{3.171}$$

The total energy change is

$$\Delta E = g_{n+1/2} \Delta u_{n+1/2}. \tag{3.172}$$

When the numerical solution of eq. (3.170) reaches convergence, $g_{n+1/2=0}$, the total energy change is zero, which guarantees the energy conservation. Calculate the differentiation of eq. (3.168)

$$\delta F_{i,n+1/2} = T_{n+1/2} \delta f_{il,n+1/2} + \delta T_{n+1/2} f_{il,n+1/2}. \tag{3.173}$$

The first right term of eq. (3.173) can be derived as follows:

$$T_{n+1/2} \delta f_{il,n+1/2} = T_{n+1/2} \delta \left(\frac{1}{2} \left(f_{il,n} + f_{il,n+1}\right)\right) = \frac{1}{2} \left(\frac{T_n + T_{n+1}}{2}\right)^T \delta f_{il,n+1}. \tag{3.174}$$

The second right term of eq. (3.173) can be derived as follows:

$$\delta T_{n+1/2} f_{il,n+1/2} = \delta\left[\frac{1}{2}(T_n + T_{n+1})\right] f_{il,n+1/2} = \frac{1}{2}\delta T_{n+1} f_{il,n+1/2}. \tag{3.175}$$

Based on eqs. (3.108) and (3.174), the linear tangent stiffness matrix can be derived from the first right term of eq. (3.173) as

$$K_{tl} = \frac{1}{2}\left(\frac{T_n + T_{n+1}}{2}\right)^T K_l T_{n+1}. \tag{3.176}$$

Based on eqs. (3.108) and (3.175), the geometric stiffness matrix can be derived from the second right term of eq. (3.173) as

$$K_{t\sigma} = \frac{1}{2}\left[\bar{N}_{n+1/2}/l_{n+1} \cdot p_{n+1} p_{n+1}^T + \left(\bar{M}_{1,n+1/2} + \bar{M}_{2,n+1/2}\right)/l_{n+1}^2 \cdot \left(r_{n+1} p_{n+1}^T + p_{n+1} r_{n+1}^T\right)\right]. \tag{3.177}$$

Then, the global tangent stiffness matrix is obtained as

$$K_{n+1/2} = K_{tl} + K_{t\sigma}. \tag{3.178}$$

It can be seen that the derivation of this method is suitable for any type of structural element. Similarly, the tangent stiffness matrix of the three-dimensional shell element based on the approximate energy conservation method can be derived using the developed eqs. (3.81) and (3.136),

$$K_{n+1/2} = K_{tl} + K_{t\sigma} = \frac{1}{2}\left(\frac{T_n + T_{n+1}}{2}\right)^T K_l T_{n+1} + \frac{1}{2} K_{\sigma m}, \tag{3.179}$$

where the calculation of $K_{\sigma m}$ needs the displacement and internal force at $n+1/2$ step. It can be seen that the tangent stiffness matrix at this time is no longer a symmetric matrix. If the degree of freedom of the structure is large, the solution requires a larger storage capacity and calculation time.

3 Solution of nonlinear dynamics – exact energy conservation algorithm

Strictly speaking, the first formula of eq. (3.171) is not an increment of strain energy from n step to $n+1$ step. Taking pure axial strain as an example, the increment of strain energy from n to $n+1$ is [3]

$$\Delta E_\varepsilon = \frac{1}{2}\left[\bar{N}_{n+1}(l_{n+1} - l_0) - \bar{N}_n(l_n - l_0)\right]. \tag{3.180}$$

Using the internal force relationship $\bar{N} = (EA/l_0)\bar{u}$, eq. (3.180) becomes

$$\Delta E_\varepsilon = \frac{1}{2}\frac{EA}{l_0}\left[(l_{n+1} - l_0)^2 - (l_n - l_0)^2\right]. \tag{3.181}$$

Introducing $\Delta l = l_{n+1} - l_n$ and eq. (3.181) is simplified as

$$\Delta E_\varepsilon = \frac{EA}{l_0}\left(l_n - l_0 + \frac{\Delta l}{2}\right)\Delta l = \bar{N}_{n+1/2}\Delta l. \tag{3.182}$$

Using eq. (3.168) and the first formula of eq. (3.171), we obtain

$$\Delta E_\varepsilon = \frac{\bar{N}_{n+1/2}}{2}(r_n + r_{n+1})^T \Delta u_{n+1/2}. \tag{3.183}$$

If there is rigid body rotation, as shown in Fig. 3.13, then obviously there is

$$\Delta l \neq \frac{(r_n + r_{n+1})^T}{2}\Delta u_{n+1/2.} \tag{3.184}$$

Thus, a correction can be made to eq. (3.184) as follows:

$$\Delta l = \frac{1}{1 + \cos\Delta\alpha}(r_n + r_{n+1})^T \Delta u_{n+1/2.} \tag{3.185}$$

Similarly, in the use of the energy conservation method, the strain caused by bending should be corrected

$$T_{n+1/2} = D\left(\frac{T_n^0 + T_{n+1}^0}{2}\right) + \bar{I} = DT_{n+1/2}^0 + \bar{I} = \bar{T}_{n+1/2} + \bar{I}, \tag{3.186}$$

where T^0 is

$$T^0 = \left(r^T - p^T / l - p^T / l\right)^T. \tag{3.187}$$

D is the diagonal matrix as follows:

$$\mathrm{dig}(D) = \left(\frac{2}{1 + \cos\Delta\alpha}, \ \frac{\Delta\alpha}{\sin\Delta\alpha}, \ \frac{\Delta\alpha}{\sin\Delta\alpha}\right). \tag{3.188}$$

There is $\bar{I} = T - T^0$ in eq. (3.186), and substituting eq. (3.186) into eq. (3.168) derives

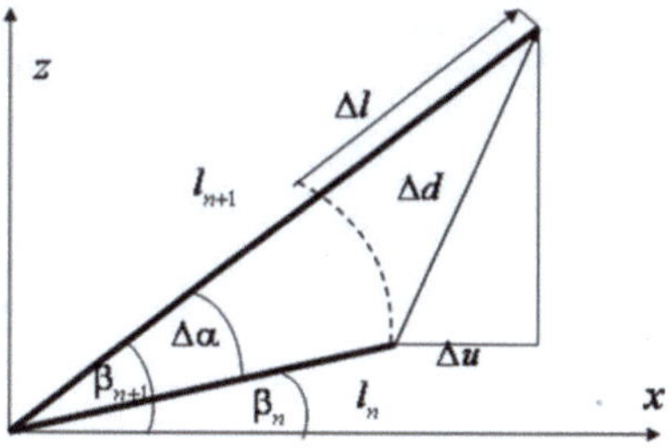

Fig. 3.13: The rotating motion of a two-dimensional beam.

$$F_{i,n+1/2} = \left(\bar{T}^T_{n+1/2} + \bar{I}^T\right) f_{il,n+1/2}. \tag{3.189}$$

Calculating the differentiation of eq. (3.189), there is

$$\delta F_{i,n+1/2} = \left(\bar{T}^T_{n+1/2} + \bar{I}^T\right) \delta f_{il,n+1/2} + \delta\left(\bar{T}^T_{n+1/2} + \bar{I}^T\right) f_{il,n+1/2}. \tag{3.190}$$

The second right term of eq. (3.190) can be derived as follows:

$$\delta\left(\bar{T}^T_{n+1/2} + \bar{I}^T\right) = \delta\bar{T}^T_{n+1/2} = \delta T^{0T}_{n+1/2} D + T^{0T}_{n+1/2} \delta D. \tag{3.191}$$

Derivation from eq. (3.108) and the first right term of eq. (3.190), we obtain

$$K_{tl} = \frac{1}{2}\left(\bar{T}^T_{n+1/2} + \bar{I}^T\right) K_l T_{n+1}. \tag{3.192}$$

The first right term of eq. (3.191) becomes

$$\delta T^{0T}_{n+1/2} D = \frac{1}{2} \delta T^{0T}_{n+1} D = \frac{1}{2} \delta T^T_{n+1} D. \tag{3.193}$$

Let $\Delta\alpha = \alpha$, and the above formula with a right multiplication factor $f_{il,n+1/2}$ can thus be written as

$$\begin{aligned} K_{t\sigma 1} = &\frac{1}{1+\cos\alpha} \frac{p_{n+1} p^T_{n+1}}{l_{n+1}} \bar{N}_{n+1/2} \\ &+ \frac{1}{2l^2_{n+1}} \frac{\alpha}{\sin\alpha} \left(r_{n+1} p^T_{n+1} + p_{n+1} r^T_{n+1}\right)\left(\bar{M}_{1,n+1/2} + \bar{M}_{2,n+1/2}\right). \end{aligned} \tag{3.194}$$

For the second right term of eq. (3.191), the differentiation of D is obtained:

$$\delta\left(\frac{\alpha}{\sin\alpha}\right) = \frac{\sin\alpha - \alpha\cos\alpha}{\sin^2\alpha} \delta\alpha, \quad \delta\left(\frac{1}{1+\cos\alpha}\right) = \frac{\sin\alpha}{(1+\cos\alpha)^2} \delta\alpha. \tag{3.195}$$

Based on the geometric relationship shown in Fig. 3.13, the following expression can be obtained:

$$\delta\alpha = \frac{-\delta(u_{2,n+1} - u_{1,n+1}) \sin\beta_{n+1} + \delta(w_{2,n+1} - w_{1,n+1}) \cos\beta_{n+1}}{l_{n+1}}. \tag{3.196}$$

Based on the previous definition, eq. (3.196) can be written as

$$\delta\alpha = \frac{1}{l_{n+1}} \begin{bmatrix} -s_{n+1} & c_{n+1} \end{bmatrix} \begin{bmatrix} \delta u_{2,n+1} - \delta u_{1,n+1} \\ \delta w_{2,n+1} - \delta w_{1,n+1} \end{bmatrix}. \tag{3.197}$$

Substitutting eq. (3.195) into eq. (3.188) results in

$$\mathrm{dig}(\delta D) = \left(\frac{2\sin\alpha}{(1+\cos\alpha)^2}\delta\alpha, \quad \frac{\sin\alpha - \alpha\cos\alpha}{\sin^2\alpha}\delta\alpha, \quad \frac{\sin\alpha - \alpha\cos\alpha}{\sin^2\alpha}\delta\alpha \right) \tag{3.198}$$

δD with a left multiplication factor $T^{0T}_{n+1/2}$ and then a right multiplication factor $f_{il,n+1/2}$ becomes

$$\begin{aligned} K_{t\sigma 2} = & \frac{\sin\alpha}{(1+\cos\alpha)^2}\frac{\bar{N}_{n+1/2}}{l_{n+1}}\left[(r_n + r_{n+1})p^T_{n+1}\right] \\ & + \frac{\sin\alpha - \alpha\cos\alpha}{\sin^2\alpha}\frac{\bar{M}_{1,n+1/2} + \bar{M}_{2,n+1/2}}{2l_{n+1}}\left[-\left(\frac{p_n}{l_n} + \frac{p_{n+1}}{l_{n+1}}\right)p^T_{n+1}\right]. \end{aligned} \tag{3.199}$$

Therefore, the tangent stiffness matrix of the exact energy conservation method is

$$K_{n+1/2} = K_{tl} + K_{t\sigma 1} + K_{t\sigma 2}. \tag{3.200}$$

This tangent stiffness matrix is still asymmetric. For the three-dimensional shell element, there is also a tangent stiffness matrix for exact energy conservation method, but the derivation is quite complicated and the calculational cost is huge.

4 Solution of nonlinear dynamics – predictor–corrector procedure based on energy conservation

The expressions of the tangent stiffness matrix of the two-dimensional beam element for the approximate energy conservation and the exact energy conversation algorithms are derived, and the approximate energy conservation tangent stiffness matrix of the three-dimensional shell element is also developed. Here, the predictor–corrector procedure is introduced in order to solve the nonlinear dynamic equilibrium equation of the structure [9, 10].

(1) Predictor step:
first, solve the initial value of the displacement increment

$$\bar{K}_{n+1/2}\Delta u = \Delta F, \tag{3.201}$$

where

$$\bar{K}_{n+1/2} = K_{n+1/2} + \frac{2}{\Delta t^2}M, \quad \Delta F = F_{e,n+1/2} - F_{i,n+1/2} + \frac{2}{\Delta t}M\dot{u}_n. \tag{3.202}$$

Thus, the displacement and velocity of $n+1$ step can be predicted by

$$\begin{cases} u_{n+1} = u_n + \Delta u \\ \dot{u}_{n+1} = 2/\Delta t \cdot \Delta u - \dot{u}_n. \end{cases} \tag{3.203}$$

Substituting the obtained velocity into eq. (3.170), obviously $g_{n+1/2} \neq 0$.

(2) Corrector step:
The residual obtained earlier is recorded as $g_{n+1/2}^{i-1}$, and thus the iterative equation is as follows:

$$g_{n+1/2}^{i-1} + \bar{K}_{T,n+1/2}^{i-1} \delta u_{n+1}^{i} = 0, \tag{3.204}$$

where $K_{T,n+1/2}^{i-1}$ is obtained from the tangent stiffness matrix derived from Section 3.5.1, and then $\bar{K}_{T,n+1/2}^{i-1}$ is obtained from eq. (3.201). Solve eq. (3.204), and update displacement and velocity at each corrector iteration

$$\begin{cases} u_{n+1}^{i} = u_{n+1}^{i-1} + \delta u_{n+1}^{i} \\ \dot{u}_{n+1}^{i} = \dot{u}_{n+1}^{i-1} + \delta \dot{u}_{n+1}^{i} \end{cases} \tag{3.205}$$

where

$$\delta \dot{u}_{n+1}^{i} = 2/\Delta t \cdot \delta u_{n+1}^{i}. \tag{3.206}$$

After the displacement and velocity are obtained, the displacement in the local coordinate system can be computed by CR theory. Therefore, the internal force at $n+1$ step can be obtained, and the internal force and then the residual $g_{n+1/2}$ at $n+1/2$ step can be calculated. By updating the tangent stiffness matrix, and the iterative calculation is repeated until the convergence criterion is satisfied.

3.6.3 Numerical analysis

1 Nonlinear static analysis of structure: displacement of a cantilever beam with concentrated load

As shown in Fig. 3.14, the cantilever beam is subjected to an end moment M. Take $l = 20$ m, $EI_z = 10$ Nm, $EA = 10$ N and $M = 0.2\pi,\ 0.4\pi,\ 0.6\pi,\ 0.8\pi,\ \pi,$ respectively. While the exact analytical solution is an arc with arc length of 20 and a central angle of $\theta = (Ml)/(EI_z) = 0.4\pi,\ 0.8\pi,\ 1.2\pi,\ 1.6\pi,\ 2\pi$.

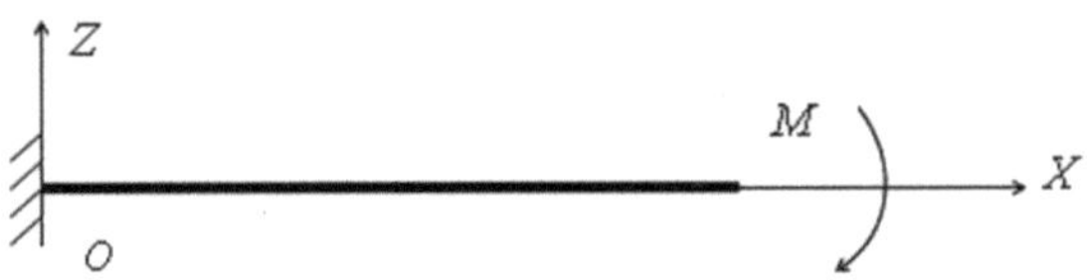

Fig. 3.14: A two-dimensional cantilever beam subjected to concentrated bending moment.

The linear finite element method, traditional nonlinear TL method, CR-based load increment method and CR-based displacement increment method are used, respectively.

The comparison of the resulting displacement u, w and analytical solution is shown in Tab. 3.1.

Tab. 3.1: Displacement of the two-dimensional cantilever beam subjected to end bend moment.

			u/l		
M	**CR load**	**CR displacement**	***T.L***	**Linear solution**	**Analytical solution**
0.2π	−0.243	−0.243	−0.134	0.0	−0.243
0.4π	−0.769	−0.768	−0.531	0.0	−0.766
0.6π	−1.154	−1.158	−1.180	0.0	−1.156
0.8π	−1.190	−1.189	−2.101	0.0	−1.189
π	−1.0	−1.0	−3.287	0.0	−1.0
			w/l		
M	**CR load**	**CR displacement**	***T.L***	**Linear solution**	**Analytical solution**
0.2π	−0.550	−0.550	−0.630	−0.628	−0.550
0.4π	−0.720	−0.720	−1.261	−1.260	−0.720
0.6π	−0.483	−0.477	−1.881	−1.88	−0.478
0.8π	−0.138	−0.136	−2.511	−2.51	−0.137
π	0.0	0.0	−3.140	−3.14	0.0

Figures 3.15–3.18 show the displacements of the cantilever beam subjected to end bend moment using different solving methods. It can be seen from the table and the figures that the linear result has no displacement in the x-direction, and the displacement in the z-direction increases linearly with the load; the traditional TL method can obtain satisfactory results when the geometric nonlinearity is not very large, but the error becomes significantly larger with the load increase, so it is not suitable for solving the nonlinear problems with large deformation; the result of the CR method is more accurate and the deformation is close to the analytical solution. From the table analysis, the accuracy of the displacement increment method is higher, but from the calculation process, the load increment method is more efficient.

2 Nonlinear dynamic analysis of structure: nonlinear dynamic response of a free beam

The cantilever beam is free swing with gravity, as shown in Fig. 3.19. Take the material parameters as $L = 1\,\text{m}$, $EI_z = 3.6 \times 10^4\,\text{Nm}$, $EA = 2.4 \times 10^5\,\text{N}$, $\rho A = 1.0\,\text{kg/m}$, and the

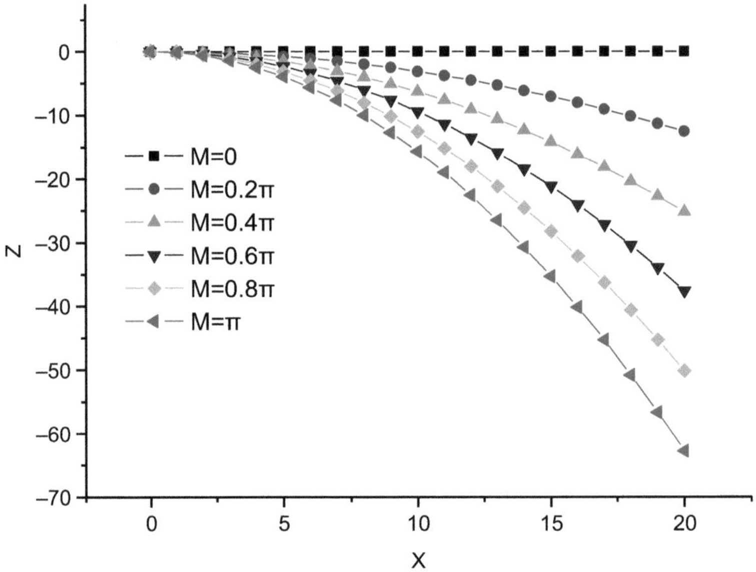

Fig. 3.15: Linear displacement solution.

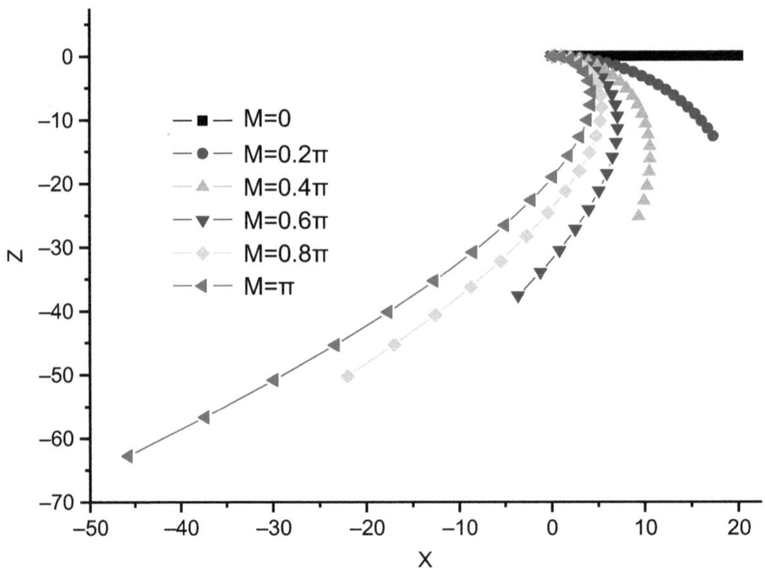

Fig. 3.16: Displacement solution via TL method.

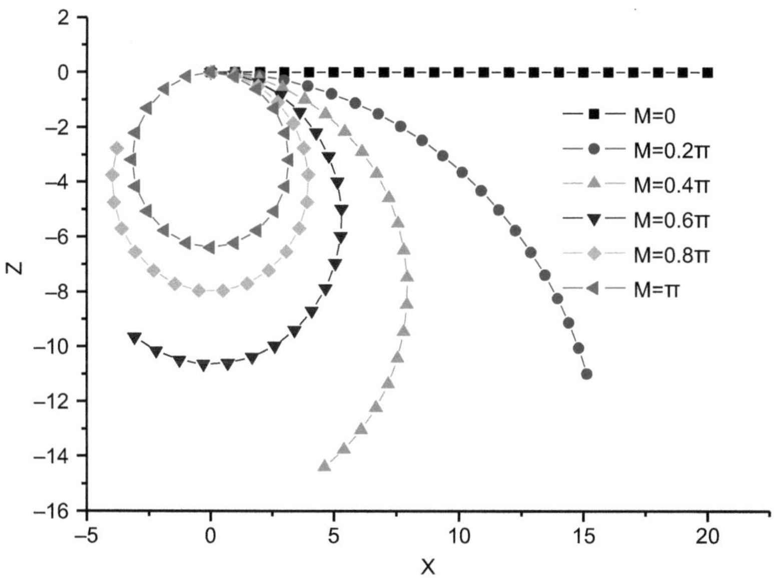

Fig. 3.17: Displacement solution via CR load increment method.

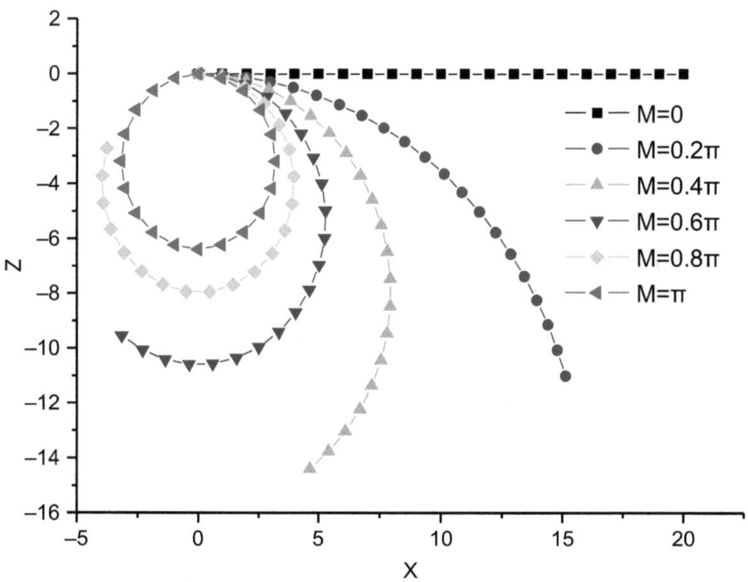

Fig. 3.18: Displacement solution via CR displacement increment method.

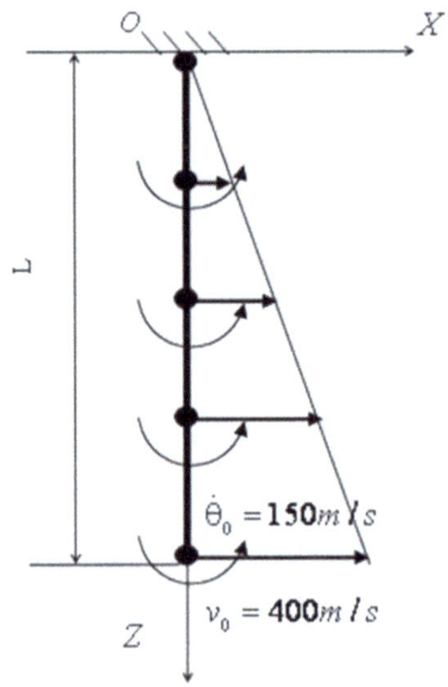

Fig. 3.19: A cantilever beam subjected to gravity.

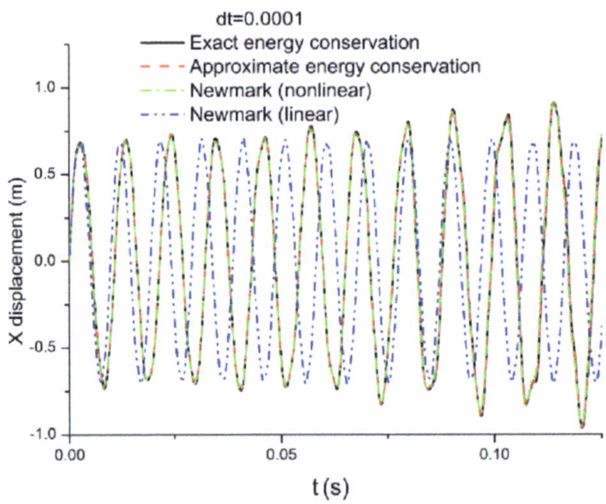

Fig. 3.20: Displacement in *x*-direction with four methods.

initial velocity of the free end of the cantilever beam as $v_0 = 400$ m/s,$\dot{\theta}_0 = 150$ m/s. In structural modeling, the concentrated mass is adopted, and four tow-dimensional beam elements are constructed. The free end response is solved by the linear Newmark method, nonlinear Newmark method, approximate energy conservation and exact energy conservation method, respectively. The time step is taken as $dt = 0.0001$ s and $dt = 0.001$ s.

Figures 3.20 and 3.21 show the free-end displacement and velocity responses calculated by the linear Newmark, nonlinear Newmark, approximate energy conservation and exact energy conservation method at $dt = 0.0001$ s, respectively. It can be seen that when the time-step is small, the linear result is obviously deviated, and the results of the three nonlinear methods are consistent. Figures 3.22 and 3.23 show the energy changes of the three nonlinear methods. Since the external load remains constant, the energy obtained by the exact energy conservation method is kept at zero, which however has a small amplitude (10^{-2}) by the approximate energy conversation method. The response obtained by the nonlinear Newmark method is consistent, but the amplitude of the energy change is close to the magnitude of 10^4.

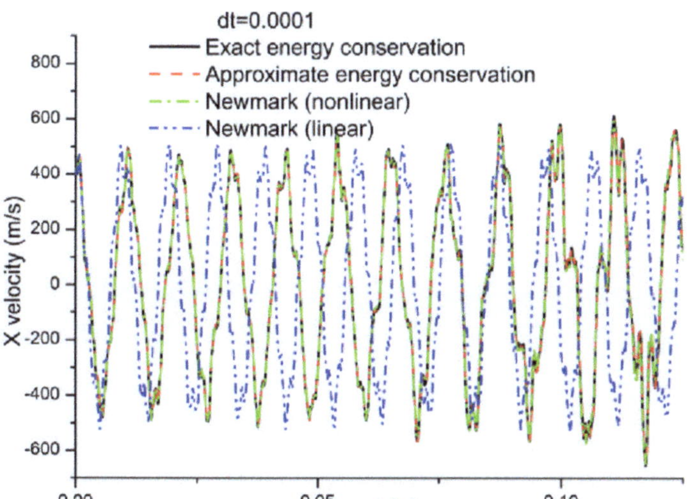

Fig. 3.21: Velocity in *x*-direction with four methods.

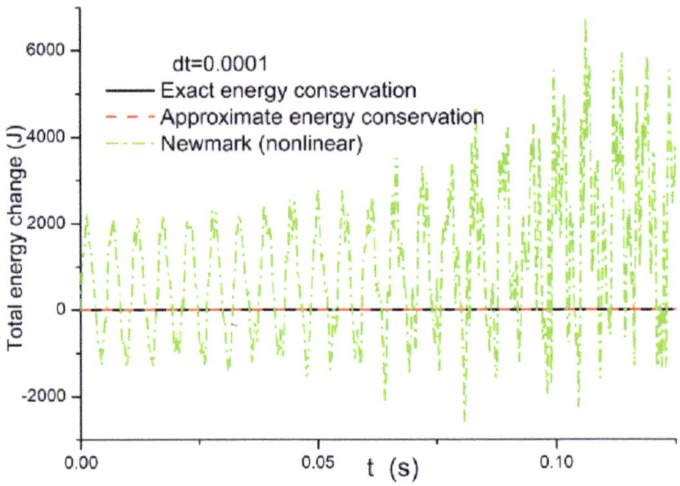

Fig. 3.22: Total energy change with three nonlinear methods.

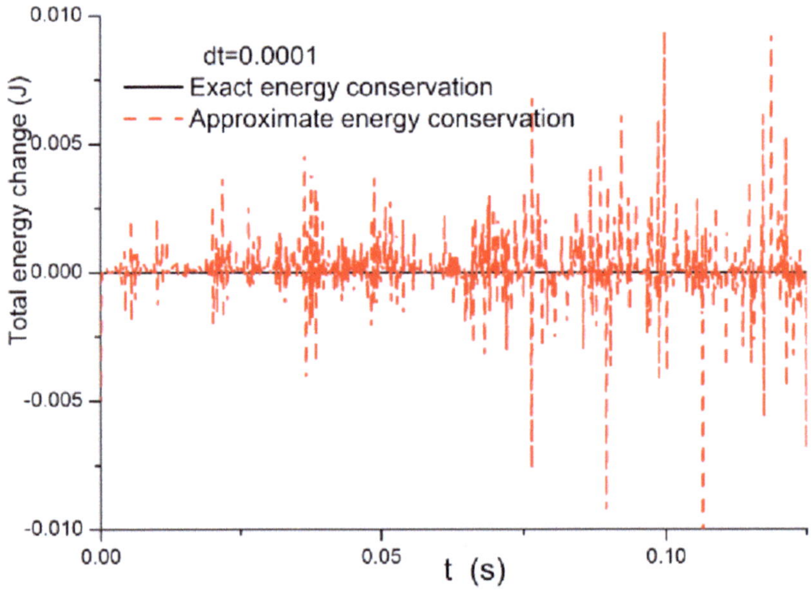

Fig. 3.23: Total energy change with two energy conversation methods.

Bibliography

[1] Battini JM. A modified corotational framework for triangular shell elements. Computer Methods in Applied Mechanics and Engineering. 2007, 196, 1905–1914.

[2] Battini JM. Co-Rotational Beam Elements in Instability Problems. Technical Reports from Royal Institute of Technology Department of Mechanics SE-100 44 Stockholm. 2002. 1.

[3] Crisfield MA, Galvaneito U. An energy-conserving co-rotational procedure for the dynamics of planar beam structures. International Journal for Numerical Methods in Engineering. 1996, 39, 2265–2282.

[4] Crisfield MA. Nonlinear Finite Element Analysis of Solids and Structures, Advanced Topics. Wiley, Chichester; 1997. 2.

[5] Hancock GJ, Clarke MJ. Study of incremental-iterative strategies for non-linear analyses. International Journal for Numerical Methods in Engineering. 1990, 29, 1365–1391.

[6] Pacoste C, Battini JM. Co-rotational beam elements with warping effects in instability problems, Comput. Methods in Applied Mechanics. 2002, 191(17/18), 1755–1789.

[7] Pacoste C. Co-rotational flat facet triangular elements for shell instability analysis. Computer Methods in Applied Mechanics and Engineering. 1998, 156, 75–110.

[8] Shieh MS, Yang YB. Solution method for nonlinear problems with multiple critical points. AIAA Journal. 1990, 28, 2110–2116.

[9] Suleman A, Relvas A. Application of the corotational structural kinematics and Euler flow to two-dimensional nonlinear aeroelasticity. Computers and Structures. 2007, 85, 1372–1381.

[10] Suleman A, Relvas A. Fluid–structure interaction modelling of nonlinear aeroelastic structures using the finite element corotational theory. Journal of Fluids and Structures. 2006, 22, 59–75.

[11] Vu-Quoc L, Tan X G. Optimal solid shells for non-linear analysis of multilayer composites. 2. Dynamics. Computer Methods in Applied Mechanics and Engineering, 2003, 192 (9–10), 1017–1059.

Chapter 4
Interpolation and moving-grid technique for CFD/CSD coupling program

4.1 Introduction

Computational fluid dynamics (CFD)/computational structure dynamics (CSD) coupling method is an effective way to solve the nonlinear aeroelastic problem, in which the unsteady aerodynamic force is computed by CFD technology, and the responses of the elastic structure are calculated by CSD technology. However, the two domains have a different description, the fluid mechanics system is always described by the Eulerian method, and the solid system is described by the Lagrange method.

In the Lagrangian description (i.e., material description or L description), the computational grid is fixed on the body and moves with the body. That is, the grid points and the material points always coincide in the deformation of the body. Consequently, there is no relative motion between the material points and the mesh points (i.e., migration motion or convective motion). It not only simplifies the solution of the governing equation, but also describes the moving interface of the body accurately, and tracks the trajectory of the particle. However, when it comes to the large deformation case, the calculation may fail if the mesh is distorted with the moving body.

In the Euler description (i.e., spatial description or *E* description), the mesh is fixed in space, that is, in the deformation of the body, the distortion can be avoided as the computational mesh stays invariant. Nevertheless, a complicated mapping technique should be introduced to describe the moving interface, which may lead to a rather serious error. In addition, when the Galerkin method is employed in discretization, the coefficient matrix of the finite element equation is asymmetric and it is possible to obtain the oscillatory solution due to the influence of the transfer term.

Both the Lagrangian and Eulerian descriptions have their own advantages and defects. Actually, these two descriptions are the extreme cases of the arbitrary Lagrangian Euler (ALE) method (ALE descriptions), it means, when the speed of the mesh equals to the speed of the body, it degenerates into Lagrangian description; on the contrary, when the mesh is fixed in space, it is considered as Eulerian description. If the two descriptions can be combined appropriately to take their own advantages and avoid the two intrinsic shortcomings, a large number of engineering problems can be solved effectively. For this purpose, Noh [27] proposed an ALE description, in which the computational mesh can move in any form in space, it is independent of material coordinates and space coordinates. Thus, the moving interface of the body

https://doi.org/10.1515/9783110576689-004

can be accurately described and the shape of the element can be maintained by defining the appropriate form of mesh motion. Noh successfully applied the finite difference method with the ALE method to solve two-dimensional (2D) fluid problems with moving boundary.

The description of the interface in the CFD/CSD coupling method for nonlinear aeroelastic numerical solution belongs to the ALE method. The two different domains, fluid and structure, will interact and influence each other on their common interface. As a result, such CFD/CSD coupling system is highly nonlinear, and the two fields are coupled by the moving interface with continuity and compatibility conditions.

In order to conserve the mass, momentum and energy of the two fields, the data exchange on the coupled interface should satisfy three conditions: (1) stress balance condition, that is, on the CFD/CSD coupled common interface, the stress on the structural mesh should be equal to the stress on the aerodynamic mesh; (2) displacement compatibility condition, that is, there is a unique displacement on the CFD/CSD coupled common interface; (3) velocity compatibility condition, there is a unique velocity on the CFD/CSD coupled common interface unit. The equations of the three conditions are expressed as follows:

$$\begin{cases} \sigma_S \cdot n = -pn \\ x = u \\ \dot{x} = \dot{u} \end{cases}, \tag{4.1}$$

where σ_S and p represent the stress on the structural mesh and the pressure on the aerodynamic mesh, respectively. $x, \dot{x}$ and $u, \dot{u}$ represent the displacements and velocities on the aerodynamic surface and structural surface, respectively.

Generally, both fluid and structure domains have their own rules for mesh generation because the two domains are controlled by totally different physical principles. That is, the difference between the Lagrangian system and Eulerian system. The computational meshes of CFD and CSD are generally inconsistent: (1) in CFD solver, the mesh is related to the gradient of physical quantities, where the gradient of physical quantities varies greatly, the meshes are relatively dense, and vice versa; whereas in CSD solution, the mesh is composed of uniform three-node or four-node elements; (2) the attributes of the two mesh systems are also different: CFD analysis is concerned with the flow around the body exposed in the flow field. For example, the flow around a rigid airfoil depends on the profile of the airfoil and the control field; instead, the CSD solution focuses on the surface loads and on how these loads affect the internal structure of the airfoil. The mesh in the CSD solution depends on the surface and interior of the airfoil, and the location of the nodes is determined by the structural element. Therefore, it is very necessary to exchange the data on the common boundary of the two mesh systems.

In CFD/CSD coupling procedure, since the common interface is the channel for information exchange between fluid and structure, the earlier three conditions can be satisfied only by making sure that the parameters on the common boundary are properly discretized and correctly exchanged, thereby the momentum and energy of the coupling system can be conserved. Generally, the exchanged data from structure to the fluid domain is the displacement of the boundary, and the transferred information from fluid to structure is the load on the common surface. Therefore, the first two equations in eq. (4.1) should be satisfied in the data exchange method. Since the velocity can be considered as the derivative of the displacement, the velocity compatibility condition on the boundary can be satisfied if the displacement compatibility condition is satisfied as well as it has a discrete solution scheme with compatible time precision.

There are two kinds of data transfer on the common boundary between CFD and CSD grid systems: one is interpolating the displacements calculated by CSD solver to CFD grid, which can be thus taken as the dynamic boundary conditions for CFD solution (displacement and velocity on the boundary); and the other is transforming aerodynamic loads computed by the CFD code to CSD mesh points, which will form the equivalent node load for the dynamic analysis of the structure. Both the transformations should consider the criteria: (1) accuracy, to reflect the real-time state of pressure and displacement information; (2) smoothness, to maintain the continuity of physical attributes in the computational domain; (3) stability, to make sure that the data exchange will not cause the divergence of the coupling calculation; (4) ease of use, to be easily programmed; (5) and efficiency, to obtain steady results with less Central Processing Units (CPU) time and small storage.

At present, the most popular interpolation methods can be categorized as two groups [14, 33, 24, 17, 8, 7, 21, 10]: one is the surface fitting method, which utilizes a global function with weighting coefficients to obtain the displacements of unknown points, for example, infinite-plate splines (IPS) method, which is dominantly used for the displacement transfer; and the other is surface tracking method, which utilizes shape functions of the finite element by projecting fluid nodes to the nearest structural elements, such as constant volume transformation (CVT) method. Currently, the surface tracking method is mostly used for the load transfer.

According to the basic theory of the ALE method, this chapter focuses on the description of space coordinate system motion as well as mesh movement. Various interpolation methods and moving-grid algorithms are discussed, the mapping method for integrated transformation of the aerodynamic load and structural displacement is developed, and a general interface mapping matrix is designed.

4.2 Interpolation methods

Currently, several popular interpolations in CFD/CSD coupling procedure are [33, 4, 30, 15]: (1) IPS; (2) finite-plate splines (FPS); (3) multiquadric biharmonic (MQ); (4) thin-plate splines (TPS); (5) inverse isoparametric mapping (IIM); (6) nonuniform B-splines (NUBS); and (7) spatial interpolation of load parameters.

4.2.1 Infinite-plate splines method

The IPS method [33, 4] is a mathematical tool for interpolating binary functions. It is based on a superposition of the solutions for the partial differential equation of equilibrium for an infinite plate. It was originally developed for the interpolation and slope calculation of wing disturbances in aeroelastic calculations. It is one of the most popular methods of interpolation used in some commercial programs. The surface spline equation is

$$W_j(x,y) = a_0 + a_1 x_j + a_2 y_j + \sum_{i=1}^{N} F_i r_{ij}^2 \ln r_{ij}^2, (j=1, \ldots\ldots N), \tag{4.2}$$

with three supplement equations $\sum_{i=1}^{N} F_i = \sum_{i=1}^{N} x_i F_i = \sum_{i=1}^{N} y_i F_i = 0$. There are $N+3$ unknowns a_0, a_1, a_2, F_1, F_2, ..., F_N. Equation (4.2) can be written in a matrix form, and linear equations with unknowns are obtained as follows:

$$\begin{bmatrix} w_1 \\ w_2 \\ w_3 \\ \vdots \\ w_N \\ 0 \\ 0 \\ 0 \end{bmatrix} = \begin{bmatrix} 1 & x_1 & y_1 & r_{11}^2 \ln r_{11}^2 & \vdots & \ddots & r_{1N}^2 \ln r_{1N}^2 \\ 1 & x_2 & y_2 & r_{21}^2 \ln r_{21}^2 & \vdots & \ddots & r_{2N}^2 \ln r_{2N}^2 \\ 1 & x_3 & y_3 & r_{31}^2 \ln r_{31}^2 & \vdots & \ddots & r_{3N}^2 \ln r_{3N}^2 \\ \cdots & \cdots & \cdots & \cdots & \vdots & \ddots & \cdots \\ 1 & x_N & y_N & r_{N1}^2 \ln r_{N1}^2 & \cdots & \ddots & r_{NN}^2 \ln r_{NN}^2 \\ 0 & 0 & 0 & 1 & 1 & \ddots & 1 \\ 0 & 0 & 0 & x_1 & x_2 & \ddots & x_N \\ 0 & 0 & 0 & y_1 & y_2 & \ddots & y_N \end{bmatrix} \begin{bmatrix} a_0 \\ a_1 \\ a_2 \\ \vdots \\ F_{N-3} \\ F_{N-2} \\ F_{N-1} \\ F_N \end{bmatrix}, \tag{4.3}$$

where $r_{ij}^2 = (x_i - x_j)^2 + (y_i - y_j)^2$. Using solutions of eq. (4.3) and substituting eq. (4.2), the corresponding value of any point can be obtained by the interpolation expression. Here some additional descriptions are needed:

(1) When $r=0$, $\ln r^2$ does not exist, and $\lim_{r\to 0} r \ln r^2 = 0$. Equation (4.2) can be derived as follows: $\partial W(x,y) \big/ \partial x = a_1 + 2\sum_{i=1}^{N} F_i (1 + \ln r_i^2)(x - x_i)$.
(2) This method can be extended to three-dimensional (3D) cases, and eq. (4.2) can be transformed into the following forms:

$$\begin{gathered} W(x,y) = a_0 + a_1 x + a_2 y + a_3 z + \sum_{i=1}^{N} F_i r_i^2 \ln r_i^2 \\ \sum_{i=1}^{N} F_i = \sum_{i=1}^{N} x_i F_i = \sum_{i=1}^{N} y_i F_i = \sum_{i=1}^{N} z_i F_i = 0, \end{gathered} \tag{4.4}$$

where $r_i^2 = (x - x_i)^2 + (y - y_i)^2 + (z - z_i)^2$.

(3) In order to avoid the nonsingular matrix in the 2D case, all structural points cannot be collinear or overlapping; in the 3D case, all structural points should not be coplanar.
(4) Scaling ratio, that is, if the grid points are located in a narrow region, it is required to transform them linearly into a rectangular region to reduce errors and facilitate calculation.
(5) Symmetry, that is, if there is symmetric or antisymmetric relation for one or more planes, the mirror relationship can be used to improve computational efficiency and accuracy. For example, if $W(x,y)$ is symmetrical about $x=0$, eq. (4.2) can be replaced by

$$W(x,y)=a_0+a_2y+\sum_{i=1}^{N}F_i\left(r_i^2\ln r_i^2+\bar{r}_i^2\ln\bar{r}_i^2\right), \tag{4.5}$$

where $\bar{r}_i^2=(x+x_i)^2+(y-y_i)^2$. Then let $\sum_{i=1}^{N}F_i=\sum_{i=1}^{N}y_iF_i=0$, $W_j=W(x_j,y_j)$, and there will be $N+2$ equations.
(6) There is no need for the load to pass through these grid points if the elastic force is proportional to the value of the smooth interpolation plane. The plane can be smoothed by elastic extension, which can be expressed as follows:

$$W_j=a_0+a_1x_j+a_2y_j+\sum_{i=1}^{N}F_ir_{ij}^2\ln r_{ij}^2+C_jF_j, \tag{4.6}$$

where the unit of the coefficient C_j is the square of the length, which is equal to $16\pi D/K_j$. D is the stiffness of the curved surface. K_j is the elastic coefficient associated with the jth point. The original spline can be obtained when $C_j=0(K_j=\infty)$, and the least squares fitness surface are derived when $C_j\to\infty(K_j\to 0)$.
(7) The smoothness can also be done by the distributed loads. If $r^2\ln r^2$ is replaced by $r^2\ln(r+\varepsilon)^2$, a new surface through N points can be generated. When $\varepsilon\to 0$, these loads tend to be point loads. Thus, the surface can be differentiable everywhere, which will partly simplify the programming.
(8) IPS method is a scalar method, which is thus unable to solve the displacements in other directions with a given displacement in a certain direction.

4.2.2 Finite-plate splines method

In FPS method, a given plane is depicted by a balanced plate element. The shape function of the plate element connects the displacement of the CSD nodes with the CFD grid points [33, 4]. It is easy to build a realistic missile body model as it is based on the finite element method. However, a $3m\times 3n$ conversion matrix (m is the number of mesh points for fluid and n is the one of structure) is required, which needs large CPU time and storage. Therefore, it is more convenient to use IPS method in a large or supercomputer.

4.2.3 Multiquadric-biharmonic method

MQ method can represent an irregular surface and perform interpolation of various topographies [33, 4, 34]. Generally, the quadratic surface can be obtained by rotating hyperbola on two thin surfaces. The addition of biharmonic is to smooth the solution of the governing equation. One-dimensional (1D) interpolation equation is determined by

$$W(x) = \sum_{i=1}^{N} \alpha_i \left[(x - x_i)^2 + r^2 \right]^{1/2} \tag{4.7}$$

The MQ method is stable and consistent with respect to the parameter that controls the shape of the basis functions. A large r gives a plane shape function, whereas a small r gives a narrow cone function. For nonzero r, MQ method provides an infinitely differentiable function that preserves monotonicity and convexity. The performance is improved by later development: permitting r to vary with the basis function; scaling or rotating the independent variables for some applications where the magnitudes of the variables change widely and applying it in overlapping subdomains. Currently, this method has been applied to solve aeroelastic problems.

4.2.4 Thin-plate splines method

TPS method [33, 4, 34] provides a means to characterize an irregular surface by using minimum energy function. One-dimensional interpolation function is as follows:

$$W(x) = \sum_{i=1}^{N} \alpha_i |x - x_i|^2 \log|x - x_i|. \tag{4.8}$$

For a 1D case, elementary cubic splines can be interpreted as equilibrium positions of a beam undergoing bending deformation. For a 2D surface, these splines are determined from the minimization of the bending energy of a thin plate. Since these types of splines are invariant with rotation and translation, they are very powerful tools for the interpolation of flexible surfaces. This method is not limited to the 2D problems, but is extendible to the 3D cases. At present, this method has been used in the aeroelastic solution.

4.2.5 Inverse isoparametric mapping method

The IIM method [33, 4] is also based on the finite element analysis. In order to interpolate coordinates and displacement vectors, the same shape function is utilized in the same variable element. It cannot be applied in extrapolation. IIM is also used in aeroelastic analysis currently.

4.2.6 Nonuniform B splines

In the NUBS method [33, 4], a tensor product of two-spline scan is used to represent a surface in 3D space. The polynomial B splines are recommended to be used to satisfy the surface blending requirements in aeroelastic applications. A surface can be represented by the tensor product of two B splines:

$$W_{kt}(x,y) = \sum_{i=1}^{m-1}\sum_{j=1}^{n-1} P_{ij}B_{ik}(x)B_{jt}(y), \tag{4.9}$$

where W is the surface deformation at any point (x, y). P_{ij} are the multiplying coefficients to make these splines fit the control point data. B_{ik} and B_{jt} are the components of B spline in the x and y directions, respectively. Currently, this method has not been applied in the coupling of fluid and structure.

4.2.7 Spatial interpolation of load parameters

In this method, the interface of the different 3D physical fields are projected into a 2D parameter space, and the fitting surface is constructed in the parameter space to fit the parameters of the source nodes, and then the values of any point can be obtained by substituting the parameter coordinates of the target nodes into the fitting surface [24]. The advantage of this method is that the original 3D coordinates can be simplified into two coordinates in the parameter space. Moreover, it can transfer all continuous variables between the different fields and can avoid the influence of the geometric shape of the model on the interpolation accuracy.

4.3 Coupling interface design

The previous section introduced the most popular displacement interpolation methods. In CFD/CSD coupling system, besides the displacement exchange, the load transfer on the interface should also be considered. Therefore, this section introduces the mapping method for the integration of both aerodynamic load and structural displacement, in order to deal with the data transfer on the coupled interface.

4.3.1 Interface mapping based on constant volume transformation method

1. Displacement conversion

CVT [14, 30] was proposed by Goura et al. This method uses volume conservation before and after deformation to calculate the interpolation. It includes three processing steps of projecting, evolving and recovering the surface. First, identifying for each aerodynamic grid point $\vec{a}(t)$, the nearest triangular element in structural mesh. Here an adaptive search algorithm is developed to find the triangular element of an adjacent CSD mesh based on the minimum energy [6, 19]. The searching process is as follows: for each CFD grid point, consider its adjacent structural triangular mesh, as shown in Fig. 4.1. The black dot is the CFD grid point $\vec{a}(t)$, and the white points are its four adjacent structural mesh points. Therefore, there are four structural triangles.

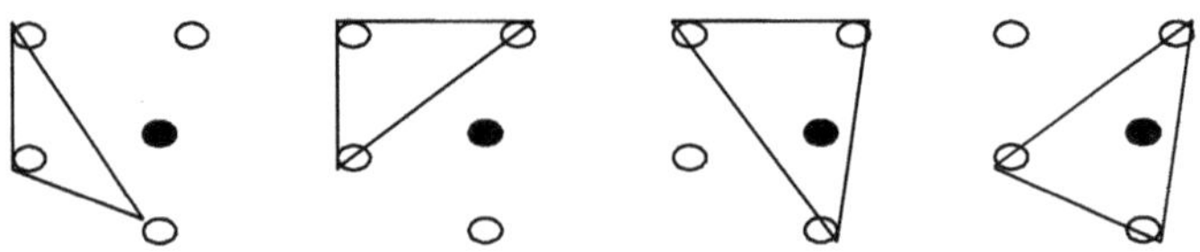

Fig. 4.1: Structural triangles adjacent to the CFD grid point.

When identifying the nearest structural triangle, those triangles that do not contain CFD points can be discarded. First, the area coordinate is used to determine whether the point $\vec{a}(t)$ is located in the structural triangle (note that the CFD grid points are not necessarily located on the same surface of the structural triangle). The area coordinate definition is shown in Fig. 4.2. The point $\vec{a}(t)$ and structural triangle vertices form three triangles, whose areas are A_1,A_2 and A_3, respectively, and $A = A_1 + A_2 + A_3$. The area coordinates α, β and γ are defined as

$$\begin{aligned} \alpha &= A_1/A \\ \beta &= A_2/A. \\ \gamma &= A_3/A \end{aligned} \tag{4.10}$$

If the sum of the area coordinate reaches $1.0 \pm \varepsilon$ (ε is the error precision), it can be considered that the point $\vec{a}(t)$ is located inside the structural triangle. It is observed from Fig. 4.1 that triangles ① and ② which do not include the CFD grid point $\vec{a}(t)$ can be omitted, and both triangles ③ and ④ include the point $\vec{a}(t)$. Here the distance L_i, which is from the point $\vec{a}(t)$ to the vertex of the structural triangle, is introduced as

$$L_i^m = \sqrt{(x_s^m - x_a)^2 + (y_s^m - y_a)^2 + (z_s^m - z_a)^2} \quad (i = 1, 3), \tag{4.11}$$

where (x_a, y_a, z_a) is the coordinate of a CFD mesh point $\vec{a}(t)$. (x_s^m, y_s^m, z_s^m) is the coordinate of the vertex s of the structural triangle m. The maximum distance of the triangle m can be obtained as

$$L_{\max}^m = \max(L_1^m, L_2^m, L_3^m). \tag{4.12}$$

It can be seen from eq. (4.12) that the triangle which has the minimum $L_{\max}^m$ is the smallest structural triangle corresponding to the CFD grid point. In Fig. 4.3, for triangles ③ and ④, $L_{\max}^4 < L_{\max}^3$, therefore, the structural triangle ④ can be selected as the structural triangle for the next transformation.

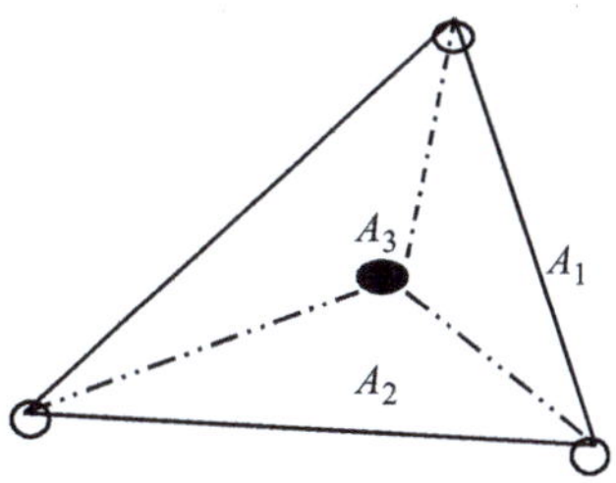

Fig. 4.2: Area coordinate definition.

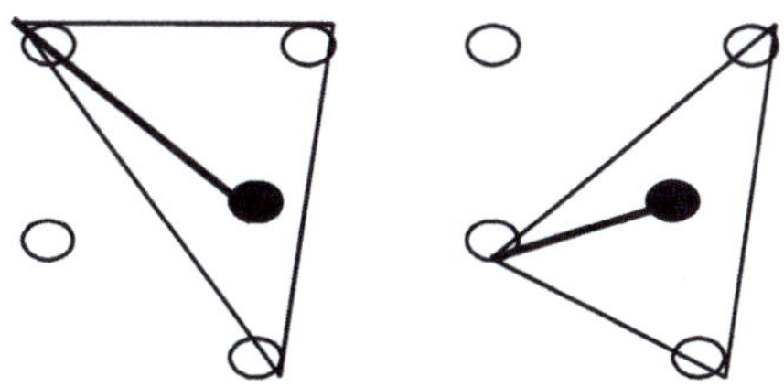

Fig. 4.3: Structure triangle selection.

Here, the CVT interpolation process is expressed by scalar equations. $\vec{s}_1$, $\vec{s}_2$ and $\vec{s}_3$ denote the vertex of the nearest triangular element for each aerodynamic grid point $\vec{a}(t)$, as shown in Fig. 4.4.

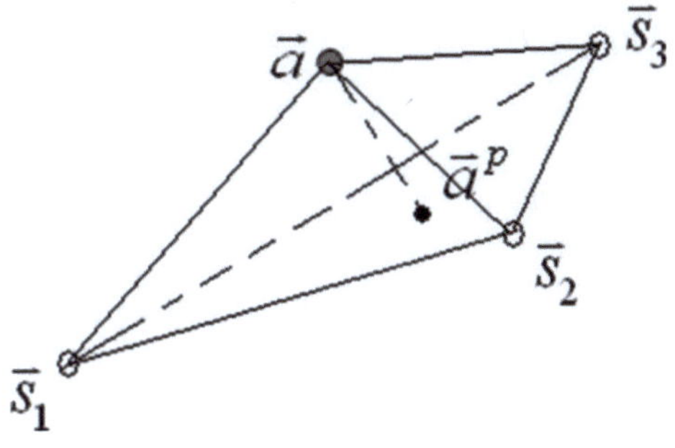

Fig. 4.4: Schema of the constant volume transformation method.

(x_1, y_1, z_1), (x_2, y_2, z_2) and (x_3, y_3, z_3) are the coordinates of the three vertices $\vec{s}_1$, $\vec{s}_2$ and $\vec{s}_3$. (x, y, z) represents the coordinate of an arbitrary point $\vec{s}$ in the triangular element. The plane equation of the structural triangular element is written as follows:

$$\begin{cases} x = \alpha x_1 + \beta x_2 + \gamma x_3 \\ y = \alpha y_1 + \beta y_2 + \gamma y_3 , \\ z = \alpha z_1 + \beta z_2 + \gamma z_3 \end{cases} \tag{4.13}$$

where $\alpha + \beta + \gamma = 1$, and the normal vector of the plane $\vec{N} = (\vec{s}_2 - \vec{s}_1) \times (\vec{s}_3 - \vec{s}_1)$ can be written as follows:

$$\vec{N} = \begin{vmatrix} \vec{i} & \vec{j} & \vec{k} \\ x_2 - x_1 & y_2 - y_1 & z_2 - z_1 \\ x_3 - x_1 & y_3 - y_1 & z_3 - z_1 \end{vmatrix} = l\vec{i} + m\vec{j} + n\vec{k}. \tag{4.14}$$

In addition, defining (x_a, y_a, z_a) as the coordinate of the aerodynamic grid point, then the volume of the tetrahedron element formed from the structural triangular element and the aerodynamic point can be obtained by $V = 1/3SH$. S is the area of the structural triangular element, and H is the distance from the aerodynamic point to the plane of the triangular element. They are expressed as

$$S = \frac{1}{2}|(\vec{s}_2 - \vec{s}_1) \times (\vec{s}_3 - \vec{s}_1)| = \frac{1}{2}\left|\vec{N}\right| = \frac{1}{2}\sqrt{l^2 + m^2 + n^2} = \frac{1}{2}\sqrt{l^2 + m^2 + n^2}, \tag{4.15}$$

$$H = \frac{l(x_a - x_1) + m(y_a - y_1) + n(z_a - z_1)}{\sqrt{l^2 + m^2 + n^2}}. \tag{4.16}$$

Substituting eqs. (4.15) and (4.16) into the volume equation, it can be simplified as

$$V = \frac{1}{6}(l(x_a - x_1) + m(y_a - y_1) + n(z_a - z_1)). \tag{4.17}$$

Supposing that the orthogonal projection of the aerodynamic point $\vec{a}(t)$ onto the plane of the triangular element is $\vec{a}^p$, and its coordinate is (x_p, y_p, z_p), then the linear equation of aerodynamic point $\vec{a}(t)$ and its projection point $\vec{a}^p$ is

$$\begin{cases} x = x_a - \mu l \\ y = y_a - \mu m , \\ z = z_a - \mu n \end{cases} \tag{4.18}$$

where μ is a parameter. l, m and n can be obtained from eq. (4.14). Connecting eqs. (4.13) and (4.18) and $\alpha + \beta + \gamma = 1$, new equations are formed as

$$\begin{cases} \alpha x_1 + \beta x_2 + \gamma x_3 + \mu l = x_a \\ \alpha y_1 + \beta y_2 + \gamma y_3 + \mu m = y_a \\ \alpha z_1 + \beta z_2 + \gamma z_3 + \mu n = z_a \\ \alpha \cdot 1 + \beta \cdot 1 + \gamma \cdot 1 + \mu \cdot 0 = 1 \end{cases}. \tag{4.19}$$

α, β, γ and μ can be obtained by solving eq. (4.19), and then substitute them into eqs. (4.13) or (4.18) to get $\vec{a}^p$. After the deformation of the structural mesh, the structural positions are changed to $\vec{s}_1{}', \vec{s}_2{}', \vec{s}_3{}'$, and the new aerodynamic point is set as $\vec{a}'(t)$. Using the principle of volume conservation and invariance of the relative coordinates of projection points on the plane of structural triangular elements (i.e., α, β, γ are fixed at their initial values), the position (x'_a, y'_a, z'_a) of the deformed aerodynamic grid point $\vec{a}'(t)$ can be determined from

$$\begin{cases} x'_a = \alpha x'_1 + \beta x'_2 + \gamma x'_3 + \mu' l' \\ y'_a = \alpha y'_1 + \beta y'_2 + \gamma y'_3 + \mu' m' \\ z'_a = \alpha z'_1 + \beta z'_2 + \gamma z'_3 + \mu' n' \end{cases}. \tag{4.20}$$

The normal vector of the new triangular plane is

$$\begin{vmatrix} \vec{i} & \vec{j} & \vec{k} \\ x'_2 - x'_1 & y'_2 - y'_1 & z'_2 - z'_1 \\ x'_3 - x'_1 & y'_3 - y'_1 & z'_3 - z'_1 \end{vmatrix} = l'\vec{i} + m'\vec{j} + n'\vec{k}, \tag{4.21}$$

and

$$V' = \frac{1}{6}\left(l'(x_a - x'_1) + m'(y'_a - y'_1) + n'(z'_a - z'_1)\right), \tag{4.22}$$

$$V = V' \tag{4.23}$$

μ' can be obtained by substituting eq. (4.20) into eqs. (4.22) and (4.23), and new aerodynamic point coordinate can also be obtained by substituting μ' back into eq. (4.20). The interpolation calculation is completed. It should be noted that CVT interpolation is a vector method, the oriented volume can be used in the calculation. That is, when the aerodynamic point is above the structural triangular element, the volume is positive and vice versa; CVT method is also able to deal with the 2D interpolation. Compared with the IPS, the practical advantages of this method are as follows: (1) it does not require to solve large algebraic equations and large matrix inversion, thus, it has high computational efficiency as smaller amount of CPU time and storage are taken in the interpolation; (2) it relies on local information, and can deal with complex geometry and discontinuous structure with the developed adaptive local searching algorithm, and can be easily applied to multiblock and parallel computation in CFD/CSD coupling system.

2. Load conversion

Load conversion is to convert the aerodynamic load on CFD grid points into CSD mesh to form equivalent nodal load in structural computation. First, according to the pressure distribution, the aerodynamic load in three directions on the CFD grid point j are obtained and expressed as

$$\begin{Bmatrix} G_x^j \\ G_y^j \\ G_z^j \end{Bmatrix} = S^j p^j \begin{Bmatrix} n_x^j \\ n_y^j \\ n_z^j \end{Bmatrix}, \tag{4.24}$$

where G_x^j, G_y^j and G_z^j are the forces on the aerodynamic grid point in the directions of x, y and z, respectively. S^j is the area of the pressure p^j. n_x^j, n_y^j and n_z^j are the components in the directions of x, y and z of the unit normal vector of the point j.

Then, the load on each CFD grid point j should be converted to the CSD mesh. Here the same searching algorithm described in Section 4.3.1.1 is used here to select the structural triangular element for each CFD grid point. The loads on CSD nodes in the triangle can be calculated as follows:

$$\begin{Bmatrix} \vec{F}_{i_1}^j \\ \vec{F}_{i_2}^j \\ \vec{F}_{i_3}^j \end{Bmatrix} = \begin{vmatrix} \alpha & 0 & 0 \\ 0 & \alpha & 0 \\ 0 & 0 & \alpha \\ \beta & 0 & 0 \\ 0 & \beta & 0 \\ 0 & 0 & \beta \\ \gamma & 0 & 0 \\ 0 & \gamma & 0 \\ 0 & 0 & \gamma \end{vmatrix} = \begin{Bmatrix} G_x^j \\ G_y^j \\ G_z^j \end{Bmatrix}, \tag{4.25}$$

where $\vec{F}_{i_1}^j$, $\vec{F}_{i_2}^j$ and $\vec{F}_{i_3}^j$ are the loads transferred from CFD grid point j to the three vertices in the structure triangle, respectively. α, β and γ are the area coordinates of the CFD grid point j in the structural triangle, respectively. i_1, i_2 and i_3 are the serial numbers of nodes in the structural triangle, respectively. Since the area coordinate is equivalent to the shape function in structural elements, there is no requirement to deal with additional moments and torques for load conversion. For each structural node i, the total force vector $\vec{F}_i$ can be calculated as

$$\vec{F}_i = \sum_{j=1}^{j_a} \vec{F}_i^j \quad (i = i_1, i_2, i_3, \ldots, i_s), \tag{4.26}$$

where i_s is the number of the CSD mesh points on the common boundary and j_a is the number of CFD grid points on the common boundary. The force vector $\vec{F}_s$ of the finite element model can be obtained by

$$\vec{F}_s = \left\{ \vec{F}_{i_1} \quad \vec{F}_{i_2} \quad \dots \quad \vec{F}_{i_S} \right\}^T. \tag{4.27}$$

3. Coupled interface design

For all the aerodynamic grid points and structural mesh points on the common boundary, the transformation of the load information can be described as

$$\begin{cases} \vec{F}_s = S_1 \vec{F}_a^{'} \\ \vec{F}_a^{'} = \vec{F}_a \end{cases}, \tag{4.28}$$

where $\vec{F}_s$ and $\vec{F}_a$ are the force vectors at all structural mesh points and all aerodynamic grid points, respectively. $\vec{F}_a^{'}$ represents the force vector at the projection point of the aerodynamic grid point in the structural triangular element. S_1 is the load transformation matrix. For the displacement transformation, it can be described as

$$\begin{cases} \vec{u}_a^{'} = S_2 \vec{u}_s \\ \vec{u}_a = \vec{u}_a^{'} + A\vec{N}_s^{'} \end{cases} \tag{4.29}$$

where $\vec{u}_a$ and $\vec{u}_s$ are the displacement vectors at all aerodynamic grid points and all structural mesh points, respectively. $\vec{u}_a^{'}$ is the displacement vector at the projection point of the aerodynamic grid point in the structural triangular element. A is a parametric matrix, $\vec{N}_s$ is the vector composed of normal vectors of the structural triangular elements corresponding to all the aerodynamic grid points, and S_2 is the displacement transformation matrix.

For the CFD grid point j, only one structural triangle element is determined by the searching algorithm, and then the structural nodes i_1, i_2 and i_3 can be determined. According to the earlier two transformations, there are

$$\begin{Bmatrix} \vec{F}_{i_1} \\ \vec{F}_{i_2} \\ \vec{F}_{i_3} \end{Bmatrix} = \begin{bmatrix} \alpha \\ \beta \\ \gamma \end{bmatrix} \{\vec{F}_j\}, \tag{4.30}$$

and

$$\{\vec{u}_j\} = [\alpha \ \beta \ \gamma] \begin{Bmatrix} \vec{u}_{i_1} \\ \vec{u}_{i_2} \\ \vec{u}_{i_3} \end{Bmatrix} + \mu_m \begin{Bmatrix} n_x \\ n_y \\ n_z \end{Bmatrix}, \tag{4.31}$$

where α, β, γ and μ_m are defined in eqs. (4.10) and (4.20), respectively. Introducing

$$S_1^j = \begin{bmatrix} \alpha \\ \beta \\ \gamma \end{bmatrix}, \quad S_2^j = [\alpha\ \beta\ \gamma]\ (j = 1, \ldots, j_{\max}), \tag{4.32}$$

the superscript j represents the jth local submatrix, which corresponds to the CFD grid point j. It can be obviously seen from eq. (4.32) that $S_1^{jT} = S_2^j$. The two matrices S_1 and S_2 described in eqs. (4.28) and (4.29) can be considered as a combination of S_1^j and S_2^j. Thus, there is a relationship that $S_1 = {S_2}^T$; as a result, the load and displacement conversion can be handled by a spline matrix and its transposition.

On the common boundary, the force relationship between the CSD node i and the CFD grid point j can be expressed as

$$\vec{F}_{si} = \sum_{j=1}^{j=ja} \vec{F}_{aj} N_i(\chi_j), \tag{4.33}$$

where subscripts s and a represent the structure and aerodynamic field, respectively. $N_i(\chi_j)$ represents the area coordinates of the CFD grid point j in the CSD triangular element. Taking into account of $\sum_{i=1}^{i=i_s} N_i = 1$, on the whole common boundary, there are

$$\sum_{i=1}^{i=i_s} \vec{F}_{si} = \sum_{i=1}^{i=i_s} \sum_{j=1}^{j=ja} \vec{F}_{aj} N_i(\chi_j) = \sum_{j=1}^{j=ja} \vec{F}_{aj}. \tag{4.34}$$

Equation (4.34) shows that the force balance is satisfied on the boundary. The total energy of all the CFD grid points on the common boundary is

$$W_a = \sum_{j=1}^{j=ja} \vec{F}_{aj}^T \vec{u}_{aj}. \tag{4.35}$$

Substituting eq. (4.34) into (4.35), the total energy can be obtained:

$$W_a = \sum_{j=1}^{j=ja} \left(\vec{F}_{aj}^T \cdot \sum_{i=1}^{i=i_s} \left(N_i(\chi_j) \vec{u}_{si} + H_i(\chi_j) \right) \right), \tag{4.36}$$

where $H_i(\chi_j)$ represents the projection coordinate distance of the CFD grid point j on the CSD node i. It is formed by the second item on the right side of eq. (4.28). From eq. (4.34), the total energy can be derived as

$$W_a = \sum_{i=1}^{i=i_s} \vec{F}_{si}^T \vec{u}_{si} + \sum_{i=1}^{i=i_s} \vec{F}_{si}^T \cdot \sum_{i=1}^{i=i_s} H_i(\chi_j) = W_s + W_e, \tag{4.37}$$

where W_s represents the total energy at the CSD mesh points on the common boundary. W_e represents the energy difference caused by the noncoincidence of the boundary between the aerodynamic grid and the structural mesh. Thus, if each aerodynamic grid point falls in the plane of the structural triangular element, there is $W_e = 0$, and $W_a = W_s$. It means that the energy on the common boundary is exactly conservative. These two transformations can be written in matrix form as

$$\begin{cases} \vec{F}_s = S\vec{F}_a \\ \vec{u}_a = S^T\vec{u}_s + A\vec{N}_s \end{cases}, \tag{4.38}$$

where S is the developed mapping matrix. A is the diagonal matrix composed of parameters μ_m and $\vec{N}_s$. It represents the vectors composed of the normal vectors of all the CSD triangular elements on the common interface. It should be noted that the mapping matrix S is only related to the initial geometric characteristics of the local CFD grid and CSD mesh on the common boundary. Consequently, it has the capability to deal with arbitrary shape geometry and if the matrix is generated in the pretreatment stage, it can be used repeatedly in the iterative process of the coupling calculation. As a result, it greatly improves the efficiency of the data exchange on the whole coupling interface.

4.3.2 Interface mapping based on boundary element method

If there is a normal distance between the structural mesh and aerodynamic grid surfaces on the boundary, the energy cannot be conserved during the transformation by the interface mapping matrix derived from the earlier CVT method, that is, in eq. (4.37) $W_e \neq 0$, and a certain amount of energy will be lost. This section will introduce an energy conservation interpolation method – boundary element method (BEM) [8, 22, 13], which can also deal with the displacement interpolation and the load interpolation.

BEM method was proposed by Professor Brebbia in 1978, and the boundary integral equation is established. With the preliminary development of the theoretical system of BEM method, it is considered as a numerical method by many researchers. It is only required to discretize the solution on the boundary instead of dividing the domain in the BEM method, which improves the computational efficiency and reduces the discretization error. It has been developed considerably and has been listed as a common numerical method with the finite element method, finite difference method and finite volume method. Currently, the BEM method is widely used in solid mechanics, dynamics, fluid mechanics and heat conduction.

1. Boundary integral equation

The equilibrium of a linear elastic continuum can be described as follows [5]:

$$\frac{\partial^2 u_i}{\partial x_j \partial x_j} + \left(\frac{1}{1-2v}\right)\frac{\partial^2 u_j}{\partial x_i \partial x_j} = 0, \quad i,j = 1,2,3 \tag{4.39}$$

where u is the displacement vector, x is Cartesian coordinate and v is Poisson's ratio. Using Betti reciprocal work principle and the Somigliana identity, the displacement $u_i(p)$ at any interior point p in the domain caused by traction $t(q)$ and displacement $u(q)$ at a point q on the boundary Γ can be obtained by

$$C_{ij}(p)u_i(p) + \int_\Gamma T_{ij}(p,q)u_j(q)d\Gamma(q) = \int_\Gamma U_{ij}(p,q)t_j(q)d\Gamma(q), \tag{4.40}$$

where $T_{ij}(p,q)$ and $U_{ij}(p,q)$ are the Kelvin solutions of the traction force and displacement, respectively. In the 3D case,

$$\begin{cases} T_{ij}(p,q) = \frac{-1}{8\pi(1-v)r^2}\cdot\left[\frac{\partial r}{\partial n}\left((1-2v)\delta_{ij} + 3\frac{\partial r}{\partial x_i}\frac{\partial r}{\partial x_j}\right) + (1-2v)\left(\frac{\partial r}{\partial x_j}n_i - \frac{\partial r}{\partial x_i}n_j\right)\right] \\ U_{ij}(p,q) = \frac{1}{16\pi\mu(1-v)r}\cdot\left[(3-4v)\delta_{ij} + \frac{\partial r}{\partial x_i}\frac{\partial r}{\partial x_j}\right] \end{cases}. \tag{4.41}$$

In the 2D case,

$$\begin{cases} T_{ij}(p,q) = \frac{-1}{4\pi(1-v)r}\cdot\left[\frac{\partial r}{\partial n}\left((1-2v)\delta_{ij} + 2\frac{\partial r}{\partial x_i}\frac{\partial r}{\partial x_j}\right) + (1-2v)\left(\frac{\partial r}{\partial x_j}n_i - \frac{\partial r}{\partial x_i}n_j\right)\right] \\ U_{ij}(p,q) = \frac{1}{8\pi\mu(1-v)}\cdot\left[(3-4v)\ln\left(\frac{1}{r}\right)\delta_{ij} + \frac{\partial r}{\partial x_i}\frac{\partial r}{\partial x_j}\right] \end{cases}, \tag{4.42}$$

where μ is the modulus of elasticity. r is the distance from the point q to the point p. n is the normal vector of point q, and n_i is the component of n in the i direction. In eq. (4.40), $C_{ij}(p)$ is a geometric function: when p locates in Ω, $C_{ij}(p) = 1$. When p is on Γ, $C_{ij}(p)$ is a function of the local shape of the boundary. In order to avoid singular integral, the rigid displacement is taken into account in the solution of these coefficients.

2. Discretization of the boundary element

It is required to discretize the boundary into finite elements to implement the boundary integral equation. In each element, a constant element, linear element, quadratic element, cubic or higher order element can be applied to describe the geometric variables and derivatives. For example, in the 3D case, two intrinsic coordinates ξ_1 and ξ_2 are defined, then the geometric variables on the boundary can be described

$$x = \sum_{n=1}^{N}\phi_n(\xi_1,\xi_2)x_n^e, \quad x_{,\xi} = \sum_{n=1}^{N}\phi_{n,\xi}(\xi_1,\xi_2)x_n^e, \tag{4.43}$$

where n and N represent the number of the element and the total number of the elements on the boundary, respectively. x and $x_{,\xi}$ are the global Cartesian coordinates and their derivatives at point (ξ_1, ξ_2), respectively. x^e is the node value of the boundary element in the global coordinates, and ϕ is the shape function. Similarly, the kernel integrals displacement and traction on the boundary can be described by the shape function as

$$\begin{cases} H_{ij}(p,q) = \int_{-1}^{1}\int_{-1}^{1} T_{ij}(p,q)\phi_n(\xi_1,\xi_2)J(\xi_1,\xi_2)d\xi_1 d\xi_2 \\ G_{ij}(p,q) = \int_{-1}^{1}\int_{-1}^{1} U_{ij}(p,q)\phi_n(\xi_1,\xi_2)J(\xi_1,\xi_2)d\xi_1 d\xi_2 \end{cases}, \tag{4.44}$$

where $J(\xi_1, \xi_2)$ is the Jacobian transformation from global to local coordinate. The integral of $(-1,1)$ can be obtained by Gauss integral method. The boundary integral equation can be described in the local coordinate as

$$C_{ij}(p)u_i(p) + \sum_{m=1}^{M} H_{ij}^m(p,q)u_j(q) = \sum_{m=1}^{M} G_{ij}^m(p,q)t_j(q), \tag{4.45}$$

where M is the total number of the boundary element. $t_j(q)$ and $u_j(q)$ are independent of integrals as they are known quantities or can be obtained by applying boundary conditions. Then, eq. (4.45) can be written in matrix form:

$$C_{ij}u_i + H_{ij}u_j = G_{ij}t_j. \tag{4.46}$$

For the points locating in the interior and on the boundary, the integrals can be expressed as

$$u_i + H_{bi}u_b = G_{bi}t_b, \qquad H_{bb}u_b = G_{bb}t_b, \tag{4.47}$$

where H and G are the matrices composed of $H_{ij}^m(p,q)$ and $G_{ij}^m(p,q)$, respectively. Subscripts b and i stand for the boundary and interior values, respectively. If the displacements u_b on the entire boundary are known, t_b on the boundary can be obtained by solving eq. (4.47), so that the displacement u_i at interior points can be obtained by a transformation matrix B as

$$u_i = Bu_b, \ B = G_{bi}G_{bb}{}^{-1}H_{bb} - H_{bi}. \tag{4.48}$$

Equation (4.48) can be considered as the displacement transformation between the boundary point and the internal point in the domain. In aeroelastic computation, the boundary points are usually regarded as CFD surface grid, and the internal points are treated as the CSD mesh nodes. The displacements on the CSD mesh nodes are often known from the structural solver, it is required to invert the matrix B to solve the displacement on the aerodynamic grid points. However, in most cases, the matrix B is asymmetric and irreversible as the number of internal points is far coarser than the number of points on the boundary. Here the minimization of the strain energy method [9] proposed by Chen and Jadic is employed to determine the universal displacement transformation matrix.

3. Minimum strain energy method

Structural strain energy can be described by

$$W = \int_{\Gamma} t_k u_k d\Gamma. \tag{4.49}$$

The displacement and traction force described by shape function are integrated over the boundary elements. Then the strain energy is written in matrix form as

$$W = u_b^T N t_b, \tag{4.50}$$

where N is the area integral matrix which contains the integral of the shape function $\phi(\xi_1, \xi_2)$ of the boundary elements. Then the strain energy can be derived from eq. (4.47)

$$W = u_b^T R u_b, \tag{4.51}$$

where $R = NG_{bb}^{-1}H_{bb}$. For the given displacement at the structural grid points, the strain energy W should be minimum in order to avoid the undue deformation of the structure. It is a constrained quadratic minimization problem and an objective function is defined as

$$F = u_b^T R u_b - \lambda^T (B u_b - u_i), \tag{4.52}$$

where λ is a vector containing the Lagrangian multipliers. Then two differential equations can be derived by calculating the derivatives of the objective function with respect to u_b and λ, respectively,

$$\begin{cases} \partial F/\partial u_b = R u_b + R^T u_b - B^T \lambda \\ \partial F/\partial \lambda = u_i - B u_b. \end{cases} \tag{4.53}$$

By applying the Lagrangian technique, eq. (4.53) is equal to 0. After rearrangement, the desired universal spline matrix associated with the displacement of the CFD grid and CSD grid is obtained

$$S = \left(R + R^T\right)^{-1} B^T \left[B\left(R + R^T\right)^{-1} B^T\right]^{-1}. \tag{4.54}$$

Thus

$$u_b = S u_i \Rightarrow \vec{u}_a = S \vec{u}_s. \tag{4.55}$$

Once the matrix S is obtained, it can be used for the two-way conversion of the interface information. One is using eq. (4.55) to convert the structural displacement through S to aerodynamic grid points; the other is applying the transpose of the spline matrix to transform the forces from the aerodynamic to the structural grid points:

$$\vec{F}_s = S^T \vec{F}_a, \tag{4.56}$$

where $\vec{F}_s$ and $\vec{F}_a$ represent the loads on the structural points and aerodynamic points, respectively. Moreover, the conservation requirements of work between two transformations can be satisfied:

$$W_s = \vec{F}_s^{\,T} \vec{u}_s = \left(S^T \vec{F}_a\right)^T \vec{u}_s = \vec{F}_a^{\,T} (Su_s) = \vec{F}_a^{\,T} \vec{u}_a = W_a. \tag{4.57}$$

4. Intermediate BEM method

There are two shortcomings when BEM method is used to deal with information conversion in CFD/CSD coupling system: (1) the aerodynamic grid is regarded as the boundary in BEM, and the structural mesh element should be located inside the aerodynamic grid. Singularity will occur if the two sets of the grid points are approaching each other; in addition, the BEM will fail if the aerodynamic grid does not fully contain the structural grid points or some of the structural points are located outside the aerodynamic grid. (2) In CFD solver, the surface grid of the aircraft may be constructed as a different type of grid system, such as structured grid, unstructured grid, nested grid or multiblock grid, and it requires fine and dense grid quality. Consequently, it is required to establish a different type of BEM model for the different aerodynamic grid styles, and it is a very challenging task to invert the huge and asymmetric matrix G_{bb}.

Here an intermediate BEM model is applied to establish a closed boundary Γ_m [7], which is constructed as a virtual platform by embracing all the CFD and CSD grid points. The mapping matrix between the two different grid systems can be related by the intermediate boundary. First, the CSD grid points and the constructed virtual boundary element are connected with the spline matrix S_{mi} by eq. (4.54). Then the transformation matrix B_{bm} from intermediate boundary to the CFD grid point is established by eq. (4.48). Therefore, eq. (4.55) can be written as

$$\vec{u}_a = S\vec{u}_s, \quad S = B_{bm} S_{mi}. \tag{4.58}$$

The intermediate BEM method can effectively deal with more complex geometries, such as aerodynamic unstructured grid, nested grid and multiblock grid by virtually constructing a simple BEM boundary to overcome the shortcomings of the direct BEM method. The most important advantage is that the designed spline matrix can deal with the two-way energy conservation interpolation of the structural displacement and aerodynamic load, and the construction of the matrix only needs the initial structural and aerodynamic grid information, so it can be reused in the whole aeroelastic simulation.

4.3.3 Numerical examples

In most of the CFD/CSD coupling applications, the aerodynamic grid points are located outside the structural grid, the displacements of the CFD surface have to be replaced or complemented by extrapolation.

1. A 2D rigid body rotation deformation

Figure 4.5 shows a 2D interpolation problem. The structural points are located inside the ring, initially in a horizontal state, and the aerodynamic points are located on the ring. When the structure rotates 0.1 rad clockwise, the aerodynamic point displacements extrapolated by BEM, CVT and IPS methods are compared with the displacements of the rigid body rotation. As shown in Fig. 4.5, at a small rotation angle, the displacement results obtained by BEM and CVT interpolation are consistent with those obtained by the rigid body rotation, while the IPS method has a slight error. Figure 4.6 shows the extrapolation results when the structural points rotate clockwise by 0.5 rad. It can be seen that the interpolation errors of IPS method become larger, while the results obtained by CVT method are still consistent with the rigid rotation, and the interpolation accuracy of BEM method is higher in the direction of the extension of the structural grid lines and lower in the direction perpendicular to the structural grid lines. It is concluded that CVT method has

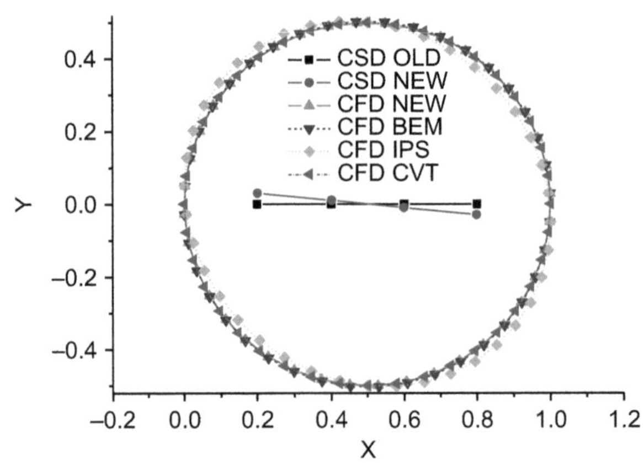

Fig. 4.5: Two-dimensional rigid body rotation extrapolation (0.1 rad).

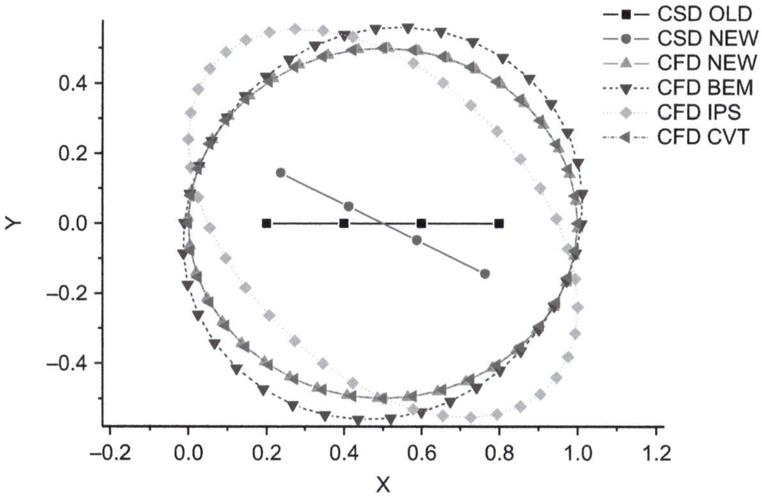

Fig. 4.6: Two-dimensional rigid body rotation extrapolation (0.5 rad).

the highest accuracy and IPS method has the lowest accuracy when describing rigid body deformation, and BEM algorithm needs more detailed structural deformation information to obtain a high accuracy.

2. A two-dimensional elastic translational deformation

Figure 4.7 shows an extrapolation example of a 2D elastic deformation. The structural grid points are located in the central position, and the deformation is given as the rigid translation, and the aerodynamic grid points are placed around the structure and extrapolated by the three methods: BEM, CVT and IPS. It can be found that the IPS method only has normal deflections, which results in exaggerated shear deformation of the aerodynamic grid. The CVT method has a rigid representation for external points as shown in Fig. 4.8, and the distance between the two aerodynamic points on the upper side is fully extended, while the two aerodynamic points on the lower side are rigidly attached. In aeroelastic solution, it may cause distortions or discontinuities if the distance between the aerodynamic points is very small for the case of the structure with large deformation. Nevertheless, the BEM method generates a moderate and smooth deformation rather than the rigid attachment, which represents physically reasonable shapes.

3. A three-dimensional missile body interpolation

One part of a missile body is selected as the interpolation example. The structural mesh points are 21 × 11 and the aerodynamic grid points are 101 × 51. Figures 4.8 and 4.9 show the structural and aerodynamic grid, respectively. Figures 4.10 and 4.11 show

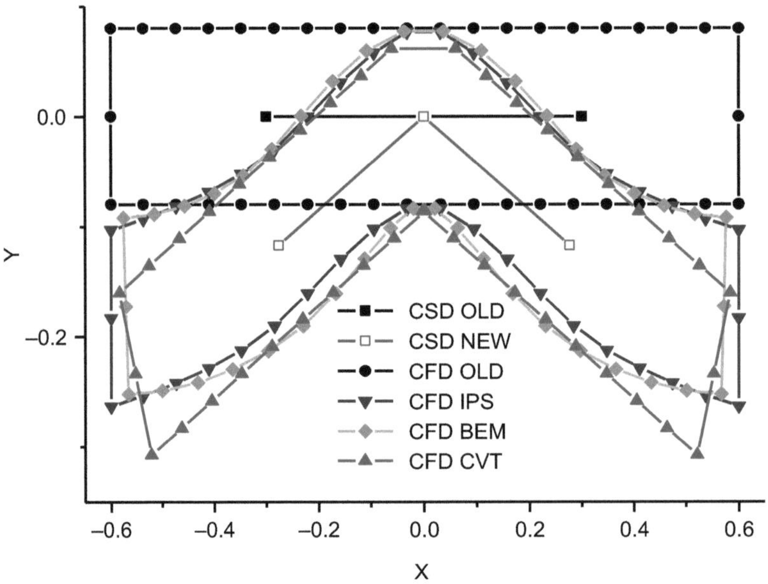

Fig. 4.7: A two-dimensional elastic extrapolation.

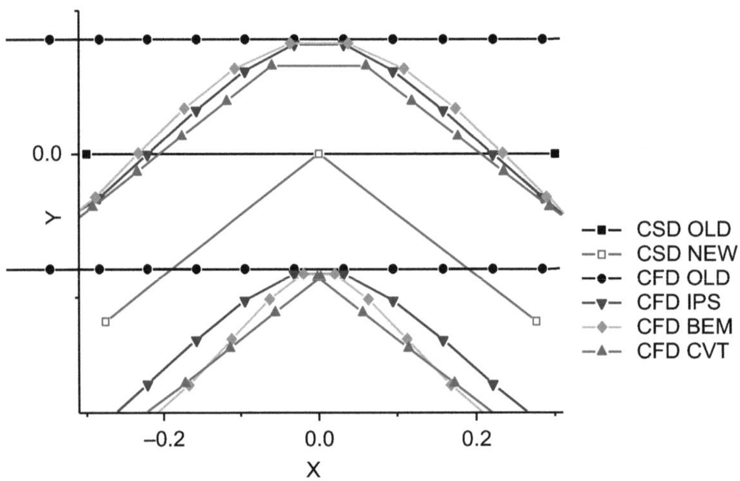

Fig. 4.8: Enlarged view of a two-dimensional elastic extrapolation.

the deformed structural and aerodynamic grid under the given geometric function, respectively. Figures 4.12–4.14 are the aerodynamic meshes and the enlarged images obtained by the BEM, CVT and IPS interpolation methods, respectively. It can be seen that the BEM method obtains a smooth deformation with high mesh quality, but there

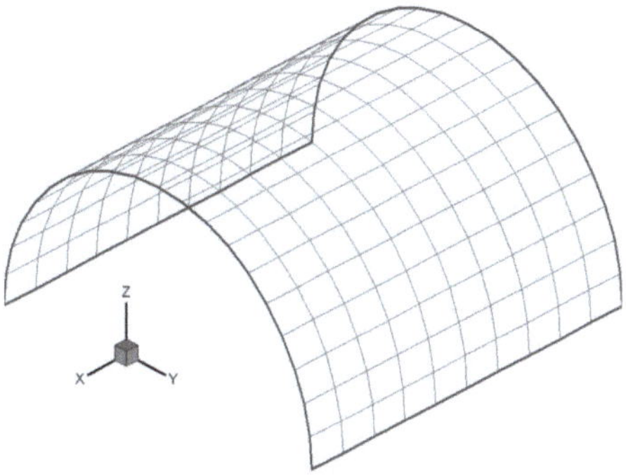

Fig. 4.9: Initial structural grid.

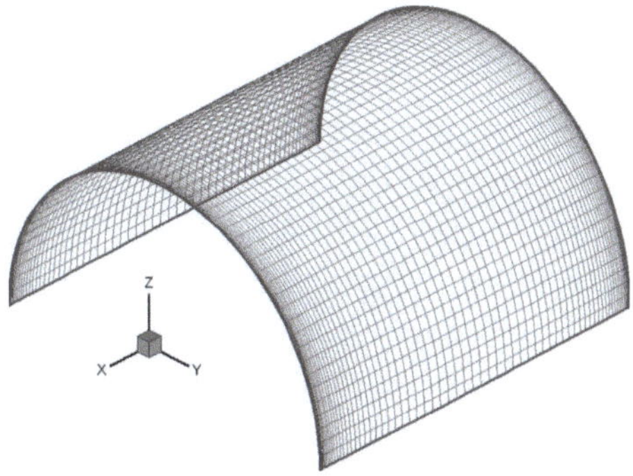

Fig. 4.10: Initial aerodynamic gird.

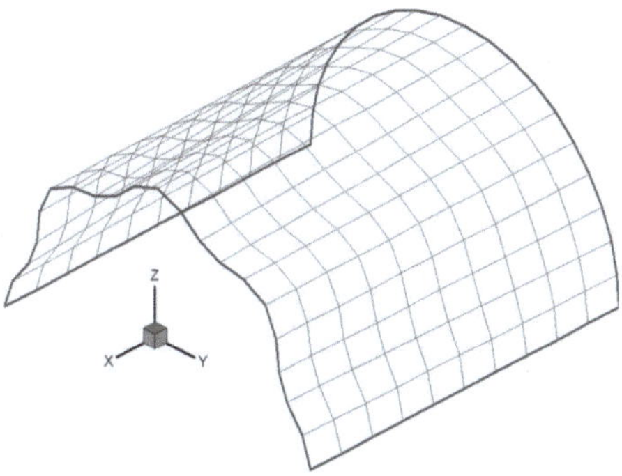

Fig. 4.11: Deformed structural grid.

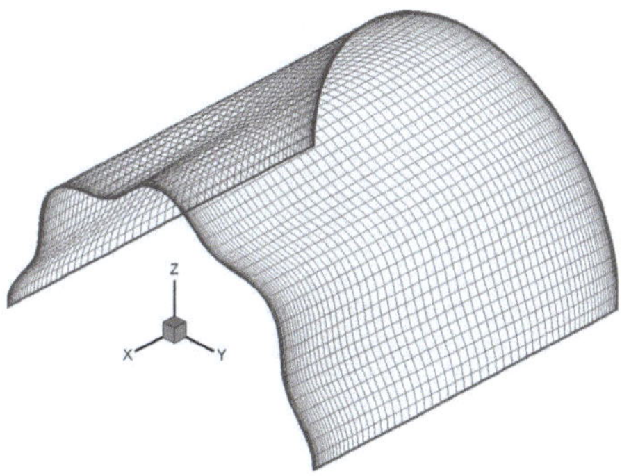

Fig. 4.12: Deformed aerodynamic gird.

occurs little wave curve on the boundary instead of a straight line. The deformed mesh obtained by the CVT method is almost the same as the standard mesh. For the IPS interpolation, there are some obvious wave lines near the boundary. Table 4.1 compares the errors of the three interpolation methods. It can be seen that the CVT method has a highest accuracy in interpolating the displacement of the missile body, because the structure mesh is relatively uniform, and the area change between the initial position and deformed position is small, which ensures the volume conservation. The maximum error of the BEM interpolation occurs on the boundary, which is attributed to the

Tab. 4.1: Error comparison of the several interpolation methods for the missile body deformation.

Interpolation method	BEM	CVT	IPS
Maximum absolute error	2.1726e-002	1.0835e-002	6.3516e-002

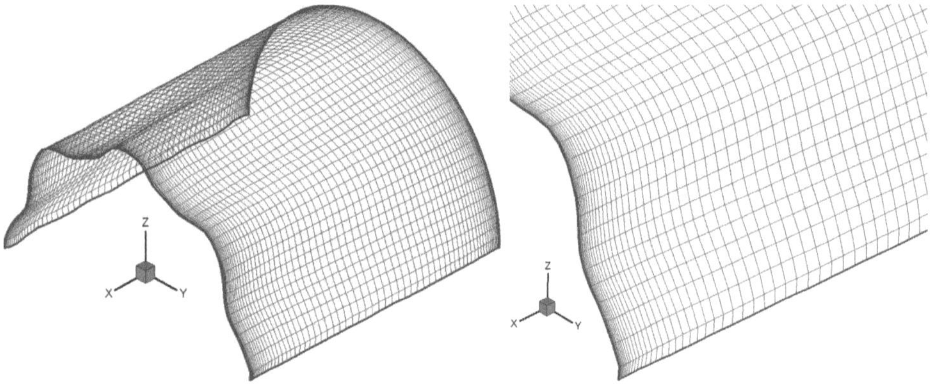

(a) (b)

Fig. 4.13: (a) Aerodynamic grid by BEM; (b) enlarged view of the BEM interpolation.

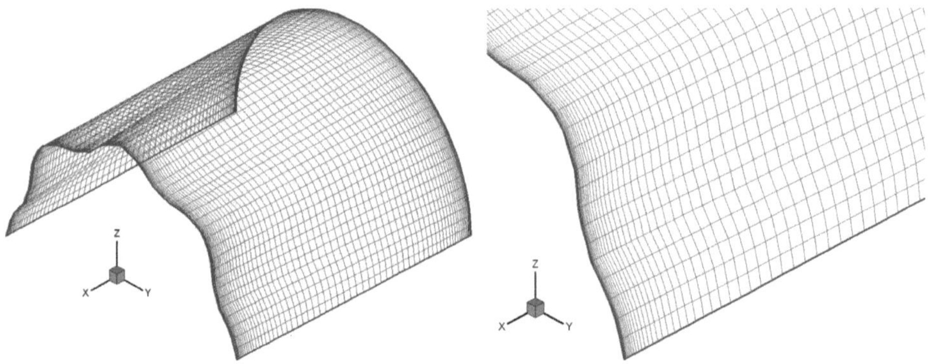

(a) (b)

Fig. 4.14: (a) Aerodynamic grid by CVT; (b) enlarged view of CVT interpolation.

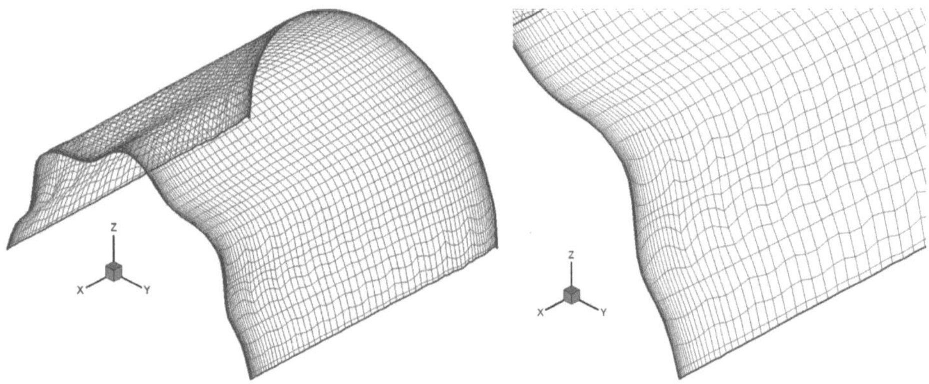

(a) (b)

Fig. 4.15: (a) Aerodynamic grid by IPS; (b) enlarged view of the IPS interpolation.

singular integral terms on the boundary, and it is required to adjust the parameters of the indirect BEM. The IPS method is prone to errors where the curvature varies greatly in the case of the 3D deformation of nonplate configuration.

4. Interpolation of structural vibration modes of AGARD445.6 wings

AGARD445.6 wing (composed of NACA 65A004 airfoil) is a standard model for validating the flutter simulations in most publications. The wing aspect ratio is 1.65, the tip-root ratio is 0.66, the half-span is 0.762 m, the root chord length is 0.5587 m, and the quarter-chord sweep angle is 45°. It is introduced to illustrate the practical significance of the interface. The finite element model of the structure is constructed with 400 anisotropic 4-node shell elements with 441 structural mesh points, and a 3D aerodynamic surface mesh contains 2940 points.

The three mapping algorithms BEM, CVT and IPS are used to interpolate the fifth modal deflection of the wing to the aerodynamic grid, which is the third bending mode with relatively large flexible deformation in the z direction. Results are compared with each other in Figs. 4.16–4.19. It is obvious that the CFD grid still maintains good quality and reflects the real CSD deformation in large-scale case transferred by the BEM method. As the CSD grid is planar, the IPS method, based on displacements normal to the plat surface, transfers the smoothing result as well

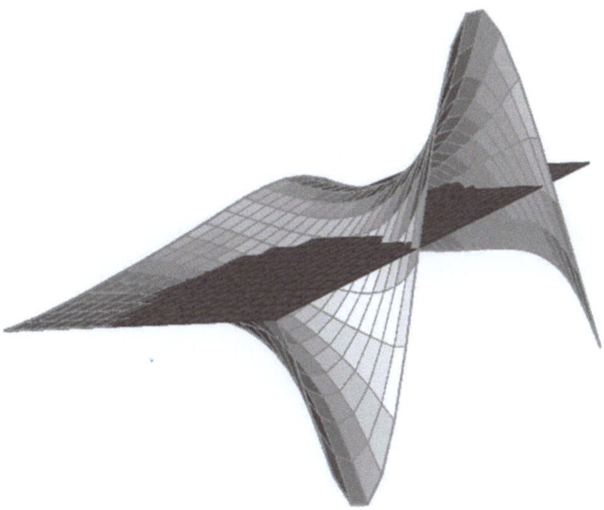

Fig. 4.16: Fifth-order mode deformation of the structure.

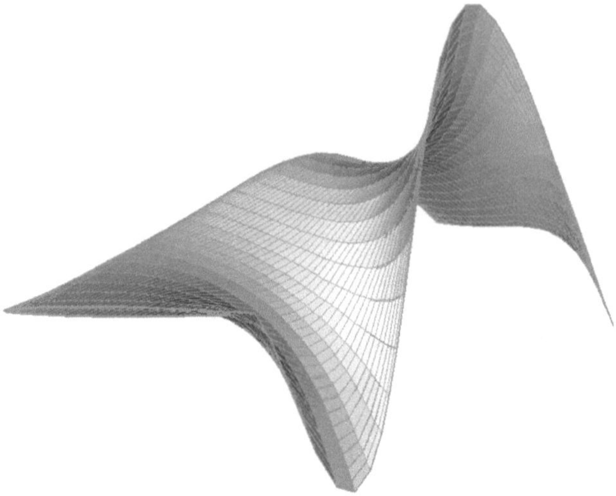

Fig. 4.17: Aerodynamic deformation by BEM.

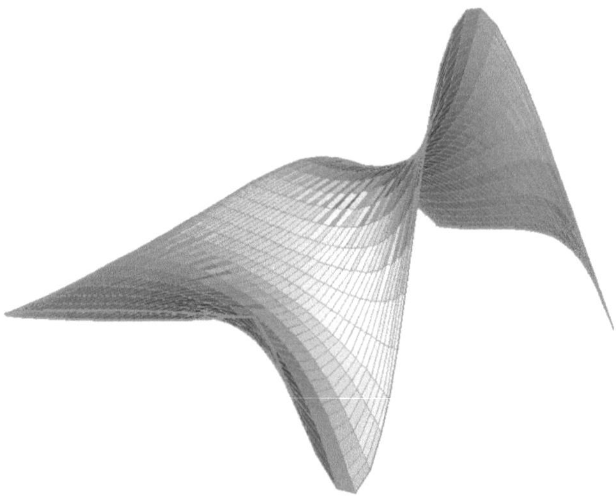

Fig. 4.18: Aerodynamic deformation by CVT.

as the deformation by the BEM method. Otherwise, the CVT result represents some small discontinuities on the surface, which is due to the fact that the aerodynamic mesh points in CVT method are only related to the three nearest structural mesh points, and there are rigid connections in the interpolation process.

Then, the three interpolations of the BEM, CVT and IPS methods are utilized to simulate the aeroelastic response of the AGARD445.6 wing. The structure and aerodynamic model are the same as before. The freestream conditions are set to $M = 1.141$,

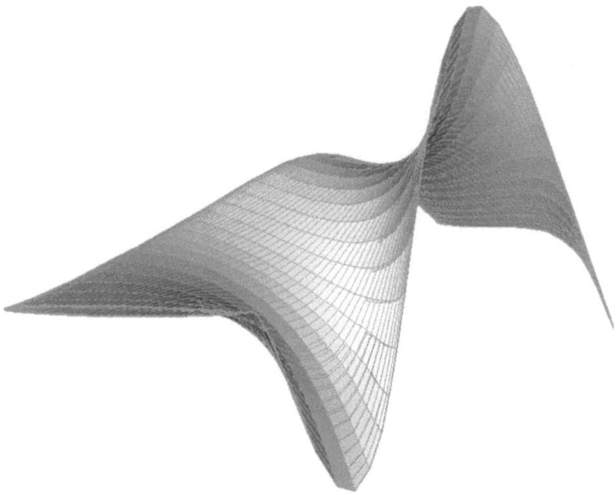

Fig. 4.19: Aerodynamic deformation by IPS.

and the ratio of the freestream velocity to the experimental flutter velocity is $V_\infty/V_f = 1.5$. The flow field is solved by N-S equation, the structure is solved by a linear method, the coupling method is applying the improved coupled method (as shown in Chapter 5) and the dimensionless time step is 0.15. Figure 4.20 shows the first-order generalized displacement time history calculated by the BEM, CVT and IPS methods. It can be seen that the three curves are identical to each other at the initial time steps. With the increase of time, there are some differences between the transformation methods. The amplitude of the CVT result tends to augment after some steps.

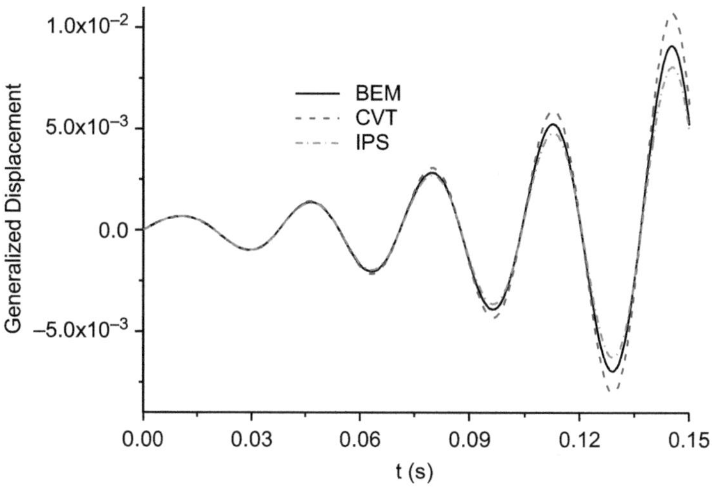

Fig. 4.20: First-order generalized displacement response.

The IPS method has a reversed effect. However, the BEM approach has a midterm value between CVT and IPS. Figure 4.21 shows the total force histories on the structural nodes, the variation trend of the three methods is similar to that of the first-order generalized displacement. Figure 4.22 shows the total load transfer error between the CFD and CSD subsystems. The error of the BEM method is obviously larger and the amplitude increases with time. The CVT and IPS maintain the strict conservation of the load transfer. Figure 4.23 shows the energy transfer error between the CFD and CSD subsystems. It is obvious that the BEM method keeps exact energy conservation, the IPS method has the largest error and the CVT method has the smaller error of energy transfer, but it also shows that the amplitude increases with time.

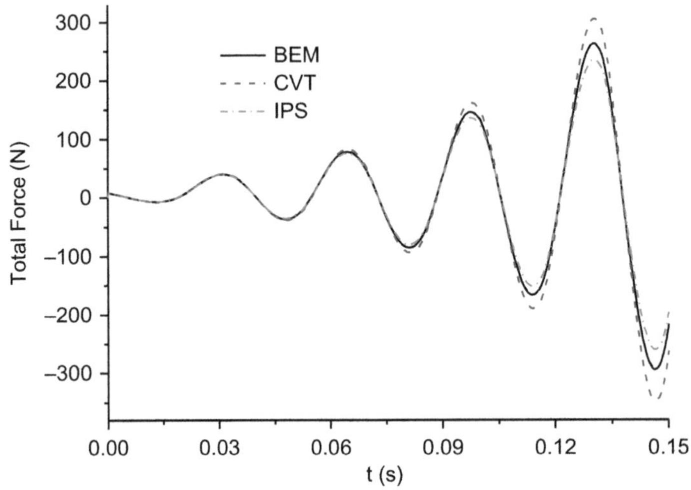

Fig. 4.21: Total load response.

The AGARD445.6 wing is regarded as a flat plate in structure modeling, and the surface mesh is generated according to the 3D configuration of the wing in aerodynamic modeling, so there is a large distance in the normal direction as the coupling boundaries of the two systems are not coincident. Because the local information with the shape function that is similar to the structural finite element is applied in the CVT method, it ensures the accuracy of the load transfer; since the IPS method is not a bidirectional interpolation method, but a pure displacement interpolation technique. In order to achieve two-way interpolation, a surface tracking method similar to CVT is used to map the load interpolation. Consequently, the strict load conversion relationship is also obtained. In order to maintain the energy transfer conservation between the two noncoincident boundaries, the BEM method loses the accuracy of the load transfer on the premise of obtaining more accurate displacement

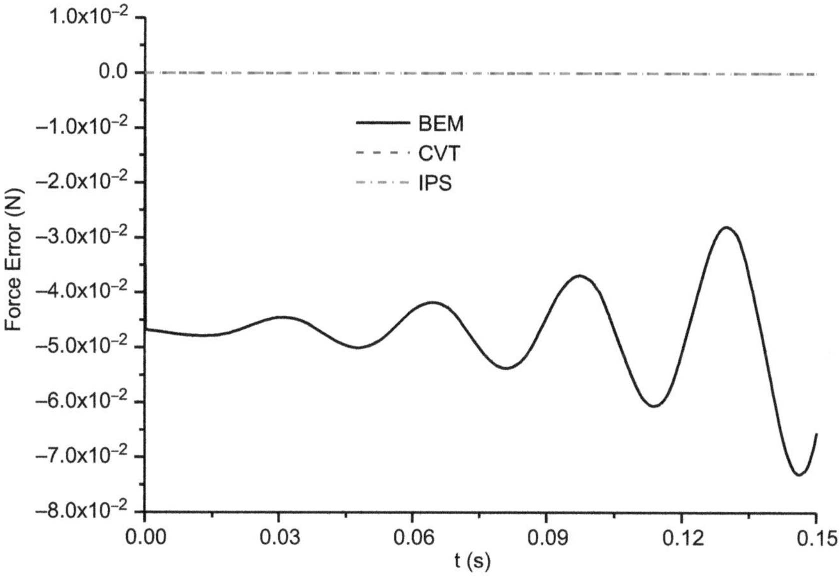

Fig. 4.22: Total load transfer error.

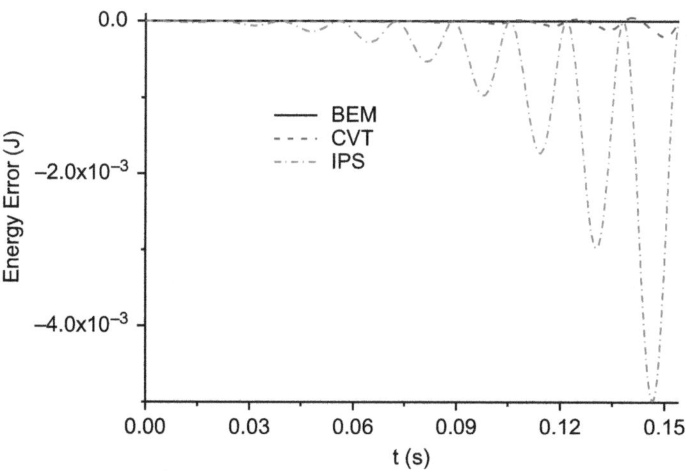

Fig. 4.23: Energy transfer error.

interpolation. The total amount of the load transfer shows great errors, but the trend of load distribution is consistent with the others. From point of view of the energy transfer errors, the BEM method is strictly satisfied; CVT method has some errors in energy transfer due to the noncoincidence of the two field boundaries; IPS method cannot deal with two-way interpolation, so it cannot guarantee energy conservation. It can be concluded that the CVT method maintains a certain accuracy in both the load and displacement conversion, but it cannot guarantee strict energy conservation; the BEM method can obtain more accurate displacement interpolation and ensure energy conservation, but lose some load conversion accuracy; the IPS method can only be used for displacement interpolation.

4.4 Moving-grid technique

In the simulation of unsteady flow, fluid-structure coupling calculation in the time domain, and aerodynamic shape optimization, the fluid grid should be adjusted continuously according to the change of the aircraft shape. For example, in flutter analysis, once the structure deforms at each time step, it is necessary to take into account the deformation for CFD calculation timely, and the moving-grid technique is required to adapt to the movement of the body. Therefore, it is necessary to study the effectiveness and efficiency of mesh regeneration. For the moving-grid algorithm, it is a challenging work to prevent the overlapping of boundary mesh points and the losing of the mesh points. One feasible method is to use interpolation or iteration method to redistribute the grid according to the distance ratio to the boundary or the original sparse ratio based on the initial grid data, which not only generates the mesh quickly but also ensures that the mesh can reflect the changes of the body surface timely.

At present, there are many mesh deformation methods, such as algebraic method [23], transfinite interpolation method (TFI) [32, 16, 25], elastomer method [37], spring analogy method [18, 26, 20, 3, 11], Delaunay map method [36] and radial basis function method.

With the development of CFD technology and moving-grid technique, the moving-grid must satisfy the following conditions:

1) Robustness. Moving-grid technology must be stable in dealing with any complex multiblock grid, including the suitable grid for N-S solver in CFD.
2) Accuracy. The quality of the grid deformation must ensure the acceptability of the fluid solver. The smoothness and orthogonality of all initial meshes should be maintained. In addition, one important characteristic on the grid interface is to synchronously deform the points in this region instead of recalculating the interpolation coefficients.
3) Ease of use. In theory, users do not need to understand the format of mesh deformation, nor do they need to make too many input commands. Nor should users bother with the special movement of the mesh points at the boundary or corner.

4) Efficiency. In the tightly coupled analysis, the fluid and structure models are iterated in advance. In time domain accuracy analysis with the tightly coupled program, it is required to be highly efficient for the mesh deformation as the mesh should be updated at each time step.
5) Parallelizable. The process should be integrated with the parallel computing code without incurring additional communication costs and idle processing time.

This section mainly introduces the algebraic method, TFI and radial basis function (RBF). Then, they are applied to some representative examples in 2D and 3D cases.

4.4.1 Geometric interpolation method

First, the 2D case $P(x_{i,j}, y_{i,j})$ is considered for the mesh point, once the points on the inner boundary have the displacement in a time step, the surrounding points will deform accordingly. The points on the inner boundary, that is, $j = 1$ are taking as the reference points to interpolate the new position of the surrounding points by the following process:

Let $P(x_{i,j}, y_{i,j})$ denotes the initial coordinate, and $P'(x'_{i,j}, y'_{i,j})$ is the deformed coordinate, when all the points on the inner boundary have displacements, such as rotation or translation, the angle of the displacement at the point $P(x_{i,1}, y_{i,1})$ on the inner boundary can be determined by the two adjacent points as

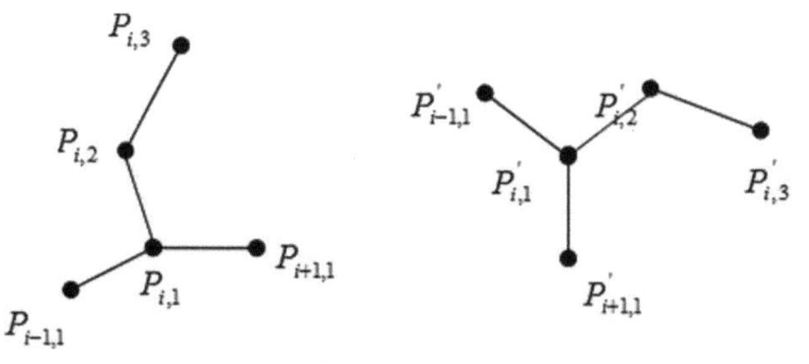

Fig. 4.24: Grid points before and after the transformation.

$$\cos(\theta_{i,1}) = \frac{(x_{i+1,1} - x_{i-1,1})(x'_{i+1,1} - x'_{i-1,1}) + (y_{i+1,1} - y_{i-1,1})(y'_{i+1,1} - y'_{i-1,1})}{\sqrt{(x_{i+1,1} - x_{i-1,1})^2 + (y_{i+1,1} - y_{i-1,1})^2}\sqrt{(x'_{i+1,1} - x'_{i-1,1})^2 + (y'_{i+1,1} - y'_{i-1,1})^2}}, \tag{4.59}$$

$$\sin(\theta_{i,1}) = \frac{(x_{i+1,1}-x_{i-1,1})(y'_{i+1,1}-y'_{i-1,1})-(y_{i+1,1}-y_{i-1,1})(x'_{i+1,1}-x'_{i-1,1})}{\sqrt{(x_{i+1,1}-x_{i-1,1})^2+(y_{i+1,1}-y_{i-1,1})^2}\sqrt{(x'_{i+1,1}-x'_{i-1,1})^2+(y'_{i+1,1}-y'_{i-1,1})^2}}. \tag{4.60}$$

For example, for the point $P(x_{i,1}, y_{i,1})$, the two adjacent points $P(x_{i-1,1}, y_{i-1,1})$, $P(x_{i+1,1}, y_{i+1,1})$ are introduced to form a triangular element, and the rigid deformation are considered for all meshes in the element. Intermediate variables are introduced as

$$x_{i,j}^{\text{ref}} = x'_{i,1} + (x_{i,j}-x_{i,1})\cos(\theta_{i,1}) - (y_{i,j}-y_{i,1})\sin(\theta_{i,1}), \tag{4.61}$$

$$y_{i,j}^{\text{ref}} = y'_{i,1} + (x_{i,j}-x_{i,1})\sin(\theta_{i,1}) + (y_{i,j}-y_{i,1})\cos(\theta_{i,1}), \tag{4.62}$$

when j > 1,

$$S_{i,j} = \sum_{k=2}^{j}\sqrt{(x_{i,k}-x_{i,k-1})^2+(y_{i,k}-y_{i,k-1})^2}, \tag{4.63}$$

when j = 1, $S_{i,j}=0$,

$$b_{i,j} = 3\left(\frac{S_{i,j}}{S_{i,j\max}}\right)^2 - 2\left(\frac{S_{i,j}}{S_{i,j\max}}\right)^3, \tag{4.64}$$

Then the new mesh points are obtained by

$$x'_{i,j} = b_{i,j}x_{i,j} + (1-b_{i,j})x_{i,j}^{\text{ref}}, \tag{4.65}$$

$$y'_{i,j} = b_{i,j}y_{i,j} + (1-b_{i,j})y_{i,j}^{\text{ref}}. \tag{4.66}$$

Obviously, on the inner boundary ($j=1$) $b_{i,j}=0$, and on the outer boundary ($j=j_{\max}$) $b_{i,j}=1$, so the outer boundary is invariant while the inner boundary is deformed as it did originally. Thus, the initial grid is readjusted by the coefficient factor $b_{i,j}$ to adapt to the deformed body surface in the new time step.

In the 3D case, each mesh point is represented by $P_{i,j,k}$, where $k=1$ represents the body surface and $k=k_{\max}$ represents the outer boundary; then the governing equation for the 3D moving grid is

when $k>1$,

$$S_{i,j,k} = \sum_{t=2}^{k}\sqrt{(x_{i,j,k}-x_{i,j,k-1})^2+(y_{i,j,k}-y_{i,j,k-1})^2+(z_{i,j,k}-z_{i,j,k-1})^2}, \tag{4.67}$$

when $k = 1$, $S_{i,j,k} = 0$,

$$b_{i,j,k} = 3\left(\frac{S_{i,j,k}}{S_{i,j,k\max}}\right)^2 - 2\left(\frac{S_{i,j,k}}{S_{i,j,k\max}}\right)^3, \tag{4.68}$$

$$x'_{i,j,k} = b_{i,j,k}x_{i,j,k} + (1 - b_{i,j,k})x^{\text{ref}}_{i,j,k}, \tag{4.69}$$

$$y'_{i,j,k} = b_{i,j,k}y_{i,j,k} + (1 - b_{i,j,k})y^{\text{ref}}_{i,j,k}, \tag{4.70}$$

$$z'_{i,j,k} = b_{i,j,k}z_{i,j,k} + (1 - b_{i,j,k})z^{\text{ref}}_{i,j,k}. \tag{4.71}$$

The experiments have shown that the interpolation method is feasible for the case of small boundary displacement in the 3D case. Where the solutions of $x^{\text{ref}}_{i,j,k}$, $y^{\text{ref}}_{i,j,k}$, $z^{\text{ref}}_{i,j,k}$ are similar to the 2D case, especially for the simple rotational deformation (rotate θ counterclockwise around Z axis) can be expressed as

$$\begin{pmatrix} x^{\text{ref}}_{i,j,k} \\ y^{\text{ref}}_{i,j,k} \\ z^{\text{ref}}_{i,j,k} \end{pmatrix} = \begin{pmatrix} \cos\theta & -\sin\theta & 0 \\ \sin\theta & \cos\theta & 0 \\ 0 & 0 & 1 \end{pmatrix} \begin{pmatrix} x_{i,j,k} \\ y_{i,j,k} \\ z_{i,j,k} \end{pmatrix}. \tag{4.72}$$

4.4.2 Transfinite interpolation method

The TFI method [32, 16, 25, 12, 38] can preserve all the features of the original grid, and it is independent of the initial grid generation. Generally, the TFI method can only be used in the structured grid with high computational efficiency and ability to deal with the moderate deformation, but it is difficult to interpolate the complex topological structures with the TFI method. The main process of the 3D structured grid block interpolation is as follows:

(1) Determine the displacement of all corners in the deformed block.
(2) Linearly interpolate the displacements of the four corners to the boundary line connected with the moving body surface, and calculate the displacement of each point inside the boundary line.
(3) Calculate the displacements of the interior points on the grid section by the TFI formulation after determining the displacement of each point on the boundary.
(4) Judge the deformed grid block in the unsteady flow field by the displacements of the four corners of the moving surface, and determine the movement of the corners of each block. In practice, the grid blocks which are not connected with the moving surface are usually fixed, and those connected with the moving surface are deformed.

(5) In all boundary lines of the grid block, the node displacement in the four boundary lines connected with the body surface can be determined by the following steps:
   a) Define the coordinate vector **x** and the displacement vector **dx** of the grid points as

$$\begin{cases} \mathbf{x} = (x(\xi,\eta,\zeta), y(\xi,\eta,\zeta), z(\xi,\eta,\zeta)) \\ \mathbf{dx} = (dx(\xi,\eta,\zeta), dy(\xi,\eta,\zeta), dz(\xi,\eta,\zeta)) \end{cases}. \tag{4.73}$$

   b) Let the displacement vector $\mathbf{dx}_A$ and $\mathbf{dx}_B$ denote the two vertices A and B on the same boundary line, respectively, then the displacement vectors of any point P on the AB line can be calculated by the interpolation equation

$$\begin{cases} \mathbf{d}x = \left(1-\frac{a}{c}\right)\mathbf{d}x_A + \left(1-\frac{b}{c}\right)\mathbf{d}x_B \\ \mathbf{d}y = \left(1-\frac{a}{c}\right)\mathbf{d}y_A + \left(1-\frac{b}{c}\right)\mathbf{d}y_B \\ \mathbf{d}z = \left(1-\frac{a}{c}\right)\mathbf{d}z_A + \left(1-\frac{b}{c}\right)\mathbf{d}z_B \end{cases}, \tag{4.74}$$

   where a, b, c are the lengths of the curves, $a = \|AP\|, b = \|BP\|, c = \|AB\|$; a, b, c can be obtained from the initial mesh coordinate.
   c) If the two vertices of the boundary line do not move, the boundary line remains invariant.

(6) In-plane node displacement interpolation

The displacement vectors of all nodes on the boundary line have been obtained in the second step, and then the displacement vectors of the in-plane nodes can be interpolated by the corresponding points in four boundary lines. Taking a plane with fixed constant ζ as an example, the TFI formulation of the displacement vector **dx** of the in-plane points is

$$\begin{cases} \mathbf{d}x(\xi,\eta) = \Phi_1^0(\eta)[\mathbf{d}x_{b1}(\xi) - f_1(\xi,0)] + \Phi_2^0(\eta)[\mathbf{d}x_{b3}(\xi) - f_1(\xi,1)] + f_1(\xi_1,\eta) \\ \mathbf{d}y(\xi,\eta) = \Phi_1^0(\eta)[\mathbf{d}y_{b1}(\xi) - f_2(\xi,0)] + \Phi_2^0(\eta)[\mathbf{d}y_{b3}(\xi) - f_2(\xi,1)] + f_2(\xi_1,\eta) \\ \mathbf{d}z(\xi,\eta) = \Phi_1^0(\eta)[\mathbf{d}z_{b1}(\xi) - f_3(\xi,0)] + \Phi_2^0(\eta)[\mathbf{d}z_{b3}(\xi) - f_3(\xi,1)] + f_3(\xi_1,\eta) \end{cases}, \tag{4.75}$$

$$\begin{cases} f_1(\xi_1,\eta) = \Psi_1^0(\xi)\mathbf{d}x_{b4}(\eta) + \Psi_2^0(\xi)\mathbf{d}x_{b2}(\eta) \\ f_2(\xi_1,\eta) = \Psi_1^0(\xi)\mathbf{d}y_{b4}(\eta) + \Psi_2^0(\xi)\mathbf{d}y_{b2}(\eta) \\ f_3(\xi_1,\eta) = \Psi_1^0(\xi)\mathbf{d}z_{b4}(\eta) + \Psi_2^0(\xi)\mathbf{d}z_{b2}(\eta) \end{cases}, \tag{4.76}$$

where Ψ and Φ are shape functions in ξ and η directions, respectively, which are related to the distribution of boundary points of each block

$$\begin{cases} \Psi_1^0(\xi) = 1 - s_1(\xi) \\ \Psi_2^0(\xi) = s_3(\xi) \\ \Phi_1^0(\xi) = 1 - s_4(\eta) \\ \Phi_2^0(\eta) = s_2(\eta) \end{cases}, \tag{4.77}$$

where $s_1(\xi)$ is the stretch function on the block boundary of $\eta = 0$; $s_2(\eta)$ is the stretch function on the block boundary of $\xi = 1$; $s_3(\xi)$ is the stretch function on the block boundary of $\eta = 1$; and $s_4(\eta)$ is the stretch function on the block boundary of $\xi = 0$. The stretch function is the ratio of the distance between the corresponding point on the boundary and the starting point of the boundary line to the total length of the boundary line. The new grid point coordinates can be expressed by the interpolation

$$\mathbf{x}(\xi, \eta, \zeta) = \mathbf{x}_0(\xi, \eta, \zeta) + \mathbf{dx}(\xi, \eta, \zeta), \tag{4.78}$$

where **dx** is the displacement vector obtained by TFI method and $\mathbf{x}_0$ is the initial coordinate.

4.4.3 Radial basis function method

The moving-grid technology based on the TFI method has the characteristic of fast operation, and it can satisfy the orthogonality of the near-surface grid under small structural disturbance. However, with the increase of the deformation scale of the structural model, due to the large distortion of the near-surface grid, the TFI method cannot exactly satisfy the orthogonality requirement for the CFD calculation. An RBF [35]-based moving-grid method is proposed to solve the problem with large deformation. The main idea of this method is to construct a radial basis function sequence by using the known body surface grid deformation, and then using the radial basis function sequence to smoothly interpolate the body surface deformation into space aerodynamic grid [35]. It can be found that the RBF method is also with high accuracy.

The RBF method does not need to use the connection between grid points in the calculation, and it is very easily parallelized since the deformation for each node is completely noninterference with each other; it can also deal with the large deformation movement of the complex shape, and it has excellent versatility, robustness, as well as good quality of the deformed grid; it is programming conveniently and implementing easily; there is no requirement for the extra interpolation as the same algorithm is utilized to determine the position of the surface and the spatial points; it can be employed in structured grid, unstructured grid and hybrid grid as well; it can support arbitrary forms of deformation, rigid motion, elastic deformation and even plastic deformation. However, the computation amount of the

RBF method is proportional to $N_{vp} \times N_{sp}$. N_{vp} is the number of the unknown aerodynamic space grid points with a general magnitude of $10^5 - 10^6$, and N_{sp} is the number of the surface nodes for interpolation with a general magnitude of $10^2 - 10^3$. Consequently, its computational amount is considerable compared with the algebraic interpolation method.

At present, almost all works of literature focus on reducing the number of surface interpolation nodes to decrease the computational complexity, but few researchers consider reducing the number of the unknown space points in the calculation. Andreas [1] proposed the algorithm for restricting the interpolation domain of RBF to separate the influence of the aircraft component motion on other components.

1. Grid deformation based on radial basis function

The basic form of the radial basis function is [35]

$$F(r) = \sum_{i=1}^{N_{sp}} \omega_i \varphi(\| r - r_i \|), \tag{4.79}$$

where $F(r)$ is the interpolation function and represents the deflection of the grid in the moving-grid problem. N_{sp} represents the total number of the radial basis functions used in the interpolation, and it is equal to the number of the surface nodes in the moving-grid problem. $\varphi(\| r - r_i \|)$ is the general form of the radial basis function. r_i is the position of the ith interpolation node; r is the position vector of any spatial point and it is the position vector of the CFD space grid point;$\| r - r_i \|$ is the distance between any spatial point and the ith interpolation node; and ω_i is the weight coefficient corresponding to the ith interpolation node. There are many kinds of the radial basis function. It has been considered that the Wendland's C2 functions [35] have better computational efficiency and grid deformation quality. Therefore, they are used as the radial basis function to implement grid deformation in the following form [35, 28]:

$$\varphi(\eta) = (1 - \eta)^4 (4\eta + 1), \tag{4.80}$$

where η is the dimensionless value of $\| r - r_i \|$ with $\eta = (\| r - r_i \| / R)$, and R is the influencing radius of the radial basis function. When $\eta > 1$, let $\varphi(\eta) = 0$, which means that when the influencing distance is exceeded, the deformation is zero.

The interpolation process by applying eqs. (4.79) and (4.80) to solve the deformation of any spatial point is as follows: Given the interpolation surface node and its displacement of r and r_i are known, and the radial basis function $\varphi(\eta)$ has been determined, the only unknown is the weight coefficient ω_i associated with the interpolation node on the surface. It can be obtained by the condition that the interpolation result on the surface must be consistent with the given displacement. That is, it can be obtained by solving the following equations:

$$\Delta X_s = \Phi W_x, \tag{4.81}$$

$$\Delta Y_s = \Phi W_y, \tag{4.82}$$

$$\Delta Z_s = \Phi W_z, \tag{4.83}$$

where the subscript s represents the surface interpolation node.

$\Delta X_s = \left\{\Delta x_{(s,1)}, \Delta x_{(s,2)}, \ldots, \Delta x_{(s,N_{sp})}\right\}^T$, $\Delta Y_s = \left\{\Delta y_{(s,1)}, \Delta y_{(s,2)}, \ldots, \Delta y_{(s,N_{sp})}\right\}^T$ and $\Delta Z_s = \{\Delta z_{(s,1)}, \Delta z_{(s,2)}, \ldots, \Delta z_{(s,N_{sp})}\}^T$ represent the displacements of N_{sp} interpolation nodes on the body surface. $W_x = \left\{\omega_{(x,1)}, \omega_{(x,2)}, \ldots, \omega_{(x,N_{sp})}\right\}^T$, $W_y = \left\{\omega_{(y,1)}, \omega_{(y,2)}, \ldots, \omega_{(y,N_{sp})}\right\}^T$ and $W_z = \left\{\omega_{(z,1)}, \omega_{(z,2)}, \ldots, \omega_{(z,N_{sp})}\right\}^T$ are the unknown weight coefficients associated with each interpolation node. Each element in the matrix is the radial basis function value of the distance between any two interpolation nodes on the surface, $\Phi(j,i) = \varphi(\| r_j - r_i \|)$, $1 \le j,\ i \le N_{sp}$.

The weight coefficients can be obtained by solving eqs. (4.81) to (4.83), and then the grid deformation at any spatial position can be determined by eq. (4.79).

2. Selection of the surface interpolation nodes

As mentioned earlier, the calculation amount of the RBF method is proportional to $N_{vp} \times N_{sp}$, where N_{vp} is the number of the unknown aerodynamic space grid points, N_{sp} is the number of the surface nodes for interpolation and it is generally selected from the structural mode in the time domain simulation of the fluid-structure coupling system. The main way to reduce the total amount of calculation is to reduce the value of N_{sp}. Rendall et al. [28, 29] adopted a greedy algorithm by gradually adding interpolation nodes according to the maximum error of the position to simplify the radial basis function sequence. The basic process is as follows.

First, the initial node set $P^0 = \{p_1, p_2, \ldots, p_I\}$ is formed by arbitrarily selected I (generally I = 3) surface nodes. The set is used to interpolate the radial basis function, and the corresponding weight coefficients are obtained by solving the equations, then the grid deformations on all the surface nodes are obtained. Obviously, the initial interpolation function is exactly accurate for all nodes in P^0. However, it will cause error for the surface nodes that do not belong to P^0 According to the principle of greedy method, the surface node with maximum error should be determined and incorporated into P^0 to form the next node set P^1, and then the updated node set P^1 is used to interpolate the radial basis function. The process is repeated until the interpolation error on the surface node satisfies the given criterion. In the process of selecting surface nodes, all the surface nodes with errors greater than the average value or a given limitation can be selected into the node set in each step. Or the earlier two selecting methods can be applied alternately in each n step, that is, the node with the maximum error is selected in n step, and the nodes with larger errors are selected in n+1 step to improve the selecting speed.

Here applying the principle of the greedy method, the step-by-step approximation theory of the function space subset [35] is introduced to enhance the efficiency of data reduction. First, selecting N_0 nodes for the radial basis function interpolation and determining the error $\Delta S^{(0)}$ on all surface nodes, and then, changing the interpolation object of $\Delta S^{(0)}$ the radial basis function from the initial grid deformation to the error of the current surface nodes, and once again, using the greedy method to select N_1 surface nodes for interpolation. Repeating these steps until the residual satisfies the criterion, and finally the interpolation coefficients of the radial basis function can be obtained by superimposing the selected interpolation nodes and the weight coefficients in the n step. The efficiency of data reduction can be improved by this way on account of only solving a $N_i \times N_1$ linear algebraic equation in each step for calculating the interpolation coefficients. In each step, the value of N_i can be controlled in a small amount, thus reducing the amount of calculation. Consequently, a method of reducing the number of the nodes to be interpolated in space can be derived by extending the idea of the step-by-step approximation theory of the function space subset.

3. A method of reducing spatial interpolated nodes

At present, the researchers focus on decreasing the amount of N_{sp}, the surface interpolation nodes to reduce the amount of computation in RBF. In order to do that, the algorithm for restricting the interpolation region of RBF is established by separating the influence of the single component motion on other components. The basic idea of the algorithm is that the restricted region of the component motion is considered as a cuboid, which includes the moving component. The deformation on the moving part is given as the required value, and the deformation on the surface of the cuboid is set as zero. The deformation of the inside region between the surface and the moving part of the cuboid is employed by RBF, while there is no interpolation for the outside of the cuboid.

Here combining the idea of the step-by-step approximation theory of the function space subset with the method of restricting interpolation region proposed by Andreas, a new effective reduction scheme for N_{sp} is developed to support the large deformed motion. The main steps of the scheme are as follows:

1) Determining the interpolation region Ω_0, the nodes with the distance to the surface which are less than a limit value $R_0 = k \cdot \Delta S_{max}$ are selected as the node set, where k is a coefficient set as $k \geq 5$. Making sure that there is no body surface boundary in Ω_0, and the boundary can be determined: Scanning each point in Ω_0, if there are nodes on the elements around the point that do not belong to Ω_0, then the point is set as the boundary, otherwise it is the interior point, and the displacement on the boundary is set as 0. An initial error criterion ϵ_0 is

given, which could be larger, such as 1.0×10^{-2}. The $N^{(0)}$ interpolation nodes are selected by the condition that the maximum value of surface interpolation error is less than $\in_0$, and the surface interpolation error ΔS_0 are recorded. Taking interpolation in region Ω_0, and at this moment, N_{vp} is large and $N_{sp} = N(0)$. Due to the large value of the error criterion, the $N^{(0)}$ will be less, that is, the total amount of the calculation will be less.

2) Taking the surface interpolation error ΔS_0 as the object of the next interpolation and determining the interpolation region Ω_1. As the maximum displacement of surface interpolation is less than $\in_0$ at this time, similarly, the nodes with the distance to the surface less than $R_1 = k \cdot \Delta S_{0,\max}$ are selected. Because of $\Delta S_{0,\max} \leq \in_0$, Ω_1 can be smaller, which makes N_{vp} smaller. Now, a relatively smaller error criterion $\in_1$ is set to choose $N^{(1)}$ interpolation points. Because of the small error criterion, the selected will be larger. However, the total amount of calculation will be still relatively small.
3) Repeating the earlier steps until the error criterion meets the requirement.

It should be noted that when R_i is relatively small, the interpolation region is also small, and the outer boundary with the body surface is too close to each other. It means that the distance among some nodes in the selected interpolation node is extremely small, which will cause a very ill matrix in the solution of the interpolation coefficient. Therefore, there should be a limit of R_i. In the present algorithm, $R_i = k_1 \cdot \max(\in_{i-1}, d_{\max})$, where k_1 is an insurance factor set as 2. $\in_{i-1}$ is the error limit of the previous step, and $d_{\max}$ is the length of the longest grid line connected to the body surface. With this restriction, in the second computation of the interpolation, R_i will reach the limit, so the third step is not necessary any more. It should be also noted that when choosing interpolation nodes on the outer boundary of the interpolation region, the error on the boundary is not required to be as small as that on the body surface, but only a requirement that the error on the boundary will not greatly decrease the grid quality near the boundary is needed. As a result, the selection of the interpolation nodes on the boundary should ensure that the maximum error on the boundary is less than the k times of the minimum size of the adjacent element, thus k can be set as 0.1, and then the value of N_{sp} will not be large in the interpolation.

1) If the error limit in the last interpolation step is far less than the minimum distance between the spatial grid point and the body surface, it is required to select more interpolation points and take the interpolation on the body surface. At this situation, the surface interpolation error will further reduce since the deformation of the body surface will not cause the distortion of the spatial grid. Then,N_{vp} is equal to the number of the surface grid, as a result, the amount of the calculation will be relatively small, and the interpolation error on the whole surface can be greatly reduced.

In the earlier process, although the calculated amount to determine the interpolation limited region is large, it is implemented only once in all process as the limited region is consistent with the initially determined region in each step of the deformation. Therefore, the cost of the whole calculation for the grid deformation does not increase despite the large calculation amount of this step.

Figure 4.25 shows the flow field grid for calculating torsional deformation of the GOLAND wing by the developed RBF method.

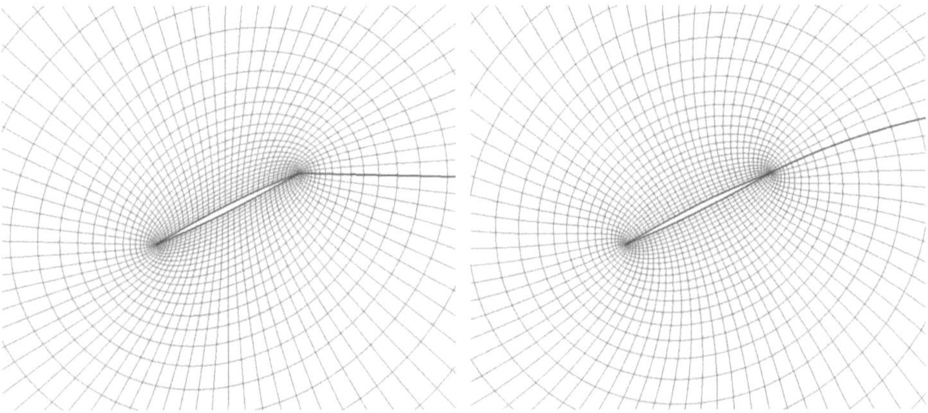

(a) Infinite interpolation method (b) Radial basis function

Fig. 4.25: Comparisons of infinite interpolation and radial basis function moving-grid method GOLAND Wing Spread Profile).

4.4.4 Shape-preserving moving-grid method

The earlier moving-grid methods have been widely used in many problems of the rigid/elastic aerodynamic grid deformation, but there are still some deficiencies in further application for aerodynamic grid deformation of the complex shapes. Especially for the rudder-body configuration, there should be a consideration that not only the rigid motion but also the elastic deformation of the moving rudder, that is, the discontinuity of the original surface deformation [1].

Aiming to deal with different complex shapes and special analysis, the projection technology based on the finite element interpolation, RBF interpolation technology and TFI moving-grid method are combined to develop a set of high-precision shape-preserving grid deforming method (SPGDM). This method belongs to a composite moving-grid method.

The TFI method and the RBF method have been introduced in Sections 4.4.2 and 4.4.3, respectively. The following will introduce the projection technique based on the finite element interpolation and the process of the SPGDM method.

1. Surface interpolation based on finite element shape function

Referring to the idea of the inverse transformation of the isoparametric element [2, 39], the local coordinates are obtained by a numerical iteration algorithm, and then a forward interpolation is performed for the physical quantities.

For the general 3D isoparametric element, the coordinate transformation is to transform each element from the global coordinate system x – y – z to the local coordinate system ξ–η–ζ. The explicit positive transformation is as follows:

$$x=\sum_{i=1}^{n}N_i(\xi,\eta,\zeta)x_i,\tag{4.84}$$

$$y=\sum_{i=1}^{n}N_i(\xi,\eta,\zeta)y_i,\tag{4.85}$$

$$z=\sum_{i=1}^{n}N_i(\xi,\eta,\zeta)z_i,\tag{4.86}$$

where x_i, y_i, z_i are the coordinates of the ith node in the element, n is the number of the nodes in the element and N_i is the shape function corresponding to the ith node. x, y, z are the coordinates of the interpolation point and ξ, η, ζ are the local coordinates of the interpolation point. The general quantity u is interpolated by the same function as

$$u=\sum_{i=1}^{n}N_i(\xi,\eta,\zeta)u_i,\tag{4.87}$$

where u_i is the physical quantity of the *i*th node.

First, the numerical iteration algorithm is used for the inverse transformation. Let $b=(x,y,z)^T, \alpha=(\xi,\eta,\zeta)^T$, then eqs. (4.84) to (4.86) can be expressed as $b=f(\alpha)$. It is a set of third-order nonlinear equations, which can be solved by the iterative method. Here, using the Newton iteration method, the solution process is as follows:

(1) The initial iteration value is given as α_0.
(2) Calculate the error vector of the *k*th step: $e_k=f(\alpha_k)-b$.
(3) If $\|e_k\|$ is small enough, go to (7).
(4) Compute Jacobian matrix of the equations: $J_k=\left(\frac{\partial(x,y,z)}{\partial(\xi,\eta,\zeta)}\right)_k$.
(5) Solve the linear equations: $J_k\Delta\alpha=e_k$.
(6) Update the iteration value: $\alpha_{k+1}=\alpha_k+\Delta\alpha$, $k=k+1$, and go back to (2).
(7) End the iteration: $\alpha=\alpha_k$.

The value of u on the other set of the grid points can be obtained by substituting the obtained α (i.e., ξ, η, ζ) into eq. (4.87). In addition, the least square method is employed in the step (5) by premultiplying $J_K{}^T$ on both sides of the equation. Moreover, the second-order element is suggested to be utilized to improve the accuracy and smoothness of surface interpolation.

The inverse transformation method presented earlier converges very quickly when the interpolation points are within the grid element or are not far away from each other. However, it is not easy to converge if the interpolation point is far away from the grid element. Therefore, the grid element where the interpolation point is located should be determined first in order to ensure the correctness and efficiency of the inverse transformation.

2. Shape-preserving moving-gird method and its implementation

First, the projection base surface grid is elastically deformed by the RBF method, and the rigid motion is added according to the given law; then, by the projection method based on the finite element interpolation, the body surface grid is projected onto the base surface in the order of point–line–plane to obtain the deformed body surface grid, during which the TFI technology is employed to maintain the distribution of the original grid; Finally, with the TFI technology, the deformation of the surface grid is evenly interpolated into the whole field grid to obtain the deformed volume grid. The strategy and implementation of the combined moving-grid are illustrated by the following schematic diagrams and steps: the solid lines represent the surface boundary of the CFD grid (Fig. 4.26), and the dotted lines represent the projection base grid (Fig. 4.27). In order to ensure the coherence of the intersection, the projection base plane should be extended appropriately.

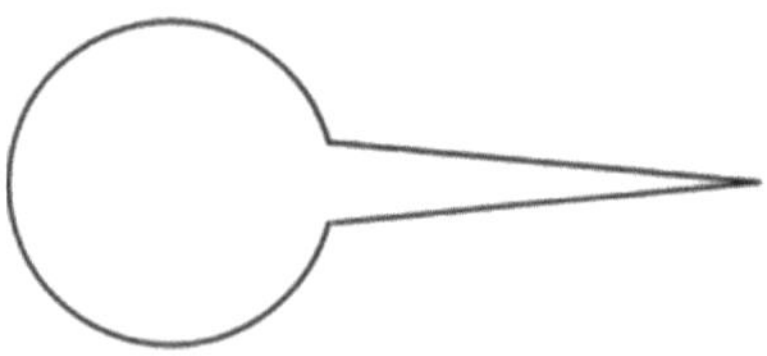

Fig. 4.26: Original CFD surface.

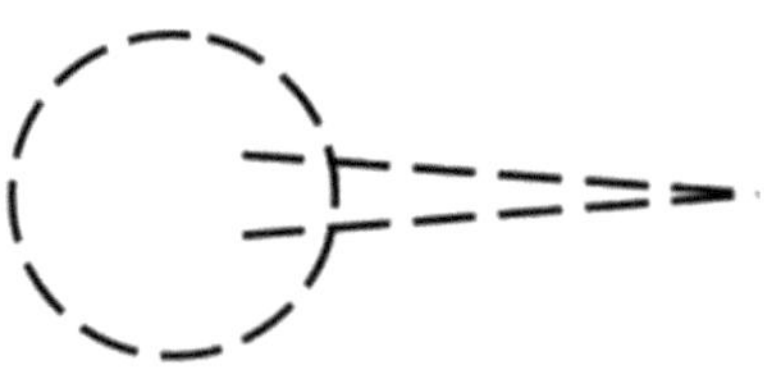

Fig. 4.27: Original projection base surface.

(1) Using RBF method, the projection base surface grid is deformed elastically and then the rigid motion is added, as shown in Fig. 4.28.
(2) By the projection method based on the finite element interpolation, the boundary of the body surface grid of the previous step is projected onto each base surface repeatedly until the intersection line position does not change anymore, as shown in Fig. 4.29.

Fig. 4.28: Deformation of the projection base surface.

Fig. 4.29: Computational intersection.

(3) By the TFI technology, the deformation of the boundary of the surface grid is evenly interpolated into the whole surface grid, as shown in Fig. 4.30.

Fig. 4.30: Body surface grid interpolation.

(4) By the projection method based on the finite element interpolation, the interior points of the body surface are projected onto each base surface to get the current surface grid as shown in Fig. 4.31.
(5) By the TFI technology, the deformation of the body surface grid is evenly interpolated into the whole field to obtain the current field grid.

Fig. 4.31: Body surface grid projection.

4.4.5 Numerical example of moving-grid techniques

1. The geometric interpolation algorithms moving-grid example

The NACA0012 airfoil is taken as an example. Figure 4.32 is the O-shape structure grid of the airfoil. The airfoil's attack angle $\pi/10$ is considered, and Fig. 4.33 shows the calculation by the geometric interpolation algorithm.

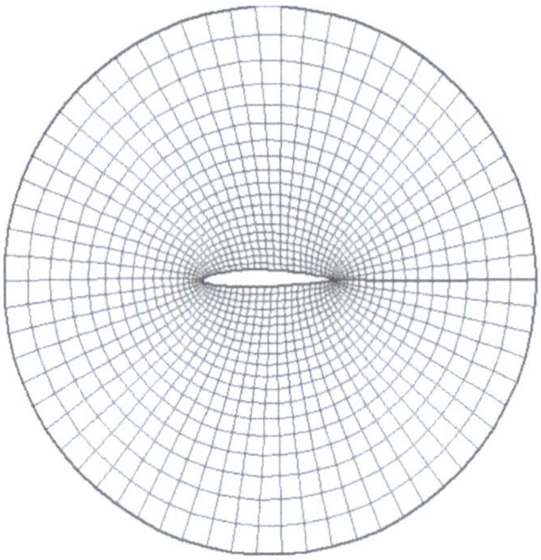

Fig. 4.32: Airfoil original grid.

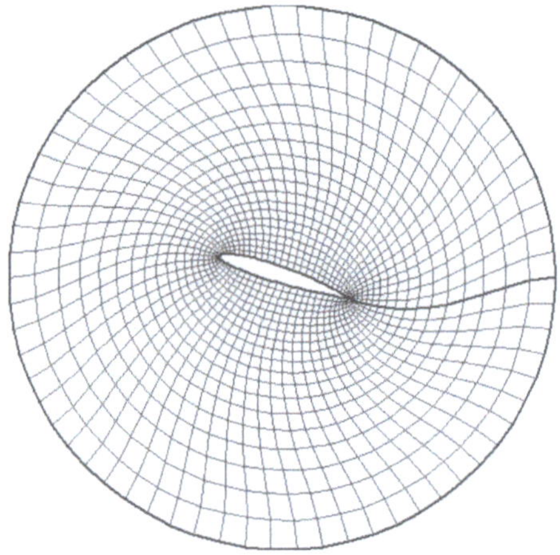

Fig. 4.33: Calculation results of moving grid.

2. The TFI moving-grid example

A 3D wing is selected as the interpolated object. First, the elliptic equation method is used to generate the C-type structured grid of the wing (Fig. 4.34(a)). The CFD–CSD coupling calculation is performed to obtain the deformed wing at a certain time, and then the 3D aerodynamic grid is reconstructed by the TFI technology (Fig. 4.34(b)).

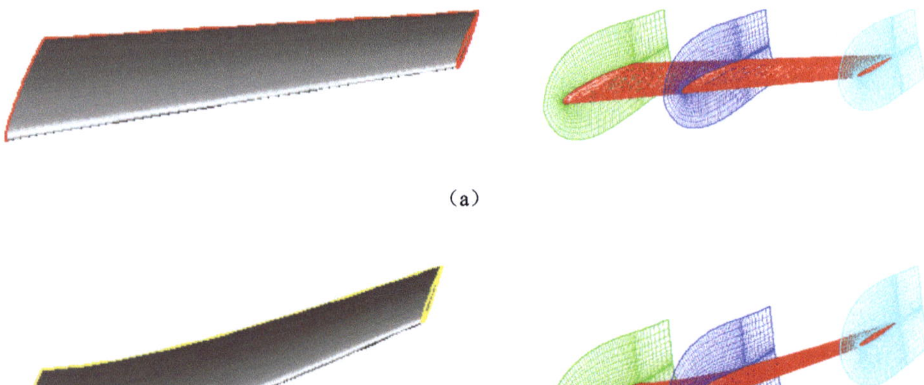

(a)

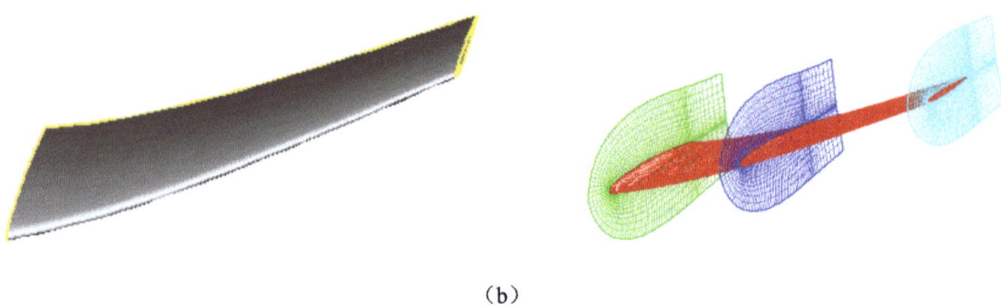

(b)

Fig. 4.34: Calculation results of wing's moving grid.

3. The RBF method in moving-grid application

1) NACA0012 airfoil moving-grid example

The 2D structured grid of NACA0012 airfoil is taken as the initial grid, and the displacement equation is give as $\Delta y = 0.03\sin(4\pi x)$ [35]. The mesh size is 323 × 81, and the error criterion is taken as 1.0×10^{-6}. The single interpolation method, as well as the multiple interpolations method, are used in the calculation. In the process of the calculation, in the first interpolation step of the multiple interpolations, the whole field mesh is directly regarded as Ω_0, as the deformation of the first step is relatively large and the far field grid themselves are sparse. In the second interpolation step, the number of the spatial grid points is reduced to Ω_1 which contains 3553 grid nodes. The error of the body surface interpolation is small enough after the second interpolation, therefore, there is no need to compute Ω_2 but the deformation of the body surface grid in the third interpolation step. The results of grid deformation are shown in Fig. 4.35.

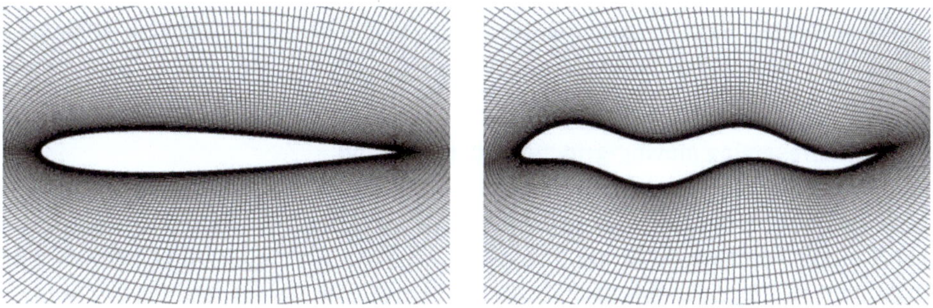

(a) original grid; (b) intepolated deformation

Fig. 4.35: Comparison of the original and deformed NACA0012 airfoil.

2) ONERA-M6 wing moving-grid example

The unstructured grid of the ONERA-M6 wing [31] is taken as the initial grid, the number of the grid elements is 1,883,014 with the 33,684 nodes, and the number of surface nodes is 33,206. The deformation mode is given as $\Delta y = 0.5bz[1 - \cos((z\pi)/b)]$, where b is the wingspan, z is the coordinate in the z-direction, and Δy is the deformation in the y-direction. The final error criterion is taken as 1.0×10^{-6}. The single interpolation method and the multiple interpolations method are used in the calculation, and the process is the same as that of the previous example. The comparison of the efficiency between the multiple interpolations and the single interpolation is shown in Tab. 4.2. It can be seen that the multiple interpolations method can greatly improve the efficiency of the grid deformation by reducing the number of spatial pending interpolated grid.

Tab. 4.2: Comparisons of the time required to complete mesh deformation of ONERA-M6 wing.

Interpolation mode	Error limit	N_{sp}	N_{vp}	$N_{sp} \times N_{vp}$	CPU time (s)
Single interpolation	1.0E-6	1137	336,684	3.83E8	6.58
Multiple interpolations ($i = 0$)	1.0E-3	28	336,684	9.43E6	0.1952
Multiple interpolations ($i = 1$)	1.0E-4	403	73,255	2.95E7	0.4013
Multiple interpolations($i = 2$, only the surface is calculated)	1.0E-6	1350	33,206	4.48E7	0.9340
Multiple interpolations (total)	1.0E-6			8.38E7	1.5305

The comparison of the initial grid and the deformed grid is shown in Fig. 4.36. Figure 4.36(a) and (b) shows the initial and deformed surface grids, respectively; Fig. 4.35(c) and (d) shows the initial and deformed grids surrounding the wing, respectively; and Fig. 4.36(e) and (f) shows the details of the initial and deformed grids near the wing tip. It can be seen that the deformed grid basically maintains the density distribution as the initial case.

4. Shape-preserving moving-grid method application

A combined rudder-body configuration is considered in the application, the aerodynamic rudder airfoil is double arc, and the relative thickness is 0.05; the plane shape is trapezoidal, the root chord length is 0.8 m, the tip chord length is 0.36 m and lead edge sweepback angle is 34.02°. The rudder-body configuration is modeled as semimodular, and the grid topology and surface grid are shown in Fig. 4.37. There are 21 grid blocks in the whole field, and the total element number is 373,480.

The structure model of the rudder is shown in Fig. 4.38. The rudder is composed of a surface and a rudder. The material parameters of the surface and shaft of the rudder are shown in Tab. 4.3. The surface is modeled by the solid elements, and the shaft is modeled by a circular cross-sectional beam with a radius of 10 mm. The shaft is located at the 1/2 chord of the root of the rudder. The fixation constraint is loaded at the root of the rudder shaft, and by RBE2, the node connecting the shaft and the surface are attached with the nearby eight points.

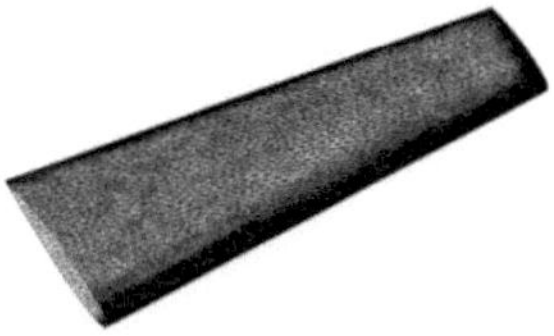

(a) Initial surface grid;

(b) Deformed surface grid;

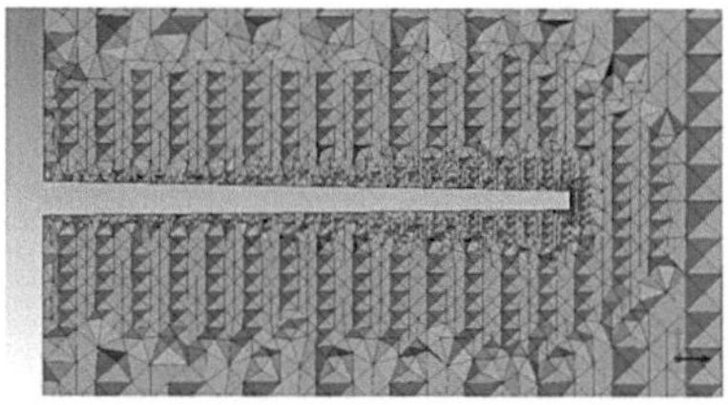

(c) Initial spatial grid surrounding the wing;

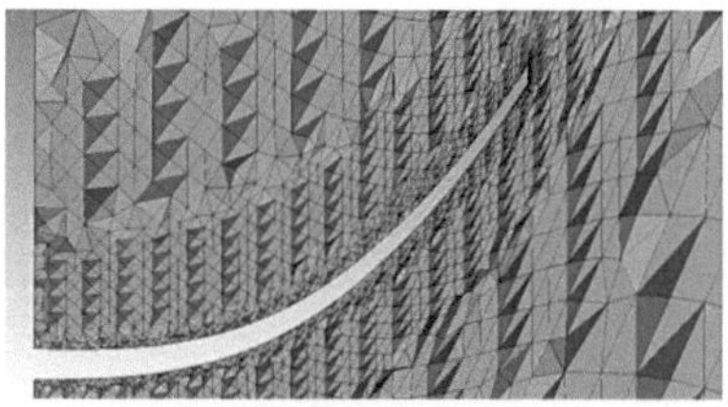

(d) Deformed grid surrounding the wing;

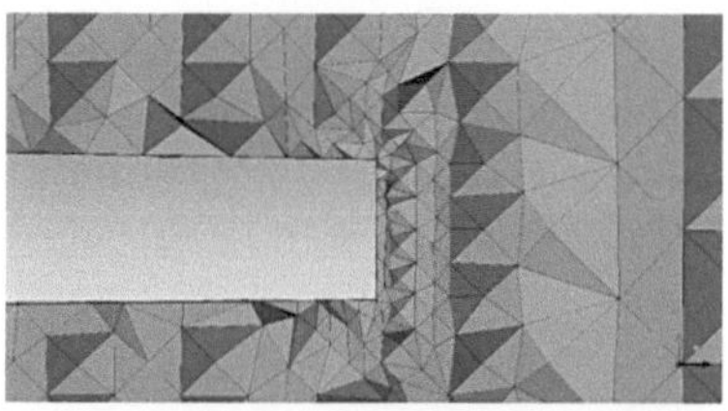

(e) Initial spatial grid near the wingtip;

(f) Deformed grid near the wingtip

Fig. 4.36: Comparisons of ONERA-M6 wing before and after mesh deformation.

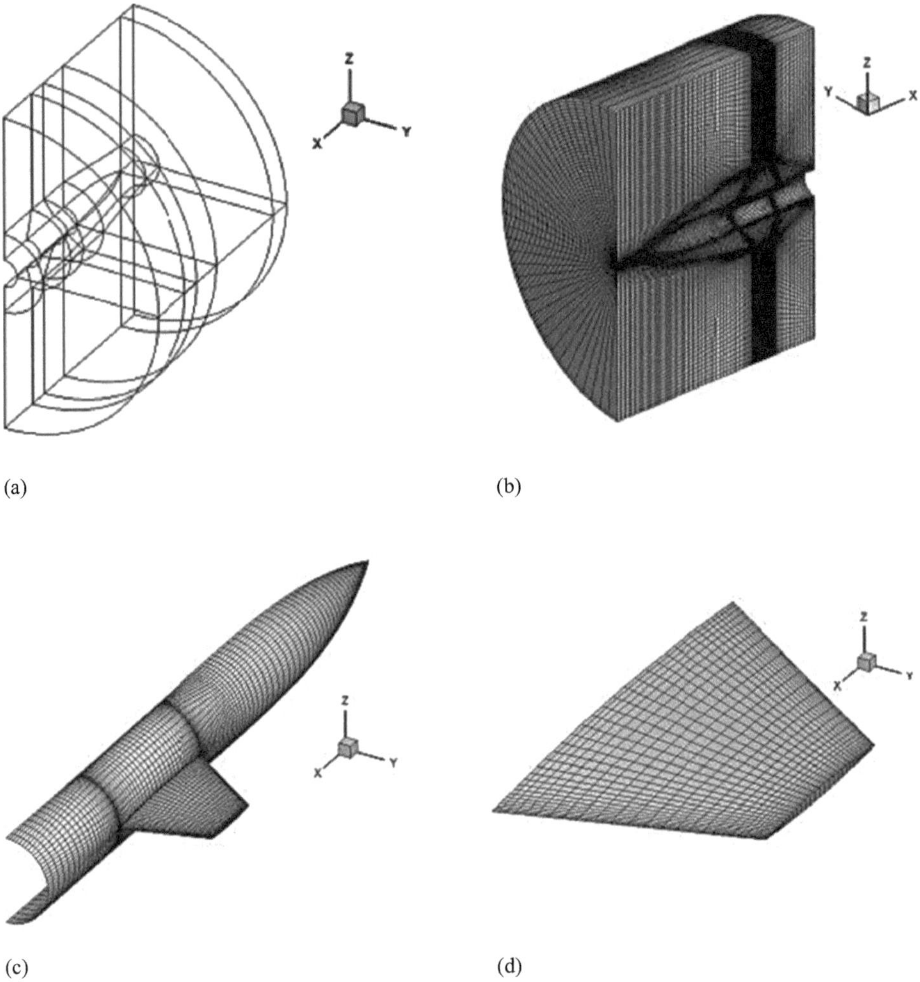

Fig. 4.37: Aerodynamic grid of the rudder body. (a) Grid topology; (b) volume grid; (c) combined surface grid; and (d) single rudder surface grid.

In order to maintain the geometric shapes of the missile body and the rudder in the coupling calculation for the rudder-body configuration, the projection base plane is selected as shown in Fig. 4.39 (the blue line is the outline of the configuration). The eight-node quadrilateral element is applied to ensure the accuracy of projection in these projections.

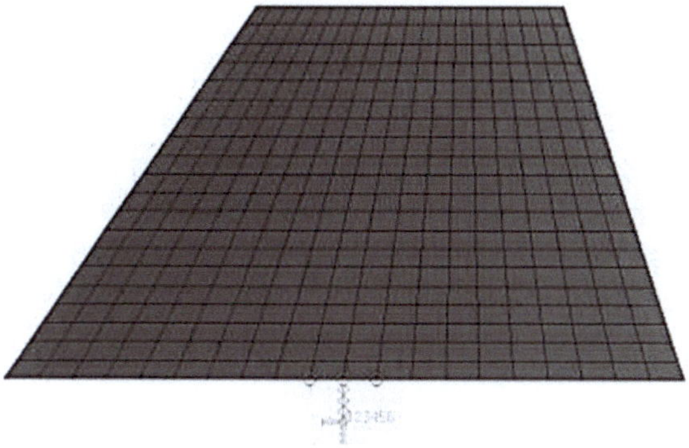

Fig. 4.38: Aerodynamic rudder structural grid.

Tab. 4.3: Rudder material properties.

Section	*E* (GPa)	μ	$\rho/(\mathrm{kg/m^3})$
Rudder	70	0.33	2800
Shaft	220	0.37	4480

The rudder deflection with 16° is imposed to the rudder surface, and the conventional TFI method and the shape-preserving moving-grid method are performed in the calculation, respectively. The deformed grid is obtained as shown in Figs. 4.40 and 4.41 for the situation of the invariant rudder shape. From Fig. 4.40, it can be clearly seen that the TFI method produces a dragging effect of the rudder surface deformation, which results in huge and unreasonable deformation of the missile body. While the shape-preserving moving-grid method keeps the invariant geometric shape for rudder surface and missile body. Table 4.4 shows that the grid points with only 7.06% in the TFI method and 100% in the SPGDM method remain on the missile body after the deformation of the aerodynamic grid nodes. Then the deformed aerodynamic grid is used to calculate the steady CFD ($Ma = 2$, $H = 3$ km). The results are shown in Fig. 4.42, and it can be found that the TFI method produces nonphysical compression and expansion zones in the damaged region of the missile body, which makes the aerodynamic characteristics not in line with the actual situation and leads to the incorrect CFD/CSD coupling results.

Bibliography

[1] Andreas KM. Aircraft control surface deflection using RBF-based mesh deformation. International Journal for Numerical Methods in Engineering. 88, 986–1007.

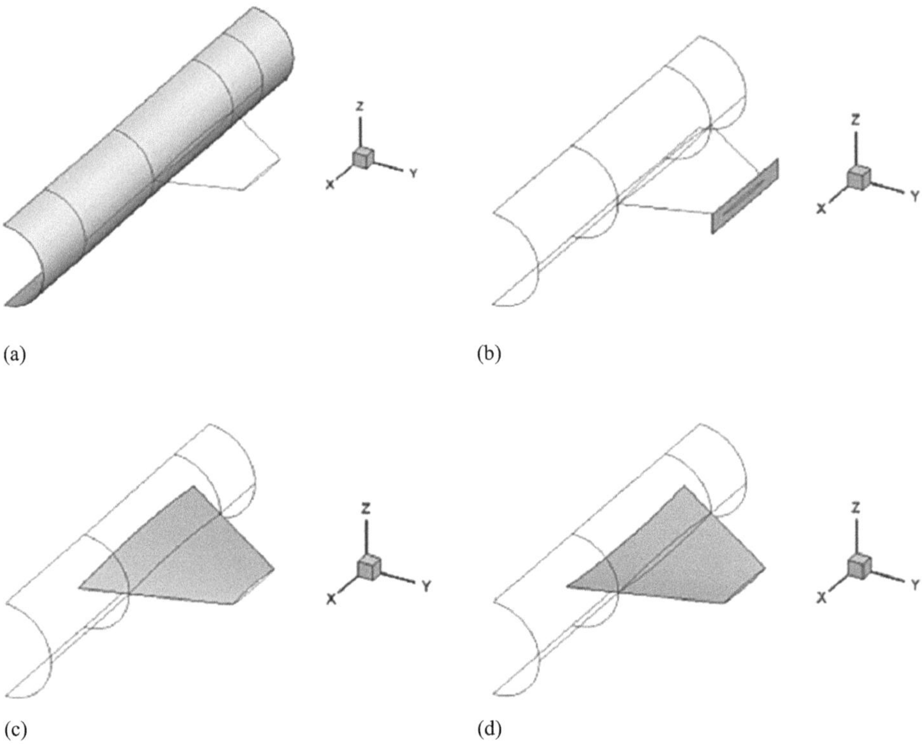

Fig. 4.39: Projection base surface grid. (a) Projection base plane of the missile body; (b) side projection base surface of the rudder; (c) upper projection base surface of the rudder; (d) lower projection base surface of the rudder.

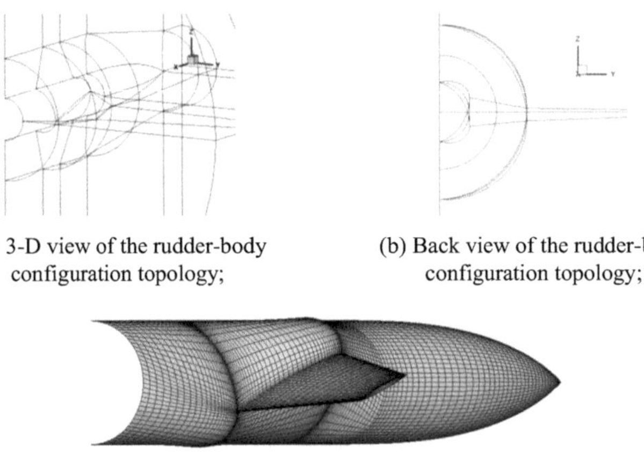

(a) 3-D view of the rudder-body configuration topology;

(b) Back view of the rudder-body configuration topology;

(c) Deformed surface

Fig. 4.40: The deformed grid of the rudder-body configuration by TFI method.

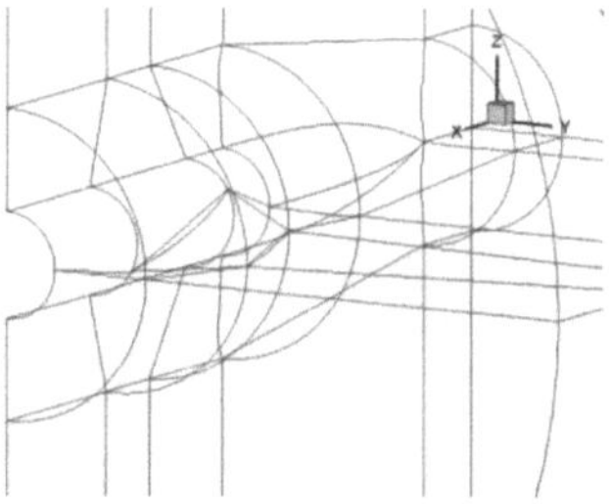

(a) 3-D view of the rudder-body configuration topology;

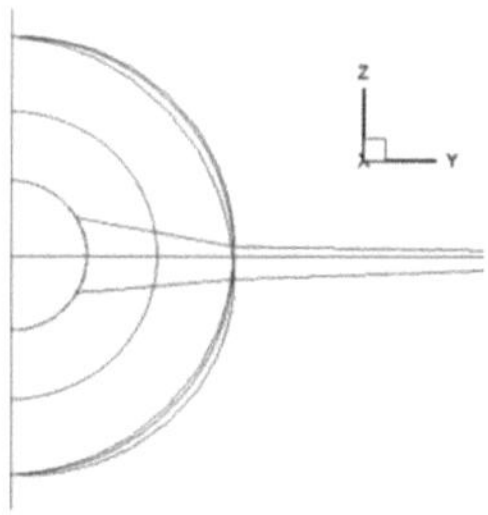

(b) Back view of the rudder-body configuration topology;

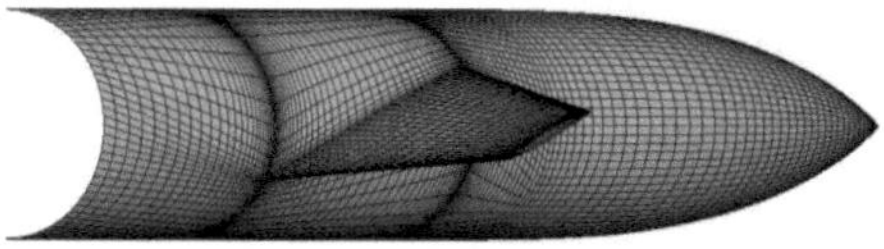

(c) Deformed surface

Fig. 4.41: The deformed grid of the rudder-body configuration by the shape-preserving moving-grid method.

Tab. 4.4: Shape preservation effect of the missile body nodes.

Type	Number of the nodes on the body	Ratio of the nodes on the body
TFI	258	7.06%
SPGDM	3654	100%

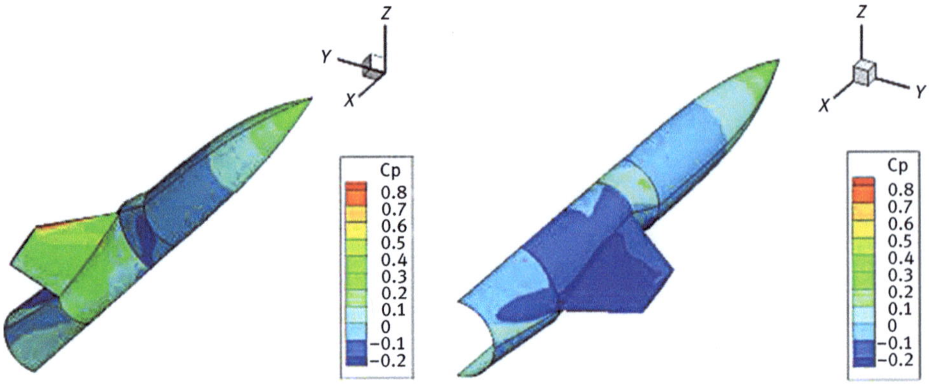

(a)

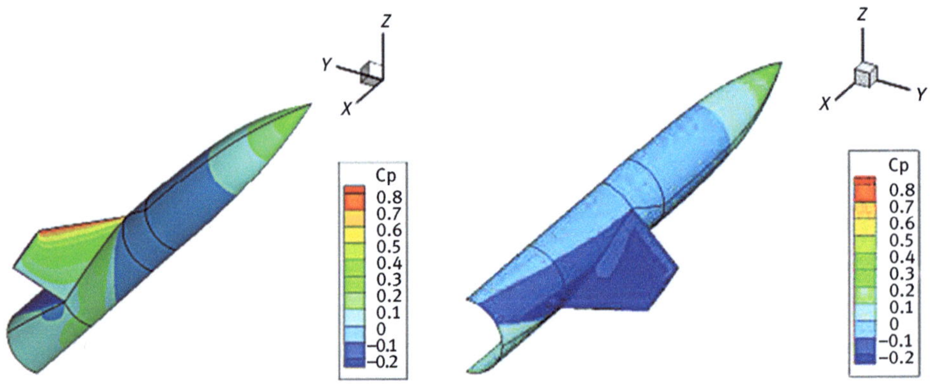

(b)

Fig. 4.42: Pressure cloud chart. (a) The TFI method. (b) Shape-preserving moving-grid method.

[2] Bao J, Yang Q, Guan F. Geometric dichotomy method for inverse isoparametric mapping in linear finite element and its correction measure. Chinese Journal of Computational Mechanics. 2010, 27(5), 770–774. (in Chinese).

[3] Batina J. Unsteady Euler algorithm with unstructured dynamic mesh for complex-aircraft aerodynamic analysis. AIAA Journal. 1994, 31(9), 1626–1633.

[4] Bhardwaj MK. A CFD/CSD interaction methodology for aircraft wings. Virginia Polytechnic Institute and state University, Ph.D. Dissertation. 1997.

[5] Brebbia CA, Dominguez J. Boundary Elements, an Introductory Course. Southampton: Computational Mechanics Inc,1992.20-300.

[6] Cebral JR Loose coupling algorithms for fluid-structure interaction dept. George Mason Univ., VA, 1996.

[7] Chen PC, Gao XW. A multi-block boundary element method for CFD/CSD grid interfacing. AIAA Paper 01-0715, 2001.

[8] Chen PC, Hill LR. A three-dimensional boundary element method for CFD/CSD grid interfacing. AIAA Paper 99-1213, 1999.

[9] Chen PC, Jadic L. Interfacing of fluid and structural models via innovative structural boundary element method. AIAA Journal. 1998, 36(2), 282–287.

[10] Claude YL, Wagdi GH. Conservative interpolation of aerodynamic loads for aeroelastic computations. AIAA Paper 2000-1449.

[11] Degand C. Moving grids for nonlinear dynamic aeroelastic simulations. University of Colorado, Ph.D. Dissertation, 2001.

[12] Dou Y Study on flutter characteristics of two-dimensional heated panel based on CFD / CSD coupling. Northwest Polytechnic University, MA.Sc. Dissertation. 2009, 3. (in Chinese)

[13] Gao XW Deforming mesh for computational aeroelasticity using a nonlinear elastic boundary element method. AIAA Paper 2001-1579, 2001.

[14] Goura GSL, Badcock KJ. A data exchange method for fluid-structure interaction problems. The Aeronautical Journal. 2001, 215–221.

[15] Guruswamy GA. New modular approach for tightly coupled Fluid/Structure analysis. AIAA paper 2004-4547, 2004.

[16] Guruswamy G. Unsteady aerodynamic and aeroelastic calculations for wings using Euler equations. AIAA Journal. 1990, 28, 461–469.

[17] Guruswamy GP. A review of numerical fluids/structures interface methods for computations using high-fidelity equations. Computers and Structures. 2002, 80, 31–41.

[18] Huo S, Wang F, Yue Z. Spring analogy method for generating of 2D unstructured dynamic meshes. Journal of Vibration and Shock. 2011, 30(10), 177—182.

[19] Hurka J, Ballmann J Elastic panels in transonic flow. AIAA Paper 2001-2722, 2002.

[20] Kholodar DB, Morton SA, Cummings RM. Deformation of unstructured viscous grids. AIAA Paper 2005-926, 2005.

[21] Kim YH, Kim JE. New Hybrid A Interpolation method using surface tracking, fitting and smoothing function applied for aeroelasticity. AIAA Paper 2005-2347, 2005.

[22] Lai KL, Tsai HM, Lum KY A CFD and csd interaction algorithm for complex configurations. AIAA Paper 2002-2715, 2002.

[23] Li J, Huang SZ, Jiang SQ, et al. Unsteady viscous flow simulations by a fully implicit method with deforming grid. AIAA 2005-1221.

[24] Li L, Lu Z, Wang J, et al. Turbine blade temperature transfer using the load surface method. Computer-Aided Design. 2007, 39, 494–505. (in Chinese).

[25] Liu F, Cai J, Zhu Y. Calculation of wing flutter by a coupled fluid-structure method. Journal of Aircraft. 2001, 38(2), 334–342.

[26] Liu J, Bai X, Guo Z. Unstructured Dynamic Grid Method and its Application in Flow Field Simulation Including Moving Interface. National University of Defence Technology Press; 2009. (in Chinese).

[27] Noh WF. CEL: A time-dependent two-space-dimensional coupled Eulerian-Lagrangian code, In: Alder B., Fernbach S., Rotenberg M., eds. Methods in Computational Physics 3, New York: Academic press; 1964.

[28] Rendall TCS, Allen CB. Efficient mesh motion using radial basis functions with data reduction algorithms. Journal of Computational Physics. 2009, 229(7), 6231–6249.

[29] Rendall TCS, Allen CB. Reduced surface point selection options for efficient mesh deformation using radial basis functions. Journal of Computational Physics. 2010, 229(8), 2810–2820.

[30] Sadeghi M, Liu F. Application of three-dimensional interfaces for data transfer in aeroelastic computations. AIAA Paper 2004-5376, 2004.

[31] Schmitt V, Charpin F, Pressure distributions on the ONER AM6-wing at transonicMMach numbers. experimental data base for computer program assessment, Report of the Fluid Dynamics Panel Working Group 04, AGARDAR138, May, 1979.

[32] Sheta E, Harrand V, Thompson D, et al. Computational and experimental investigation of limit cycle oscillations of nonlinear aeroelastic systems. Journal of Aircraft. 2002, 39(1), 133–141.

[33] Smith MJ, Hodges DH. Evaluation of computational algorithms suitable for fluid-structure interactions. Journal of Aircraft. 2000, 37(2), 282–294.

[34] Snyder RD, Hur JY, Strong DD, et al. Aeroelastic analysis of a high-altitude long-endurance joined-wing aircraft. AIAA Paper 2005-1948, 2005.

[35] Wang G, Lei B, Ye Z. An efficient deformation technique for hybrid unstructured grid using radial basis functions. Journal of Northwestern Polytechnical University. 2011, 29(5), 783—788. (in Chinese).

[36] Wu Y, Tian S, Xia J. Unstructured grid methods for unsteady flow simulation. Acta Aeronautica et Astronautica Sinica. 2011, 32(1), 15—26. (in Chinese).

[37] Zhang J, Tan J, Chu J, et al. New method for generating unstructured moving grids. Journal of Nanjing University of Aeronautics & Astronautics. 2007, 39(5), 633–636.

[38] Zhang Z Grid generation technology in computational fluid dynamics. Teaching materials for Postgraduates of Northwest University of Technology, 2010. (in Chinese)

[39] Zhou Y, Chen S, Zhang X, et al. Improved inverse isoparametric mapping method and its application to coupled analysis. Rock and Soil Mechanics. 2008, 9(11), 3170–3173. (in Chinese).

Chapter 5
CFD/CSD coupling solution technology

5.1 Introduction

In the nonlinear aeroelastic problem, the interaction between the flexible structure of the aircraft and the surrounding flow field is involved. Two different physical fields interact on the coupling interface and affect each other. On the one hand, the fluid problem itself involves a large number of nonlinear phenomena, such as complex turbulent motion, shock waves generated by high-speed motion, boundary layer separation, unsteady vortex shedding, movement and evolution as well as the unsteady fluid motion caused by the structural deformation or vibration in the case of the fluid-structure coupling problems; on the other hand, structural problems involve nonlinear large geometric deformation, the nonlinearity of elastoplastic materials and interface nonlinearities with indefinite contact surfaces. Since the aerodynamic loads and structural response cannot be solved synchronously in the same discretized form, the computational fluid dynamics (CFD) governing eq. (5.1) and the computational structural dynamics (CSD) governing eq. (5.2) should be coupled together in an appropriate way. In addition, even if the two physical fields are linear, the uncertain coupling on the common interface of the CFD and CSD will cause new nonlinear problems. Therefore, the coupling of the CFD and CSD system is not a simple superposition of these two problems. It is a highly nonlinear problem for the CFD/CSD system coupled through the continuity-compatible conditions on the interface, and it may cause the failed calculation of the entire system by the improper handling way:

$$\frac{\partial}{\partial t}\int \bar{U}\, d\Omega + \int \bar{F}\cdot ds = \frac{1}{\mathrm{Re}}\int \bar{F}^{v}\cdot d\vec{s}, \tag{5.1}$$

$$M\ddot{u} + C\dot{u} + Ku = F. \tag{5.2}$$

The meanings of the variables in the above two formulas can be found in Chapters 2 and 3, respectively. Furthermore, it is required that the coupling scheme meets the speed coordination conditions for the aerodynamic grid and structural grid on the coupling boundary, that is

$$\frac{\partial x}{\partial t} = \frac{\partial u}{\partial t}. \tag{5.3}$$

Based on the conservation of energy transfer at the two interfaces, the three CFD/CSD coupling algorithms: traditional loosely coupling [30, 29, 16, 15, 2, 22, 21, 20, 27, 17], predictive multistep coupling and tightly coupling algorithms [17, 11, 23, 1, 10, 35, 31, 32, 8, 14, 12] are analyzed in this chapter to evaluate the accuracy and coordination conditions

https://doi.org/10.1515/9783110576689-005

in the time domain. Then, a high-precision coupling scheme of CFD/CSD serial and parallel with second-order time precision and program modularity is introduced.

5.2 Accuracy analysis of the common CFD/CSD coupling algorithms

The research of the coupling algorithm focuses on the time advancing problem between the different physics fields, which includes data transfer and algorithm design. The traditional fluid-structure coupling problem has not encountered great difficulties in these aspects, because the traditional unsteady aerodynamic theory (such as Theodorson theory and doublet lattice method) and quasistatic aerodynamic theory (aerodynamic derivative method, piston theory, etc.) can express the unsteady aerodynamic loads into an explicit function of the structural parameters by an aerodynamic influence coefficients matrix [33]. However, in the solution of the CFD/CSD coupling system, such explicit function cannot be obtained to represent the unsteady aerodynamics, thus the coupling algorithm should be utilized to connect the two fields in some way. The commonly used CFD/CSD coupling algorithms are classified into the fully coupling [13, 28], loosely coupling, predictor-multiple iterative coupling and tightly coupling algorithm [17].

5.2.1 Introduction to the common coupling algorithms

1. Fully coupling algorithm

Fully coupling algorithm [13, 28] refers to solving the main governing equations of the CFD and CSD system in the same time domain as well as in the same spatial domain. The boundary condition equations, including stress balance condition, displacement and velocity-coordinated condition, should be satisfied and added to the simultaneous equations. The fully coupling calculation process is shown in Fig. 5.1.

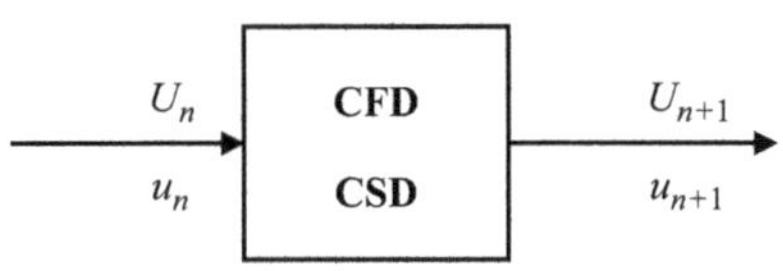

Fig. 5.1: CFD/CSD fully coupling flowchart.

The node degree of freedom vector of the coupled system can be expressed as $X = (X_f, X_s)^T$, where X_f represents the solution vector of the fluid subsystem, and X_s represents the solution vector of the structural subsystem. A simultaneous equation containing a coupled boundary condition can be written as

$$F(X) = \begin{bmatrix} F_f[X_f, \bar{u}_s(X_s)] \\ F_s[X_s, \bar{\sigma}_f(X_f)] \end{bmatrix} = 0, \tag{5.4}$$

$\bar{u}_s(X_s)$ in the fluid governing equation denotes that the fluid is affected by the displacement $\bar{u}_s$ of the structure at the coupling interface; and $\bar{\sigma}_f(X_f)$ in the structural governing equation represents that the structure is affected by the fluid force $\bar{\sigma}_f$ at the coupling interface. In this algorithm, the coupling equation is directly solved synchronously, and there are no requirements for the information transformation between the two grid systems, thus it has a high time precision. However, due to its highly nonlinear characteristics, it is very difficult to use a synchronous advancing solution in one grid system, which leads to great limitations in the practical applications. Currently, this coupling algorithm is limited to solving the two-dimensional specific problems.

2. Loosely coupling algorithm

The loosely coupling algorithm [30, 29, 16, 15, 2, 22, 21, 20, 27, 17] is to integrate the governing equations of the CFD and CSD subsystems with their own solvers in the time domain, and advance in a stagger way to obtain the response of the coupled system. This method is also referred to as a collocated serial-partitioned coupling algorithm because the structure and flow field are advancing to the same step. The typical process is shown in Fig. 5.2(a) and (b).

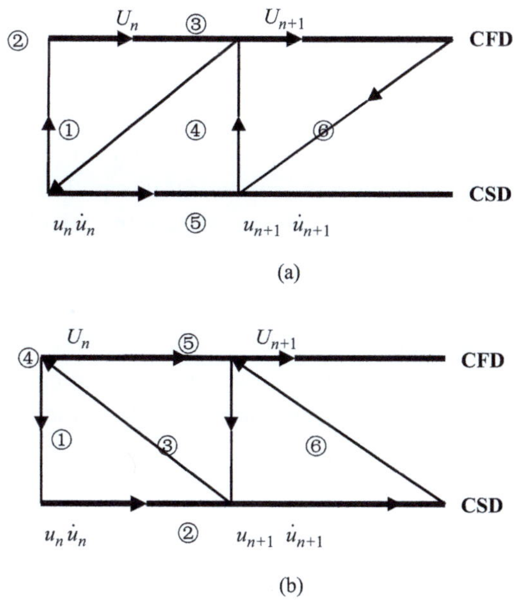

Fig. 5.2: CFD/CSD loosely coupling flowchart.

Taking Fig. 5.2(a) as an example, the thick line in the figure indicates the time integral marching of the fluid subsystem and the structure subsystem, and the thin line indicates the data transformation between the interfaces. Obviously, the time-marching process of the loosely coupling method is staggered rather than synchronous. The basic idea is to assume that the variables of the fluid and structure at n step have been obtained, in order to obtain the structure and fluid parameters at $n+1$ step, the typical process of the loosely coupling method is as follows:
(1) Transferring the displacement and motion of the boundary of the structure at n step to the fluid system using a data transformation method
(2) Updating the fluid dynamic grid by moving-grid method
(3) Integrating the fluid governing equation under the current grid state and boundary condition to obtain the flow parameters, such as the pressure distribution at $n+1$ step
(4) Converting the fluid pressure at $n+1$ step to the structural equivalent load by the data transformation method
(5) Solving the structural dynamic equation to advance the structural response to the $n+1$ step
(6) Entering the next iteration loop

Since the loosely coupling method maximizes the independence of the various disciplines, it can make use of the existing CFD and CSD codes, as long as a small number of data exchange modules are added, thereby reducing computational complexity and simplifying the implicit/explicit treatment, subiteration [6, 25] and software modularity. These advantages make it widely used in the fluid-structure coupling problems [6, 3, 24, 34, 4, 24].

However, due to the unsynchronized integral time, the loosely coupling method cannot achieve the dynamic balance of the coupling interface. That is, the $n+1$ step on the fluid interface is corresponding to the n step on the structural interface, that is, there is a time lag phenomenon, which will cause the numerical dissipation or the excess energy absorption on the interface. Therefore, it may reduce the calculation accuracy and limit the stability of the algorithm.

3. Predictor-multiple iterative coupling algorithm

In order to compensate the lagging problem caused by the unsynchronized integral time in the loosely coupling method, a staggered coupling method with predictor-multiple iterations is introduced in one time step is shown in Fig. 5.3.

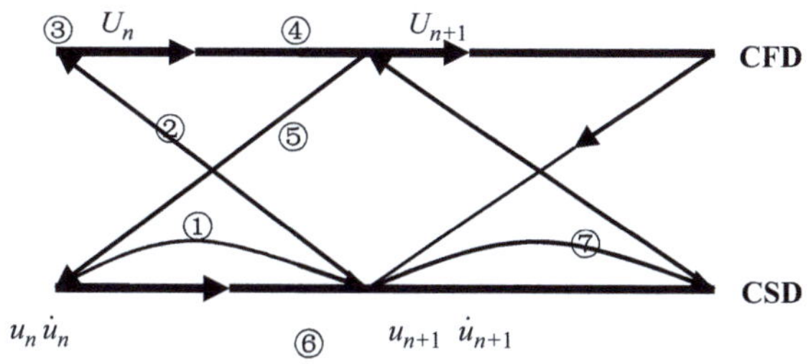

Fig. 5.3: CFD/CSD predictor-multiple iterative coupling flowchart.

In Fig. 5.3, a prediction calculation is added in a one step of the structure. The basic steps are as follows:

(1) Predicting the structural state $\tilde{u}_{n+1}$ at $n+1$ step according to the displacement and motion of the structure at n step
(2) Converting $\tilde{u}_{n+1}$ and the corresponding boundary information into the fluid system
(3) Updating the fluid dynamic grid by the moving-grid method
(4) Integrating the fluid governing equation to obtain the flow parameters at the $n+1$ step
(5) Converting the fluid pressure at the $n+1$ step into the structural equivalent load
(6) Integrating the structural dynamic equation to get the structural state at the $n+1$ step, and judging

$$\frac{\|u_{n+1}-\tilde{u}_{n+1}\|}{\|u_{n+1}\|}<\varepsilon, \tag{5.5}$$

where ε is the error criterion. If formula (5.5) is satisfied to complete the integration of the current period and start the next iteration loop; otherwise, to use u_{n+1} as the new predicting value $\tilde{u}_{n+1}$, return to (3), and repeat steps (3) to (6) until the accuracy criterion is satisfied. Obviously, as adding a step for judging whether the calculation satisfies the accuracy criterion in this algorithm, it reduces the error of the calculation from the n step to the $n+1$ step to some extent. If ε is set relatively large, this method becomes a loosely coupling method; If ε is far small, it will increase the cost of the calculation and reduce the efficiency.

4. Tightly coupling algorithm

For the convenience of the computation and programming, the tightly coupling algorithm [17, 11, 23, 1, 10, 35, 31, 32, 8, 14, 12] shown in Fig. 5.4 is often used to correct the predictor and corrector steps, and the fluid and structural systems are iteratively solved in one integrated time step until the accuracy requirements are met. Some researchers put this process in the subiteration to achieve the tightly coupling format, and fine results are obtained when the program is implemented [35]. Although a large time step can be taken in this method, the calculation efficiency is not significantly improved compared with the loosely coupling method as it is usually iterated many times in one time step to eliminate the accumulated error with the time marching.

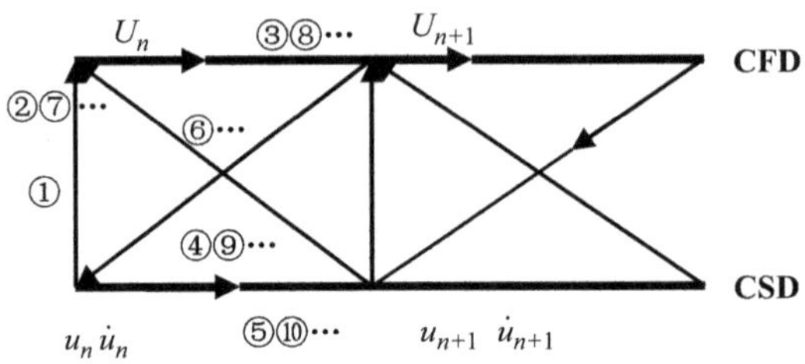

Fig. 5.4: CFD/CSD tightly coupling flowchart.

5.2.2 Accuracy analysis based on energy transfer

1. Basic assumptions of the accuracy analysis

It is assumed that the flow field solver is the second-order subiterative format for the solution of the flow field governing equation, and the structure solver selects the Newmark method with the second-order time-accurate implicit scheme. Assuming that all external forces in the structure are aerodynamic loads, then [25, 26]

$$\begin{cases} u_{n+1} = u_n + \Delta t(\dot{u}_n + \dot{u}_{n+1})/2 \\ \dot{u}_{n+1} = \dot{u}_n + \Delta t(\ddot{u}_n + \ddot{u}_{n+1})/2 \end{cases}, \tag{5.6}$$

$$M\ddot{u}_{n+1} + C\dot{u}_{n+1} + Ku_{n+1} = Fs_{n+1}. \tag{5.7}$$

By the structural energy definition [7] of $Es = \frac{1}{2}\dot{u}^T M\dot{u} + \frac{1}{2}u^T Ku$, the energy of the structural system from n step to $n+1$ step is

$$Es_{n+1} - Es_n = \frac{1}{2}\dot{u}_{n+1}{}^T M\dot{u}_{n+1} + \frac{1}{2}u_{n+1}{}^T Ku_{n+1} - \frac{1}{2}\dot{u}_n{}^T M\dot{u}_n - \frac{1}{2}u_n{}^T Ku_n. \tag{5.8}$$

From eq. (5.6), eq. (5.8) can be simplified to

$$Es_{n+1} - Es_n = \frac{1}{2}\Delta t\left(\frac{\dot{u}_n + \dot{u}_{n+1}}{2}\right)^T (M\ddot{u}_n + Ku_n) + \frac{1}{2}\Delta t\left(\frac{\dot{u}_n + \dot{u}_{n+1}}{2}\right)^T (M\ddot{u}_{n+1} + Ku_{n+1}). \tag{5.9}$$

Using the equilibrium relationship (5.7) at $n+1$ step, the above formula can be simplified as

$$Es_{n+1} - Es_n = \left(\frac{Fs_n + Fs_{n+1}}{2}\right)^T (u_{n+1} - u_n) - \Delta t \left(\frac{\dot{u}_n + \dot{u}_{n+1}}{2}\right)^T C \left(\frac{\dot{u}_n + \dot{u}_{n+1}}{2}\right). \quad (5.10)$$

Obviously, the first item on the right side of eq. (5.10) is the work done by the external force, and the second item is the energy dissipated by the damping. Since the external forces of the structure are all aerodynamic loads, the energy transferred from the aerodynamic field to the structure from n to $n+1$ step is

$$\Delta Es_{n+1} = \left(\frac{Fs_n + Fs_{n+1}}{2}\right)^T (u_{n+1} - u_n). \quad (5.11)$$

For the aerodynamic forces solved from the unsteady flow field, the work acting on the aerodynamic and structural coupling boundary from n to $n+1$ step is

$$\Delta Ea_{n+1} = \int_{t_n}^{t_{n+1}} F_a^T(t)\dot{x}(t)dt \quad (5.12)$$

where $\dot{x}(t)$ is the aerodynamic grid velocity on the coupling boundary. In the subiterative process of the flow field solution, $\dot{x}(t)$ remains invariant during $[n, n+1]$, with a definition of $\dot{x}(t) = (x_{n+1} - x_n)/\Delta t$; thus, eq. (5.12) can be written as

$$\Delta Ea_{n+1} = \int_{t_n}^{t_{n+1}} F_a^T(t)dt \cdot \dot{x}(t) = \frac{1}{\Delta t}\int_{t_n}^{t_{n+1}} F_a^T(t)dt \cdot (x_{n+1} - x_n). \quad (5.13)$$

The energy transfer is required to be conserved on the aerodynamic/structural coupling interface, which means that $\Delta Es_{n+1} = \Delta Ea_{n+1}$. Assume that (1) the structure is freely vibrating with a constant amplitude u_0 and circular frequency ω; (2) the pressure on the coupling boundary is also vibrating with a constant amplitude F_0 and the same circular frequency ω, but with a phase difference φ. Then there is

$$u(t) = u_0 \cos(\omega t), \quad (5.14)$$

$$F(t) = F_0 \cos(\omega t + \varphi). \quad (5.15)$$

Moreover, $F(t)$ and $u(t)$ should satisfy not only the structural equilibrium equation (5.2), but also the requirements as follows:

$$\begin{cases} F_0 \cos\varphi = (K - \omega^2 M)U_0 \\ F_0 \sin\varphi = \omega C U_0 \end{cases}. \quad (5.16)$$

Define:

$$\begin{cases} f_1 = u_0^T F_0 \cos\varphi \\ f_2 = u_0^T F_0 \sin\varphi \end{cases}. \tag{5.17}$$

2. Time-accuracy analysis of the traditional loosely coupling algorithm

For the traditional loosely coupling process shown in Fig. 5.2, on the coupling boundary, there is

$$x_{n+1} = u_n. \tag{5.18}$$

Substituting it into eq. (5.13):

$$\Delta Ea_{n+1} = \frac{1}{\Delta t}\int_{t_n}^{t_{n+1}} F_a^T(t)dt \cdot (u_n - u_{n-1}). \tag{5.19}$$

Substituting eqs. (5.14) and (5.15) into eq. (5.19), then:

$$\begin{aligned} \Delta Ea_{n+1} &= \frac{1}{\Delta t}\int_{t_n}^{t_{n+1}} (F_0\cos(\omega t+\varphi))^T dt \cdot u_0(\cos(\omega t_n) - \cos(\omega t_{n-1})) \\ &= \frac{1}{\omega\Delta t} F_0^T(\sin(\omega t_{n+1}+\varphi) - \sin(\omega t_n+\varphi)) \cdot u_0(\cos(\omega t_n) - \cos(\omega t_{n-1})). \end{aligned} \tag{5.20}$$

Using (5.17), the above equation can be written as

$$\Delta Ea_{n+1} = [f_1(\sin(\omega t_{n+1}) - \sin(\omega t_n)) + f_2(\cos(\omega t_{n+1}) - \cos(\omega t_n))] \cdot (\cos(\omega t_n) - \cos(\omega t_{n-1}))\frac{1}{\omega\Delta t}. \tag{5.21}$$

Define $T = 2\pi/\omega$, $h = \omega\Delta t$, the total energy transferred from the fluid to the structure during [0, NT] can be regarded as the summing of ΔEa_{n+1} in this period as follows:

$$Ea_{\mathrm{NT}} = \sum_{n=1}^{n=\frac{2N\pi}{h}} \Delta Ea_{n+1}. \tag{5.22}$$

By the formula of the trigonometric function product and the characteristic of the trigonometric function series, there is

$$\begin{aligned} \sum_{n=1}^{n=\frac{2N\pi}{h}} \cos(\omega t_{n+1})\cos(\omega t_{n-1}) &= \frac{1}{2}\sum_{n=1}^{n=\frac{2N\pi}{h}} [\cos(2\omega t_{n-1}+2h) + \cos(2h)] \\ &= \frac{1}{2}\cdot\frac{2N\pi}{h}\cdot\cos(2h) = \frac{N\pi}{h}\cdot\cos(2h). \end{aligned} \tag{5.23}$$

Similarly, there are

$$\sum_{n=1}^{n=\frac{2N\pi}{h}} \cos(\omega t_{n+1})\cos(\omega t_n) = \frac{N\pi}{h}\cdot\cos(h),$$

$$\sum_{n=1}^{n=\frac{2N\pi}{h}} \sin(\omega t_{n+1})\cos(\omega t_{n-1}) = \frac{N\pi}{h}\cdot\sin(2h),$$

$$\sum_{n=1}^{n=\frac{2N\pi}{h}} \cos(\omega t_{n+1})\sin(\omega t_n) = -\frac{N\pi}{h}\cdot\sin(h),$$

$$\sum_{n=1}^{n=\frac{2N\pi}{h}} \cos(\omega t_{n+1})\sin(\omega t_{n-1}) = -\frac{N\pi}{h}\cdot\sin(2h),$$

$$\sum_{n=1}^{n=\frac{2N\pi}{h}} \sin(\omega t_{n+1})\sin(\omega t_n) = \frac{N\pi}{h}\cdot\cos(h),$$

$$\sum_{n=1}^{n=\frac{2N\pi}{h}} \sin(\omega t_{n+1})\sin(\omega t_{n-1}) = \frac{N\pi}{h}\cdot\cos(2h),$$

$$\sum_{n=1}^{n=\frac{2N\pi}{h}} \sin(\omega t_{n+1})\cos(\omega t_n) = \frac{N\pi}{h}\cdot\sin(h),$$

$$\sum_{n=1}^{n=\frac{2N\pi}{h}} \sin(\omega t_n)\sin(\omega t_n) = \frac{N\pi}{h},$$

$$\sum_{n=1}^{n=\frac{2N\pi}{h}} \cos(\omega t_n)\cos(\omega t_n) = \frac{N\pi}{h},$$

$$\sum_{n=1}^{n=\frac{2N\pi}{h}} \sin(\omega t_n)\cos(\omega t_n) = 0.$$

Substituting eq. (5.23) into eqs. (5.21) and (5.22), there is

$$Ea_{\mathrm{NT}} = f_1\cdot\frac{N\pi}{h^2}[2\sin(h) - \sin(2h)] + f_2\cdot\frac{N\pi}{h^2}[2\cos(h) - \cos(2h) - 1]. \tag{5.24}$$

Assuming $h = \omega\Delta t \ll 1$, using the Taylor's formula, eq. (5.24) can be deduced as

$$Ea_{\mathrm{NT}} = N\pi\left[f_1 h + f_2\left(1 - \frac{7}{12}h^2\right) + o(h^3)\right]. \tag{5.25}$$

Similarly, the energy on the structure can be derived as follows:

$$\begin{aligned}\Delta Es_{n+1} &= \left(\frac{Fs_n + Fs_{n+1}}{2}\right)^T (U_{n+1} - U_n) \\ &= \left(\frac{F_0\cos(\omega t_n + \varphi) + F_0\cos(\omega t_{n+1} + \varphi)}{2}\right)^T (U_0\cos(\omega t_{n+1}) - U_0\cos(\omega t_n)) \\ &= \left(\frac{f_1(\cos(\omega t_n) + \cos(\omega t_{n+1})) - f_2(\sin(\omega t_n) + \sin(\omega t_{n+1}))}{2}\right)(\cos(\omega t_{n+1}) - \cos(\omega t_n)).\end{aligned} \tag{5.26}$$

As the total energy transferred from the fluid to the structure during [0, NT] can be regarded as the summing of ΔEa_{n+1}, then

$$Es_{\mathrm{NT}} = \sum_{n=1}^{n=\frac{2N\pi}{h}} \Delta Es_{n+1}. \tag{5.27}$$

Substituting eq. (5.26) into eq. (5.27), expanding and using the formula of the trigonometric function product and the characteristic of the trigonometric function series, we get

$$Es_{NT} = f_2 \cdot \frac{N\pi}{h}\sin(h) = N\pi\left[f_2\left(1 - \frac{h^2}{6} + o(h^4)\right)\right]. \tag{5.28}$$

Within $[0, \mathrm{NT}]$, the error in the energy transfer between the two systems is

$$\begin{aligned}\Delta E = Ea_{\mathrm{NT}} - Es_{\mathrm{NT}} &= N\pi\left[f_1 h + f_2\left(1 - \frac{7}{12}h^2\right) + o(h^3)\right] - N\pi\left[f_2\left(1 - \frac{h^2}{6} + o(h^4)\right)\right] \\ &= N\pi\left[f_1 h - \frac{5}{12}f_2 h^2 + o(h^3)\right].\end{aligned} \tag{5.29}$$

Due to $\lim\limits_{\Delta t\to 0}\frac{\Delta E}{\Delta t} = N\pi\cdot f_1\omega = N\pi\omega\cdot u_0^T F_0\cos\varphi$, then

$$\Delta E = N\pi\cdot f_1\omega\Delta t + o(\Delta t^2). \tag{5.30}$$

It can be seen that even if both the fluid and the structural subsystems reach the second-order time accuracy, the traditional loosely coupling method can only satisfy the first-order time accuracy due to the time lag in the information exchange on the coupling boundary. In order to reduce the error, it is generally necessary to take a relatively small time step, which limits the stability range and computational efficiency of the algorithm.

3. Time-accuracy analysis of the predictor-multiple iterative coupling algorithm

For the predictor-multiple iterative coupling method shown in Fig. 5.3, if the structural predictor is chosen as a second-order format:

$$u_{n+1}^{p} = u_n + \alpha_0 \Delta t \dot{u}_n + \alpha_1 \Delta t(\dot{u}_n - \dot{u}_{n-1}). \tag{5.31}$$

Then there is

$$x_{n+1} = u_{n+1}^{p}. \tag{5.32}$$

Generally, the predictor formula $\alpha_0 = 1,\ \alpha_1 = 0$ has first-order precision, and the predictor formula $\alpha_0 = 1,\ \alpha_1 = 1/2$ has second-order precision. Taking $\alpha_0 = 1,\ \alpha_1 = 1/2$ as an example, substituting it into eq. (5.13), using eqs. (5.16) and (5.17)

$$\Delta Ea_{n+1} = \frac{1}{\Delta t} \int_{t_n}^{t_{n+1}} F_a^T(t)dt \cdot (u_{n+1}^{p} - u_n)$$

$$= [f_1(\sin(\omega t_{n+1}) - \sin(\omega t_n)) + f_2(\cos(\omega t_{n+1}) - \cos(\omega t_n))] \cdot \left[\frac{1}{2}\sin(\omega t_{n-1}) - \frac{3}{2}\sin(\omega t_n)\right]. \tag{5.33}$$

In $[0, \mathrm{NT}]$, using the trigonometric function product formula and the characteristic of the trigonometric function series, via Taylor expansion, there is

$$Ea_{\mathrm{NT}} = \sum_{n=1}^{n=\frac{2N\pi}{h}} \Delta Ea_{n+1} = N\pi\left[\frac{1}{4}f_1 h^3 + f_2\left(1 + \frac{1}{3}h^2\right) + o(h^4)\right]. \tag{5.34}$$

In $[0, \mathrm{NT}]$, Es_{NT} is the same as eq. (5.28); thus, the error energy transfer between the two systems is

$$\Delta E = Ea_{\mathrm{NT}} - Es_{\mathrm{NT}} = N\pi\left[\frac{1}{4}f_1 h^3 + \frac{1}{2}f_2 h^2 + o(h^4)\right]. \tag{5.35}$$

Due to $\lim\limits_{\Delta t \to 0} \frac{\Delta E}{\Delta t^2} = \frac{1}{2}N\pi \cdot f_2\omega^2$, then

$$\Delta E = \frac{1}{2}N\pi \cdot f_2\omega^2\Delta t^2 + o(\Delta t^3). \tag{5.36}$$

Thus, the predictor-multiple iterative coupling method with the structure predictor meets the second-order time accuracy.

4. Time-accuracy analysis of the tightly coupling algorithm

If the tightly coupling process as shown in Fig. 5.4 is employed, there is

$$\lim_{p \to \infty} u_{n+1}^{p} = u_{n+1}. \tag{5.37}$$

Then, substituting it into eq. (5.13), and using eqs. (5.16), (5.17) and (5.22) for the simplification, using the trigonometric function product formula and the characteristic of the trigonometric function series, via Taylor expansion, there is:

$$Ea_{\mathrm{NT}} = N\pi\left[f_2\left(1-\frac{h^2}{12}\right)+O(h^4)\right]. \tag{5.38}$$

In $[0, \mathrm{NT}]$, Es_{NT} is still in the form of eq. (5.28), and

$$\Delta E = Ea_{\mathrm{NT}} - Es_{\mathrm{NT}} = N\pi f_2\left(\frac{h^2}{12}+o(h^4)\right). \tag{5.39}$$

Due to $\lim\limits_{\Delta t\to 0}\dfrac{\Delta E}{\Delta t^2} = \dfrac{1}{12}N\pi\cdot f_2\omega^2$, then

$$\Delta E = \frac{1}{12}N\pi\cdot f_2\omega^2\Delta t^2 + o(\Delta t^3). \tag{5.40}$$

Thus, when $p\to\infty$, the tightly coupling method has a second-order time accuracy. However, the subiteration steps will inevitably increase in order to meet the time accuracy requirements, consequently affecting its efficiency.

5.3 The design of the high-accuracy CFD/CSD coupling algorithm

5.3.1 Improved coupling algorithm design

From the above analysis, it can be concluded that the traditional loosely coupling method has only first-order time accuracy, even though the subsystem of the fluid or structure has very high precision. In order to reduce the calculation error, the time step of the coupling calculation should be generally taken quite small, which decreases the computational efficiency. In order to eliminate the error caused by such integral unsynchronization in the predictor-multiple iterative coupling method or the tightly coupling method, it is required to iterate over and over in one time step to achieve the second-order time accuracy; therefore, the number of internal iterations will increase, and there is no significant improvement in computational efficiency.

On the other hand, from eqs. (5.18) and (5.32), it cannot satisfy the continuous condition on the interface, which results in the time step of the algorithm being greatly limited in the practical calculation. It is required to meet the continuous boundary condition on the interface as follows:

$$\begin{aligned}\mathbf{x}^{n+1} &= \mathbf{u}^{n+1}\\ \dot{\mathbf{x}}^{n+1} &= \dot{\mathbf{u}}^{n+1}\end{aligned} \tag{5.41}$$

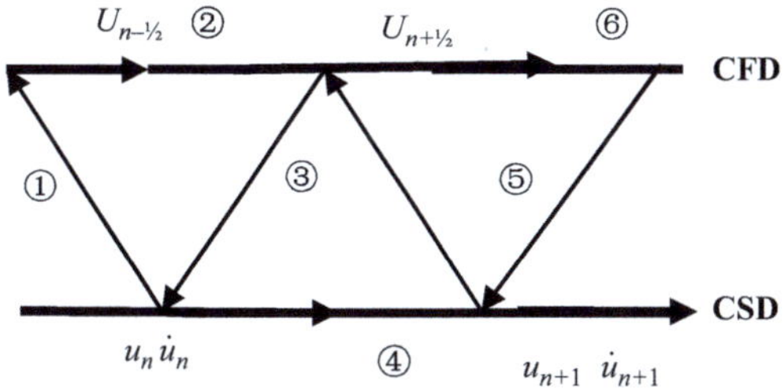

Fig. 5.5: CFD/CSD high-order serial loosely coupling flowchart.

Although the above condition is approximately valid in the tightly coupling algorithm, the computational efficiency is greatly reduced. Moreover, the geometric conservation law should be satisfied for $\mathbf{x}^{n+1}$ and $\dot{\mathbf{x}}^{n+1}$.

Farhat and coworkers [25, 26, 7] proposed a coupling scheme with second-order time accuracy, which eliminates the weakness of the calculation accuracy in the loosely coupling method, and makes the calculation efficiency higher than the tightly coupling algorithm as well [35, 8, 7, 5, 9]. The flowchart is shown in Fig. 5.5. The basic idea is as follows:

(1) Predicting the structural displacement at $n+1/2$ step by the structural state at n step:

$$u^p_{n+\frac{1}{2}} = u_n + \frac{\Delta t}{2}\dot{u}_n + \frac{\Delta t}{8}(\dot{u}_n - \dot{u}_{n-1}). \tag{5.42}$$

(2) Transferring the structural state to the aerodynamic grid, updating the fluid grid, and applying the second-order accuracy solver to predict the pressure $Fa_{n+1/2}$ at $n+1/2$ step

(3) Converting the aerodynamic load into the structural element to get the equivalent load $Fs_{n+1/2}$, and employing the second-order accuracy structure solver to solve the structural equilibrium equation at $n+1/2$ step as

$$M\ddot{u}_{n+1/2} + C\dot{u}_{n+1/2} + Ku_{n+1/2} = Fs_{n+1/2}. \tag{5.43}$$

(4) Computing the structural motion at $n+1$ step as

$$u_{n+1} = 2u_{n+1} - u_n. \tag{5.44}$$

Then, begin the next iterative calculation step. Farhat proved that the method has second-order time accuracy.

1. Serial coupling algorithm design

The process shown in Fig. 5.5 is still being taken as an example, a dual-time marching technique with a subiterative step is employed in the fluid solution, and the

Newmark method as well the energy conservation algorithm are considered in the structure solver. The basic process is as follows:

(1) Predicting the structural displacement at $n+1/2$ step by the structural motion u_n at step, and transferring them to the aerodynamic grid by the designed coupling interface mapping method in Chapter 4. If the structure solver is Newmark method, then

$$x_{n+1/2} = u_n + \Delta t/2\dot{u}_n. \tag{5.45}$$

If the structure solver is the energy conservation algorithm, then

$$x_{n+1/2} = u_n + \Delta t/4(\dot{u}_n + \Delta u_n/\Delta t). \tag{5.46}$$

(2) Updating the fluid grid by the developed moving-grid technique in the previous chapter, solving the unsteady flow field to obtain and calculate the load $Fa_{n+1/2}$ at $n+1/2$ step

(3) Converting the aerodynamic force into structural element to get the equivalent load $Fs_{n+1/2}$ by the interface mapping method. If the structure solver is the Newmark method, the structural equivalent load at $n+1$ step is predicted by

$$Fs_{n+1} = 2Fs_{n+1/2} - Fs_n. \tag{5.47}$$

(4) Solving the structural motion. If the Newmark second-order implicit format is used to calculate the linear structure problem, then there are

$$\begin{cases} \dot{u}_{n+1} = \dot{u}_n + [(1-\delta)\ddot{u}_n + \delta\ddot{u}_{n+1}]\Delta t \\ u_{n+1} = u_n + \dot{u}_n\Delta t + \left[\left(\frac{1}{2} - \alpha\right)\ddot{u}_n + \alpha\ddot{u}_{n+1}\right]\Delta t^2 \end{cases} \quad \alpha = \frac{1}{4}, \delta = \frac{1}{2}. \tag{5.48}$$

If the Newmark second-order implicit method is used to calculate the nonlinear structure problem, it is required to solve the nonlinear incremental equation as

$$\begin{cases} u_{n+1}^i = u_{n+1}^{i-1} + \Delta u^i \\ \dot{u}_{n+1}^i = a_1\Delta u^i - a_4\dot{u}_n^i - a_5\ddot{u}_n^i \\ \ddot{u}_{n+1}^i = a_0\Delta u^i - a_2\dot{u}_n^i - a_3\ddot{u}_n^i \end{cases}. \tag{5.49}$$

If the energy conservation algorithm is utilized to solve the nonlinear structural problem, the structural motion at $n+1$ step is approximated by directly solving the structural equilibrium equation at $n+1/2$ step.

2. Parallel coupling algorithm design

All the above methods can be categorized as the serial coupling method, in which the advancement of the structure is always following the fluid solution, so the total computing time required for each time step is

$$t_n^{n+1} = \left(t_n^{n+1}\right)_a + \left(t_n^{n+1}\right)_s + \left(t_n^{n+1}\right)_{a/s}. \tag{5.50}$$

The right end of the above equation indicates that the total time of the aeroelastic calculation in one time step is equal to the sum of the fluid calculation time, the structural calculation time, and the data transfer time between the two systems. Obviously, for general engineering problems, the structural solution time for the large case is no longer a nonnegligible small amount compared to the fluid solution time. Therefore, it may encounter computational efficiency problems in large engineering case by serial methods. Hence, a parallel algorithm is introduced here, as shown in Fig. 5.6.

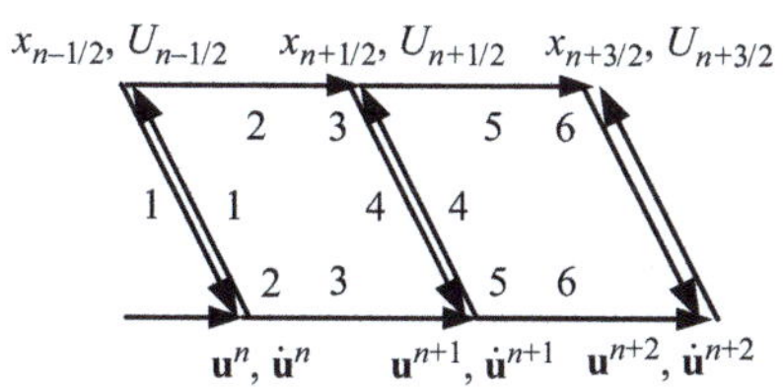

Fig. 5.6: CFD/CSD high-order parallel loosely coupling flowchart.

Its process is as follows:

In fluid solver:

(1) Predicting the structural displacement at the $n+1/2$ step by eq. (5.45) or (5.46)
(2) Transferring the structural displacement to the fluid system and updating the fluid grid
(3) Solving the fluid equation, and advancing the time from $n-1/2$ step to $n+1/2$ step to calculate the corresponding aerodynamic load

In structure solver:

(1) Interpolating the equivalent load at $n+1$ step by the structural equivalent load in the first three and the current time steps:

$$Fs_{n+1} = 4Fs_n - 6Fs_{n-1} + 4Fs_{n-2} - Fs_{n-3}. \tag{5.51}$$

(2) Solving the displacement and velocity of the structure at $n+1$ step by Newmark or energy conservation algorithm

It can be found that the method is completely parallel, and both the fluid and structure fields can be solved separately, as well as the load balance is maintained in their own subdomains.

5.3.2 Time-accuracy analysis of the improved coupling algorithm

With the serial coupling algorithm taken as an example, it is assumed that the dual-time marching technique with a subiterative step is applied in fluid solver, and the Newmark linear method is utilized in the structure solver. Then the work acting on the aerodynamic and structural coupling boundary from n step to $n+1$ step is

$$\Delta Ea_{n+1} = \Delta Ea_{n,n+1/2} + \Delta Ea_{n+1/2,n+1} = \frac{1}{2}\Delta Ea_{n-1/2,n+1/2} + \frac{1}{2}\Delta Ea_{n+1/2,n+3/2}$$

$$= \frac{1}{2\Delta t}\int_{t_{n-1/2}}^{t_{n+1/2}} F(t)^T dt(x_{n+1/2} - x_{n-1/2}) + \frac{1}{2\Delta t}\int_{t_{n+1/2}}^{t_{n+3/2}} F(t)^T dt(x_{n+3/2} - x_{n+1/2}). \quad (5.52)$$

Substituting eq.(5.45) into eq.(5.52), combining and using eqs. (5.16), (5.17) and (5.22), using the trigonometric function product formula and the characteristic of the trigonometric function series, via Taylor expansion, there is

$$Ea_{\mathrm{NT}} = N\pi\left[f_1\left(\frac{1}{24}h^3\right) + f_2\left(1 + \frac{1}{24}h^2\right) + o(h^4)\right]. \quad (5.53)$$

The energy delivered from aerodynamic field to the structure from n to $n+1$ step is

$$\Delta Es_{n+1} = \left(\frac{Fs_n + Fs_{n+1}}{2}\right)^T (u_{n+1} - u_n)$$

$$= \left(\frac{2Fs_{n-1/2} - Fs_{n-1} + 2Fs_{n+1/2} - Fs_n}{2}\right)^T (u_{n+1} - u_n) = Fs_{n+1/2}^T (u_{n+1} - u_n) \quad (5.54)$$

Within $[0, \mathrm{NT}]$, it can be derived as

$$Es_{\mathrm{NT}} = N\pi f_2\left(1 - \frac{h^2}{24} + o(h^4)\right) \quad (5.55)$$

Then there is

$$\Delta E = Ea_{\mathrm{NT}} - Es_{\mathrm{NT}} = N\pi\left(\frac{f_1}{24}h^3 + \frac{f_2}{12}h^2\right) + o(h^4). \quad (5.56)$$

Due to $\lim\limits_{\Delta t \to 0} \frac{\Delta E}{\Delta t^2} = N\pi \cdot \frac{f_2}{12}\omega^2$, then

$$\Delta E = \frac{N\pi}{12} \cdot f_2 \omega^2 \Delta t^2 + o(\Delta t^3). \quad (5.57)$$

From the above derivation, it can be concluded that although the improved serial staggered method has a similar process with the loosely coupling algorithm, it can achieve the second-order time accuracy. Consequently, it improves the calculation accuracy of the traditional loosely coupling method. Moreover, the improved algorithm satisfies the continuous conditions of displacement and velocity at the coupling boundary as

$$\begin{aligned} x_n &= \frac{x_{n+1/2}+x_{n-1/2}}{2} = \frac{u_n+\frac{1}{2}\Delta t\dot{u}_n+u_{n-1}+\frac{1}{2}\Delta t\dot{u}_{n-1}}{2} = u_n, \\ \dot{x}_n &= \frac{x_{n+1/2}-x_{n-1/2}}{\Delta t} = \frac{u_n+\frac{1}{2}\Delta t\dot{u}_n-u_{n-1}-\frac{1}{2}\Delta t\dot{u}_{n-1}}{\Delta t} = \dot{u}_n. \end{aligned} \tag{5.58}$$

For the parallel coupling algorithm, the energy delivered from the aerodynamic field to the structure is the same as the serial improved algorithm, except the energy variation of the structure. However, since eq. (5.51) is obtained by the extrapolation technology [18, 19], it is analyzed to have the same energy accuracy as the serial algorithm.

5.4 Numerical example of the CFD/CSD coupling algorithm

5.4.1 Isogai wing profile aeroelastic simulation

The two-dimensional aeroelastic model Isogai Wing is taken as an example, and the coupling algorithm in this book is performed to compare the computational efficiency. Define the dimensionless flutter speed as

$$V_f = V_\infty/(b\omega_\alpha\sqrt{\mu}) \tag{5.59}$$

where V_∞ is the free stream velocity; ω_α is the pitch frequency; μ is the mass ratio; and b is the half chord length. The fluid grid is an O-shaped structured grid (121×33), the Roe format is adopted to discretize the inviscid flux and the LUSGS-τTs implicit format is employed in the time advancement. The calculated case is $M_\infty = 0.8$, $V_f = 0.8$ and with the time step $\Delta t_f = \Delta t_s = T_a/100(T_a = 2\pi/\omega_a)$. In the calculation, the first-order mode is fixed first, and the second-order mode is imposed two cycles of sinusoidal motion with the frequency ω_α and amplitude $\pi/180$. Then the first-order mode is "released," and the response can be obtained by coupling the structural equation. The structural damping is not considered in the simulation, and the values of k and d are calculated by

$$\begin{cases} d = 0 \\ k = \sum_{i=1}^{\text{NMODE}} (\omega_i^2 - \omega_{\text{cr}}^2) U_i^T U_i \xi_{i,\text{cr}}^2 \end{cases} \tag{5.60}$$

where NMODE represents the mode number. ω_i and ω_{cr} represent the ith-order natural frequency and the flutter critical frequency of the structure, respectively. U_i and $\xi_{i,\text{cr}}$ represent the ith-order normalized mode and generalized displacement amplitude at flutter onset (because the flutter point or nearby is much more attractive). $\omega_{cr} \approx 1.0$, and in the calculation $k < 0$. The dual-time marching technique in Chapter 2, which corresponds to the subiteration step method, is applied to the solution of the unsteady flow field.

(1) The tightly coupling, traditional loosely coupling, and predictor-multiple iterative coupling ($\alpha_0 = 1,\ \alpha_1 = 0$) algorithms are applied in the solution, respectively. The time step is taken as $\Delta t = T_\alpha/1{,}000$, and the results of the first-order generalized displacement are shown in Fig. 5.7. It can be seen that the result of the traditional loosely coupling algorithm is divergent as its computing error is only first-order time accuracy. The result of the predictor-multiple iterative coupling algorithm is consistent with that of the tightly coupling method as the energy error of the method has second-order time accuracy as well as the time step is relatively small.

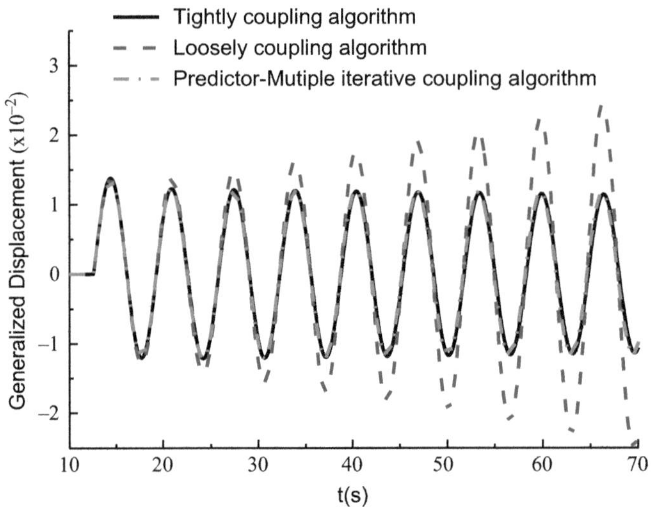

Fig. 5.7: Comparison of the three commonly used coupling algorithms.

(2) Then, the improved serial coupling, parallel coupling and tightly coupling methods are applied in the calculation, respectively. The conditions are the same as above. The first-order generalized displacement results are shown in Fig. 5.8. It can be seen that the curves obtained by the two improved methods are compared well with each other, which is consistent with the same energy error in the theoretical analysis. Figure 5.9 shows the Isogai wing airfoil flutter velocity boundary calculated by the improved parallel coupling method. It can be seen that the method is good at simulating the transonic velocity pit and the multiple flutter points, which are consistent with the reference results.

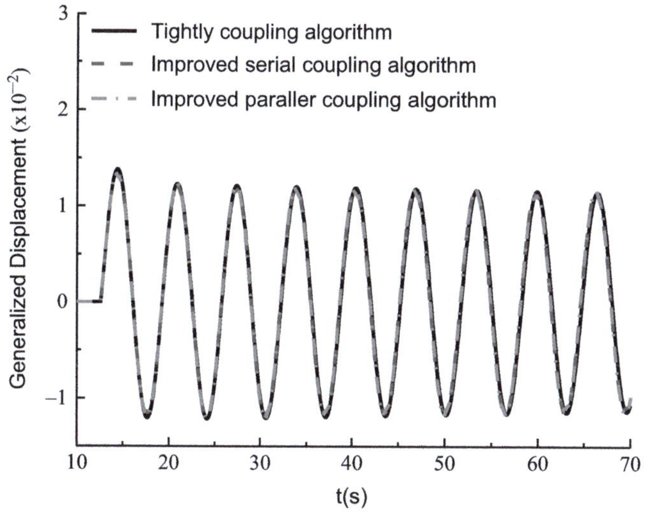

Fig. 5.8: Comparison of the improved serial and parallel coupling algorithm.

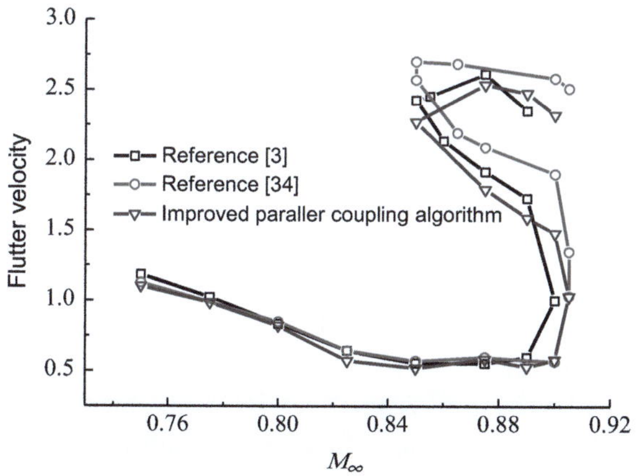

Fig. 5.9: Isogai wing flutter velocity boundary.

5.4.2 Aeroelastic response analysis of AGARD 445.6 wing

AGARD 445.6 wing is a standard model for validating the flutter calculations, and it has relatively complete wind tunnel test data. The wing composes of NACA65A004 airfoil, and anisotropic material with the longitudinal elastic modulus $E_1 = 3.1511\text{GPa}$, the transverse elastic modulus $E_2 = 0.4162$ GPa, Poisson's ratio $\nu = 0.31$, shear modulus $G = 0.4392\text{GPa}$ and density $\rho = 381.98$ kg/m^3 is used in the structure. The calculation condition is set as $M_\infty = 0.96$, and the ratio of the freestream velocity to the experimental flutter velocity is $V_\infty / V_f = 0.97$. The finite element model of the flat structure is constructed with 400 anisotropic 4-node shell element points. The aerodynamic modeling generates a surface grid with 2,940 points along the span of NACA 65A004 airfoil, and the fluid grid is constructed by $160 \times 45 \times 45$ points. The traditional loosely coupling, tightly coupling and the second-order time-accuracy serial coupling methods are applied in the simulation to verify and compare with the designed coupling method. The fluid solver is the solution of the NS equation, the structural solver still uses the linear method, and the interface interpolation applies the CVT method. The nondimensional time step is $dt = \Delta t \cdot V_\infty / L$, where the reference length L is the half length of the wing.

(1) The calculation is firstly performed by the loosely coupling algorithm as shown in Fig. 5.2, and the dimensionless time step dt is set as 0.01, 0.05, 0.1 and 0.2, respectively. The histories of the generalized displacement are shown in Fig. 5.10. It can be seen that when the time step is larger, the generalized displacement will diverge, which is due to the accumulation of errors between the fluid and the structure. When the time step is set small, the generalized displacements of $dt = 0.01$ and $dt = 0.05$ appear to be approximately equal-amplitude oscillation, and there is little difference between the two curves. It is concluded that in the loosely coupling calculation, the time step should be set as much small as possible to obtain more realistic results.

(2) The calculation is then performed by the tightly coupling algorithm as shown in Fig. 5.4, and the dimensionless time step dt is taken as 0.05, 0.1, 0.2 and 0.4, respectively. For the convenience of the calculation, the subiteration step number is taken as 10. The histories of the generalized displacement are shown in Fig. 5.11. It can be found that there is little difference between the results of $dt = 0.05$ and $dt = 0.1$. For the result of $dt = 0.2$, the amplitude of the oscillation is the same as the previous two small time steps, except that the phase varies with the time. When the time step is set to 0.4, the generalized displacement tends to converge as a large error occurs. However, when the subiteration step number is doubled as 20, the generalized displacement agrees well with the result of $dt = 0.2$ with 10 steps in the subiteration. It is concluded that although the time step can be taken larger in the tightly coupling method, the subiteration step number should be increased to obtain the accurate results. To some extent, compared with the loosely coupling method, the calculation efficiency is not significantly improved.

(3) Finally, the improved serial algorithm as shown in Fig. 5.5 is applied in the calculation. The dimensionless time step dt is taken as 0.05, 0.1 and 0.2, respectively. The histories of the generalized displacement are shown in Fig. 5.12. It can be seen that the result at $dt = 0.05$ compares well with that at $dt = 0.1$. When $dt = 0.2$, the amplitude of the oscillation increases slightly with time.

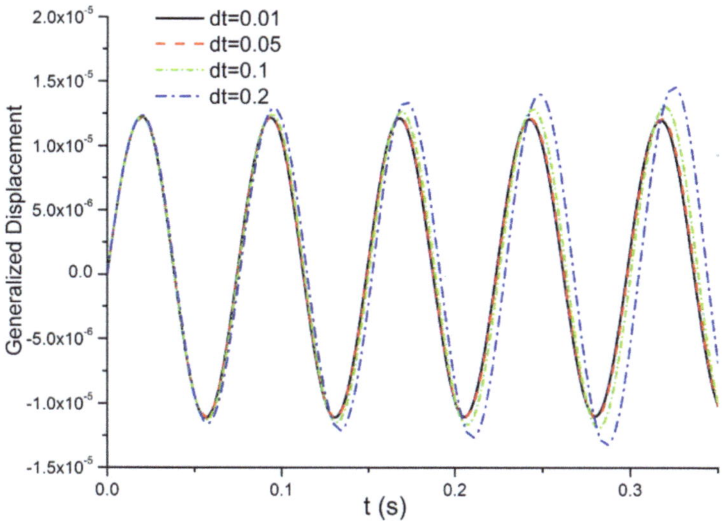

Fig. 5.10: Histories of the generalized displacement by the traditional loosely coupling calculation.

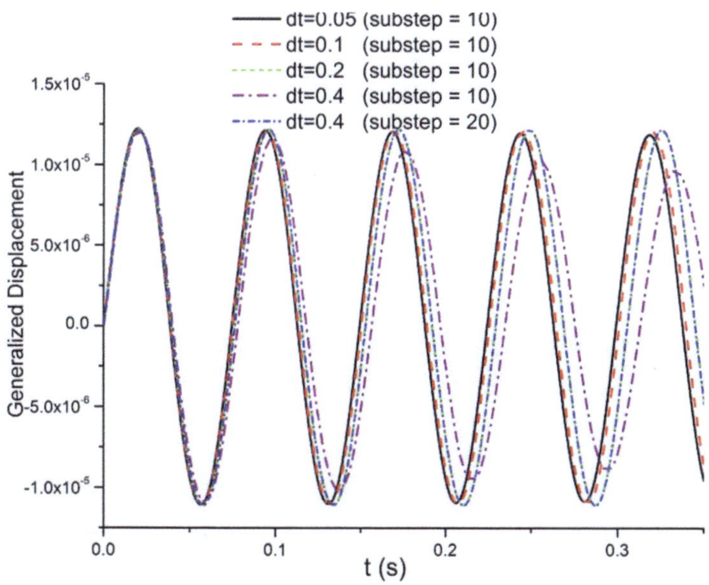

Fig. 5.11: Histories of the generalized displacement by the tightly coupled calculation.

Figures 5.13 to 5.15 show the histories of the generalized displacements calculated by the three methods at $dt = 0.05$, 0.1 and 0.2, respectively. From these figures, it can be found that when $dt = 0.05$, the curves are nearly the same with each other by

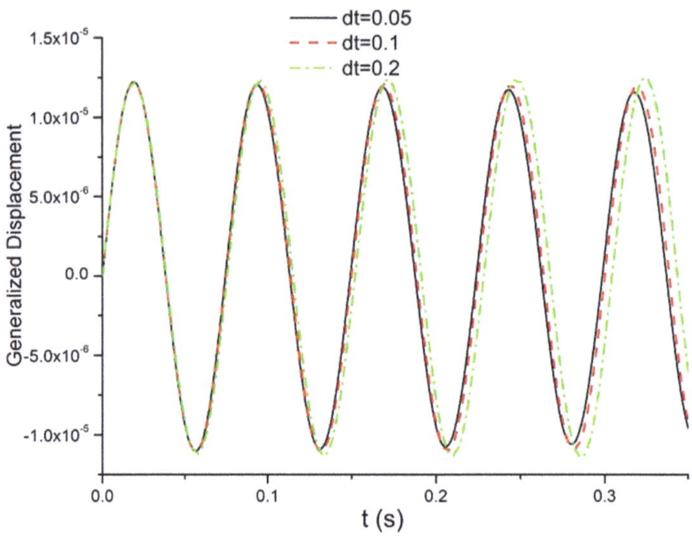

Fig. 5.12: Histories of the generalized displacement by the improved coupling calculation.

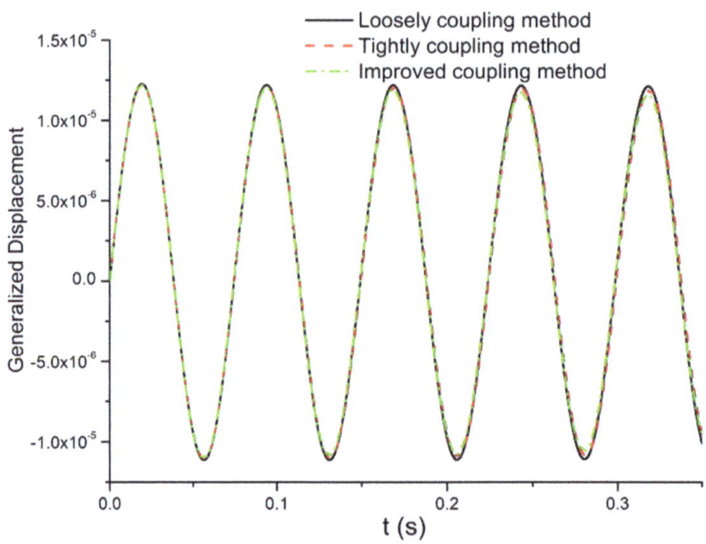

Fig. 5.13: Histories of the generalized displacement for the three methods at $dt = 0.05$.

the three methods. When $dt = 0.1$, the curves obtained by the tightly coupling method and the improved serial coupling method are consistent with each other, but the amplitude calculated by the traditional loosely coupling method increases after a period of time. When $dt = 0.2$, the result obtained by the tightly coupling

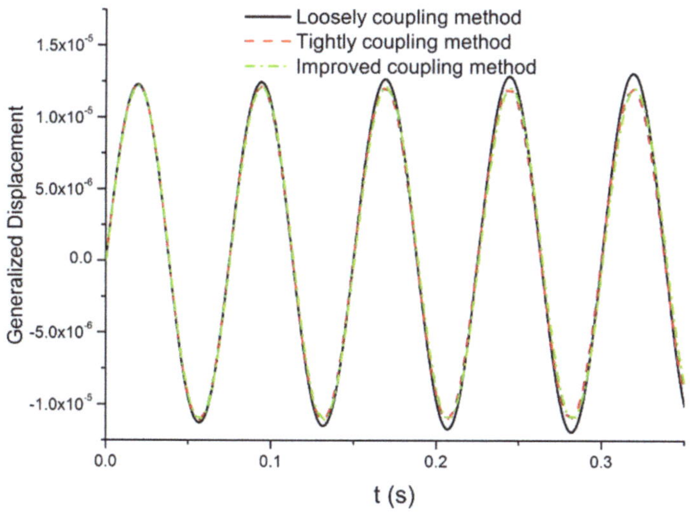

Fig. 5.14: Histories of the generalized displacement for the three methods at $dt = 0.1$.

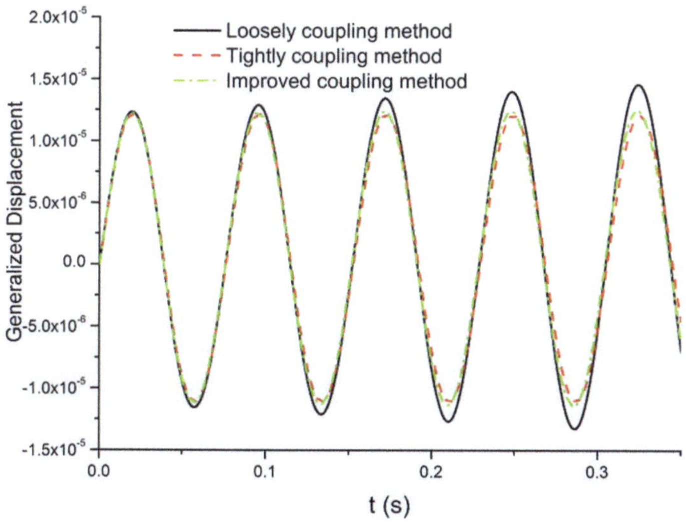

Fig. 5.15: Histories of the generalized displacement for the three methods at $dt = 0.2$.

method is still approximately equal-amplitude oscillation, and the amplitude of the curve calculated by the improved coupling method slightly increases. The curve calculated by the traditional loosely coupling method tends to diverge. It can be concluded that when the coupling time step is sufficiently small, all the three methods can obtain accurate results. When the time step is large, the error of the traditional loosely coupling method becomes significantly larger, whereas, the tightly coupling and the improved coupling method can still obtain an accurate solution. From the point of view of computational efficiency, compared with the tightly coupling method, the improved coupling method significantly improves the computational efficiency because it directly advances the subsystems instead of calculating the structural motion and transferring information in the subiteration. Moreover, there is little difference in the accuracy of calculation between the two methods.

Bibliography

[1] Attar PJ, Gordnier RE. Aeroelastic prediction of the limit cycle oscillations of a cropped delta wing. AIAA Paper 2005-1915, 2005.

[2] Bhardwaj MK A CFD/CSD Interaction Methodology for Aircraft Wings. Virginia Polytechnic Institute and state University, Ph.D. Dissertation, 1997.

[3] Bohbot J, Garnier J, Toumit S, et al. Computation of the flutter boundary of an airfoil with a parallel Navier-Stokes Solver. AIAA 2001–0572. 39th AIAA Aerospace Sciences Meeting Conference & Exhibit, January 8–11, 2001/ Reno, Nevada.

[4] Dou Y, Xu M, An X, Yao W. Flutter analysis of hypersonic rudder. Engineering Mechanics. 2009, 26(11), 232–237. (in Chinese).

[5] Farhat C, Geuzaine P, Brown G, et al. Nonlinear flutter analysis of an F-16 in stabilized, accelerated, and increased angle of attack flight conditions. AIAA Paper 2002–1490, 2002.

[6] Farhat C, Lesoinne M, Maman N. Mixed explicit/implicit time integration of coupled aeroelastic problems: Three-field formulation, geometric conservation and distributed solution. International Journal for Numerical Methods in Fluids. 1995, 21, 807–835.

[7] Farhat C, Lesoinne M. Two efficient staggered algorithms for the serial and parallel solution of three-dimensional nonlinear transient aeroelastic problems. Computer Methods in Applied Mechanics and Engineering. (182), 499–515.

[8] Farhat C, Pierson K, Degand C CFD based simulation of the unsteady aeroelastic response of a maneuvering vehicle. AIAA Paper 2000-0899, 2000.

[9] Geuzaine P, Brown G, Farhat C, Three-field-based nonlinear aeroelastic simulation technology: status and application to the flutter analysis of an f-16 configuration. AIAA Paper 2002-0870, 2002.

[10] Gordnier R, Fithen R. Coupling of a nonlinear finite element structural method with a Navier-Stokes solver. American Institute of Aeronautics and Astronautics Computers and Structures. 2003, (81), 75–89.

[11] Gordnier R, Visbal M. Computation of the aeroelastic response of a flexible delta wing at high angles of attack. Journal of Fluids and Structures. 2004, 19, 785–800.

[12] Gordnier RE, Visbal MR Development of a three-dimensional viscous aeroelastic solver for nonlinear panel flutter. AIAA Paper 2000-2337, 2000.

[13] Goura GSL, Badcock KJ, et al. A data exchange method for fluid-structure interaction problems. The Aeronautical Journal. 2001, 215–221.

[14] Guruswamy GA New modular approach for tightly coupled Fluid/Structure analysis. AIAA Paper 2000-4547, 2004.

[15] Guruswamy GP, Byun C Fluid-Structural interactions using Navier-Stokes flow equations coupled with shell finite element structures. AIAA Paper 93-3087, 1993.

[16] Guruswamy GP Coupled Finite difference/finite element approach for wing body aeroelasticity. AIAA Paper 92-4680, 1992.

[17] Hurka J, Ballmann J. Elastic panels in transonic flow. AIAA Paper 2001-2722, 2002.
[18] Jiang Y, Ye Z, Zhang W. A semi-implicit aeroelastic time domain method. Engineering Mechanics. 2012, 29(4), 66–72. (in Chinese).
[19] Jiang Y, Zhang W, Ye Z. Research on time domain coupling method of flow field and structure based on CFD Technology. Journal of Vibration Engineering. 2007, 20(4), 396–400. (in Chinese).
[20] Kamakoti R, Lian YS, Regisford S, et al. Computational aeroelasticity using a pressure-based solver. AIAA Paper 2002-0869, 2002.
[21] Kamakoti R, Shyy W, Thakur S, et al. Time dependent rans computation for an aeroelastic wing. AIAA Paper 2004-0886, 2004.
[22] Kamakoti R. Fluid–structure interaction for aeroelastic applications. Progress in Aerospace Sciences. 2004, (40), 535–558.
[23] Kilpatrick K A preconditioned krylov subspace approach to a tightly coupled aeromechanical system. The University of Tennessee, Ph.D. Dissertation, 2001.
[24] Liu F, Cai J, Zhu Y. Calculation of Wing Flutter by a Coupled Fluid-Structure Method. Journal of Aircraft. 2001, 38(2), 334–342. (in Chinese).
[25] Pipeno S, Farhat C, Larrouturou B. Partitioned procedures for the transient solution of coupled aeroelastic problems Part I: Model problem, theory and two-dimensional application. Computer Methods in Applied Mechanics and Engineering. 1995, 124, 79–112.
[26] Pipeno S, Farhat C. Partitioned procedures for the transient solution of coupled aeroelastic problems Part II: energy transfer analysis and three-dimensional applications. Computer Methods in Applied Mechanics and Engineering. 2001, 190, 3147–3170.
[27] Sheta E, Harrand V, Thompson D, et al. . Computational and experimental investigation of limit cycle oscillations of nonlinear aeroelastic systems. Journal of Aircraft. 2002, 39(1), 133–141.
[28] Smith MJ, Hodges DH. Evaluation of computational algorithms suitable for fluid-structure interactions. Journal of Aircraft. 2000, 37(2), 282–294.
[29] Soulaïmani A, Ali ABH. A parallel-distributed approach for multi-physics problems with application to computational nonlinear aeroelasticity. Canadian Aeronautics and Space Journal. 2004, 4(50).
[30] Soulaimani A, BenElHajAli A, Feng Z A distributed computing-based methodology for nonlinear aeroelasticity. AIAA Paper 2002-0868, 2002.
[31] Strganac TW, Cizmas PG, Nichkawde C, et al. . Aeroelastic analysis for future air vehicle concepts using a fully nonlinear methodology. AIAA Paper 2005-2171, 2005.
[32] Wang Z, Chen PC, Liu DD, Time domain nonlinear aeroelastic analysis for hale wings. AIAA Paper 2006-1640, 2006.
[33] Zeng X Application of model reduction in aeroelasticity. Northwest Polytechnic University, MA.Sc. Dissertation, 2008, 3. (in Chinese)
[34] Zhang Z, Yang S, Liu F, Prediction of Flutter and LCO by an Euler Method on non-moving Cartesian grids with boundary-layer corrections. AIAA 2005–833. 43rd AIAA Aerospace Sciences Meeting Conference & Exhibit, January 10–13, 2005/Reno, Nevada. (in Chinese)
[35] Zwaan R, Prananta B. Fluid/structure interaction in numerical aeroelastic simulation. International Journal of Non-linear Mechanics. 2002, 37(4-5), 987–1002.

Chapters 6
Reduced-order modeling techniques of nonlinear aeroelastic system

6.1 Introduction

In order to reduce expensive computational cost during the aeroelastic computation, it is necessary to develop a simplified model instead of a full aeroelastic model. One of them is reduced-order modeling (ROM) techniques. The ROM is a simplified mathematical model that can present the main dynamic features of a complex system completely or to a large extent. Because of its simplification, the computational cost (such as computer memory, computation time and design cycle) of the ROM is several orders of magnitude less than that the original complex system. For computational fluid dynamics (CFD), the aerodynamic ROM is an effective and high-precision surrogate model. It greatly reduces the computational time of complex systems and expands the application range of CFD methods. Table 6.1 shows the development of computational methods for unsteady aerodynamic loads.

Tab. 6.1: Development of computational methods for unsteady aerodynamic loads [14].

Time	Model
1930–1940	Theodorsen model, Grossman model and strip theory
1950–1960	Linear panel method [13]
1970–1980	State space fitting from aerodynamic loads in frequency domain [32]
1980–1990	CFD simulation of unsteady nonlinear flow [11, 33]
1990–now	Reduced-order model of unsteady aerodynamic loads using CFD [2, 16, 34, 35]

This chapter introduces reduced-order modeling of aeroelastic systems, including unsteady aerodynamic load models based on Volterra series, unsteady aerodynamic load models based on state space, ROMs of aeroelastic systems based on proper orthogonal decomposition (POD) methods and balanced POD (BPOD)-based methods. Finally, the POD-Galerkin method for nonlinear aerodynamic system is introduced.

https://doi.org/10.1515/9783110576689-006

6.2 Reduced-order model of unsteady aerodynamic force based on Volterra series

The signal-based ROM is one of the commonly used methods for unsteady aerodynamic force. It constructs a low-order model that can approximate the original system based on the input–output data of the original system and can represent the dynamic characteristics of the system [18, 24, 25, 28].

This section introduces the ROM based on the Volterra series and multiwavelet multiresolution analysis, which is commonly used for signal-based reduced-order modeling. The procedures of piecewise quadratic orthogonal multiwavelet and scaling function are explained. Furthermore, a one-dimensional discrete wavelet transform based on boundary-adaptive scaling function and wavelet function is derived. According to the definition of multiresolution analysis, the approximate Volterra series expression in the wavelet domain is given. For the second-order Volterra series, the QR Decomposition based Recursive Least Square (QRD-RLS) algorithm is used to identify the Volterra kernel in the wavelet domain. Finally, three examples are given to verify the presented techniques.

6.2.1 Multiwavelet and multiresolution analysis

This section uses the multiwavelet method to approximate the Volterra kernel based on Prazenica's work [27, 28]. Multiwavelet multiresolution analysis is defined as a sequence of subspaces $\{V_j\}_{j\in Z}$ in a square-integrable space $L^2(R)$ that satisfies the following conditions:

$$m\begin{cases} \ldots V_{-2} \subset V_{-1} \subset V_0 \subset V_1 \subset V_2 \ldots \\ \underset{j\in Z}{\cap} V_j = \{0\} \\ \overline{\underset{j\in Z}{\cup} V_j} = L^2(R) \\ f(x) \in V_j \Leftrightarrow f(2x) \in V_{j+1} \\ f(x) \in V_j \Leftrightarrow f(x-k) \in V_j \quad \forall k \in Z \\ \text{Suppose the translation of } r \text{ scale function } \{\phi_1, \ldots, \phi_r\} \text{ is} \\ \phi_s(x-k): s = 1 \cdots r\}_{k\in Z}, \text{which spans } V_0 \text{ as Riesz basis.} \end{cases} \tag{6.1}$$

Then the scaling function $\{\phi_1, \ldots, \phi_r\}$ constitutes a multiresolution analysis. Each subspace V_j constructed is a finitely generated translation-invariant space. Prazenica et al. chose a scale function in the interval on $[-1, 1]$ to construct a space V_0.

6.2.2 Piecewise quadratic orthogonal multiwavelet method

A piecewise quadratic orthogonal multiwavelet function proposed by Prazenica is used in this section. Choose the shape function of the three-node beam element eq. (6.2) as the basis function (see Fig. 6.1):

$$\begin{cases} N_1 = 2x^2 - 3x + 1 \\ N_2 = -4x^2 + 4x \qquad x \in [0,1]\,. \\ N_3 = 2x^2 - x \end{cases} \tag{6.2}$$

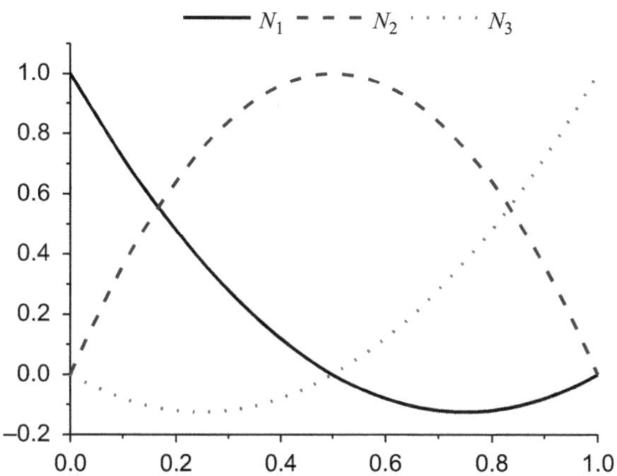

Fig. 6.1: Shape function of a three-node beam element.

Scale functions ϕ_1 and ϕ_2 are given as follows (see Fig. 6.2):

$$\phi_1 = \begin{cases} 2x^2 + 3x + 1 & x \in [-1,0] \\ 2x^2 - 3x + 1 & x \in [0,1] \end{cases}, \tag{6.3}$$

$$\phi_2(x) = -4x^2 + 4x \quad x \in [0,1]\,. \tag{6.4}$$

The finitely generated translation-invariant space is formed by the scale function and defined as V_0. It is a sequence of subspaces generated by the translation and scaling of scale function, which construct a multiresolution analysis within a square-integrable space. It should be noted that the multiresolution analysis constructed is not orthogonal. Donovan [12] proved that orthogonal multiresolution analysis can be constructed by the entanglement of scale functions. In order to construct an orthogonal multiresolution analysis from $\{V_j\}_{j\in Z}$, it is necessary to ensure that the minimum support interval of the scale function is [−1, 1]. Donovan [12] gives the following definitions to construct orthogonal multiresolution analysis from $\{V_j\}_{j\in Z}$.

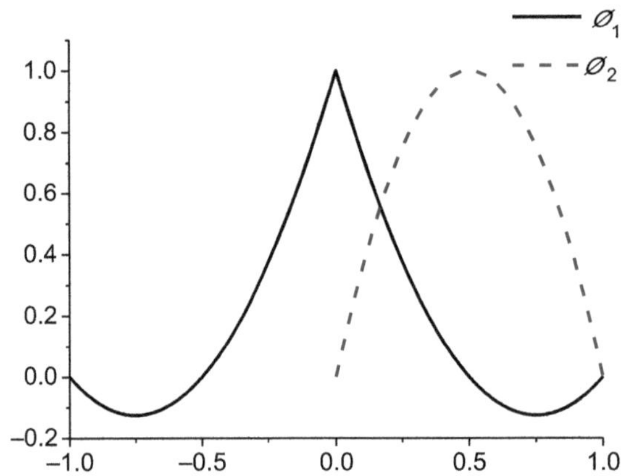

Fig. 6.2: Scale function.

Definition 1. Assume that $\{\phi_1, \ldots, \phi_n\}$ constructs a multiresolution analysis, in which the support interval of $\{\phi_1, \ldots, \phi_k\}$ is [−1, 1] and $\{\phi_{k+1}, \ldots, \phi_n\}$ has a support interval of [0, 1], then the minimum support interval is [−1,1] as the following conditions are met:

1) $\{\phi_1, \ldots, \phi_k\}$ linearly independent of the interval [0,1]
2) $\{\phi_1, \ldots, \phi_k\}$ linearly independent of the interval [−1,0]
3) $\{0\} = \text{span}(\{\phi_1(x)\chi_{[0,1]}(x), \ldots, \phi_k(x)\chi_{[0,1]}(x)\} \cap \{\phi_1(x)\chi_{[0,1]}(x-1), \ldots, \phi_k(x)\chi_{[0,1]}(x-1)\})$ where $\chi_{[0,1]}$ is the eigenfunction defined on the interval [0,1]

$$\chi_{[0,1]}(x): \begin{cases} 1 & x \in [0,1] \\ 0 & \text{else} \end{cases}. \tag{6.5}$$

Define

$$A(V_0) := \text{span}\{\phi_s : s = k+1, \ldots, n\}, \tag{6.6}$$

$$B_\sigma(V_0) := \text{span}\{\{\phi_s(\cdot - \sigma)\chi_{[0,1]} : s = 1, \ldots, k\} \cup A(V_0)\}, \tag{6.7}$$

$$C_\sigma(V_0) := B_\sigma(V_0) - A(V_0) \quad \sigma = 0, 1. \tag{6.8}$$

Theorem 1. Assume $\{V_j\}_{j \in Z}$ is a multiresolution analysis with a minimum support interval of [−1, 1] constructed by a multiscale function $\{\phi_1, \ldots, \phi_n\}$. Suppose there exists such a subspace that makes

$$(I - P_W)C_0(V_0) \perp (I - P_W)C_1(V_0), \tag{6.9}$$

where Pw is an orthogonal mapping to W. Assuming that $\{w_1, \ldots, w_k\}$ is the basis function of space W, then function $\{\phi_1, \ldots, \phi_n, w_1, \ldots, w_k\}$ produces an orthogonal multiresolution analysis $\{\tilde{V}_j\}_{j\in Z}$ tangling with $\{V_j\}_{j\in Z}$, as follows:

$$\ldots V_0 \subset \tilde{V}_0 \subset V_1 \ldots . \tag{6.10}$$

Theoretically, as long as the scale function $\{\phi_1, \ldots, \phi_n\}$ and $\{w_1, \ldots, w_k\}$ are given, orthogonal multiresolution analysis can be constructed. However, the space $W \subset A(V_1) - A(V_0)$ is often inadequate for us to obtain the expected multiwavelet function. Donovan found out that the stated problem can be solved by two consecutive tangling methods. Therefore, the construction for a quadratic orthogonal multiwavelet scaling function is resolved theoretically. Take the scale function ϕ_1 and ϕ_2 as an example. The process is as follows: the finitely generated translation-invariant space generated by the scaling function is expressed as

$$V_j := \text{span}\{\phi_{j,k}^s : s = 1, 2\}_{k\in Z}. \tag{6.11}$$

Function $\phi_{j,k}^s$ represents the translation and scaling of the scale function, defined as

$$\phi_{j,k}^s(x) = 2^{j/2}\phi^s(2^j x - k). \tag{6.12}$$

Select a function $\tilde{\phi}_3$ in V_1 (as shown in Fig. 6.3)

$$\tilde{\phi}_3(x) = \phi_2(2x) - \phi_2(2x - 1). \tag{6.13}$$

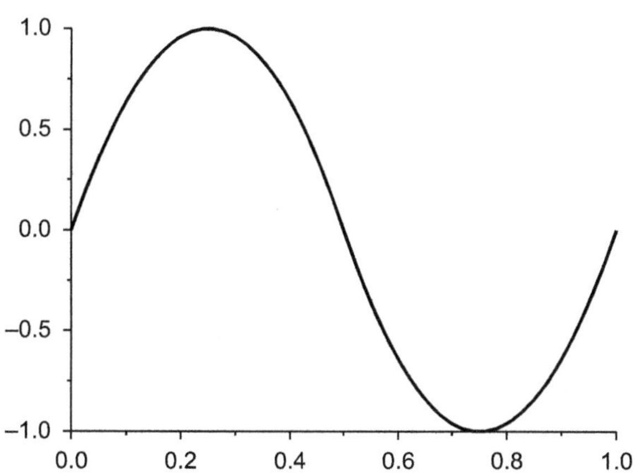

Fig. 6.3: Function $\tilde{\phi}_3$.

It can be seen that the functions $\tilde{\phi}_3$ and ϕ_2 are orthogonal, so choose $\tilde{\phi}_2 = \phi_2$ to normalize $\tilde{\phi}_2, \tilde{\phi}_3$:

$$\begin{cases} \tilde{\phi}_2 = \tilde{\phi}_2 / \left\| \tilde{\phi}_2 \right\| \\ \tilde{\phi}_3 = \tilde{\phi}_3 / \left\| \tilde{\phi}_3 \right\| \end{cases}, \tag{6.14}$$

where $\|\bullet\|$ represents the inner product of the function; the normalized function is

$$\tilde{\phi}_2 = \frac{\sqrt{15}}{2\sqrt{2}}(-4x^2 + 4x) \quad x \in [0, 1], \tag{6.15}$$

$$\tilde{\phi}_3 = \begin{cases} \frac{\sqrt{15}}{2\sqrt{2}}[-16x^2 + 8x] & x \in [0, 0.5] \\ \frac{\sqrt{15}}{2\sqrt{2}}[16x^2 - 24x + 8] & x \in [0.5, 1] \end{cases}. \tag{6.16}$$

$\tilde{\phi}_1$ is derived using Gram–Schmidt orthogonalization method

$$\begin{aligned} \tilde{\phi}_1(x) = \phi_1(x) &- <\phi_1(x), \tilde{\phi}_2(x+1) > \tilde{\phi}_2(x+1) - <\phi_1(x), \tilde{\phi}_3(x+1) > \tilde{\phi}_3(x+1) \\ &- <\phi_1(x), \tilde{\phi}_2(x) > \tilde{\phi}_2(x) - <\phi_1(x), \tilde{\phi}_3(x) > \tilde{\phi}_3(x) \end{aligned}. \tag{6.17}$$

And then, get $\tilde{\phi}_1$

$$\tilde{\phi}_1 = \begin{cases} -\frac{5}{2}x^2 - 4x - \frac{3}{2} & x \in [-1, -0.5] \\ \frac{15}{2}x^2 + 6x + 1 & x \in [-0.5, 0.0] \\ \frac{15}{2}x^2 - 6x + 1 & x \in [0.0, 0.5] \\ -\frac{5}{2}x^2 + 4x - \frac{3}{2} & x \in [0.5, 1.0] \end{cases}. \tag{6.18}$$

The scale function $\{\tilde{\phi}_1, \tilde{\phi}_2, \tilde{\phi}_3\}$ constitutes a multiresolution analysis of $\{\tilde{V}_j\}_{j \in Z}$. The scale function after the first tangling is shown in Fig. 6.4.

Hereby the first tangling process is completed and now for the second tangling. Choose the function

$$w = \tilde{\phi}_1(2x - 1) + a[\tilde{\phi}_2(2x) + \tilde{\phi}_2(2x - 1)] + b[\tilde{\phi}_3(2x) - \tilde{\phi}_3(2x - 1)], \tag{6.19}$$

which satisfies the following condition:

$$< w, \tilde{\phi}_2 > = 0. \tag{6.20}$$

Based on Theorem 1, yielding

$$(I - P_W)C_0(\tilde{V}_0) \perp (I - P_W)C_1(\tilde{V}_0). \tag{6.21}$$

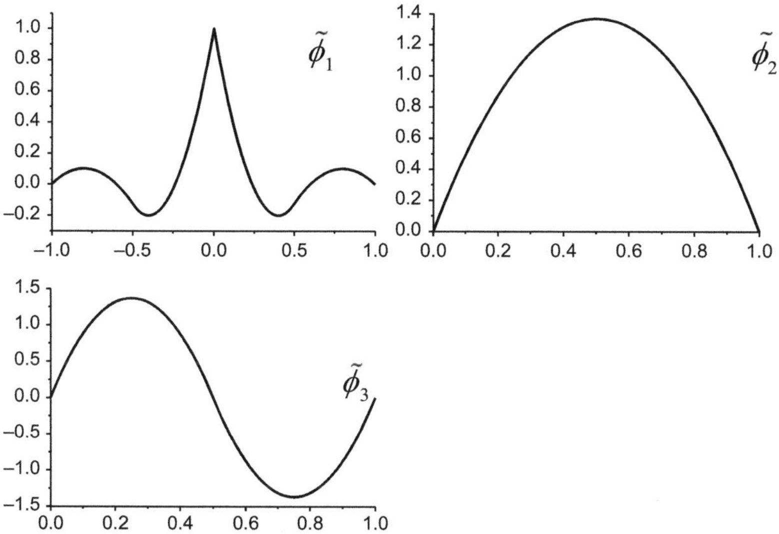

Fig. 6.4: Scaling function after the first winding.

According to the definition of the scale function $\{\tilde{\phi}_1, \tilde{\phi}_2, \tilde{\phi}_3\}$ and the spaces $C_0(\tilde{V}_0)$ and $C_1(\tilde{V}_0)$, it yields

$$\begin{cases} c_0 := C_0(\tilde{V}_0) = \tilde{\phi}_1(x)\chi_{[0,1]}(x) \\ c_1 := C_1(\tilde{V}_0) = \tilde{\phi}_1(x-1)\chi_{[0,1]}(x) \end{cases}. \tag{6.22}$$

The orthogonal condition of eq. (6.21) can be rewritten as

$$< (I - P_W)c_0, (I - P_W)c_1 > = 0. \tag{6.23}$$

Expand formula (6.23) and yield

$$< c_0, c_1 > + < P_W c_0, P_W c_1 > - < c_0, P_W c_1 > - < P_W c_0, c_1 > = 0. \tag{6.24}$$

For any function $f \in L^2(R)$ in the square-integrable space, the orthogonal mapping is

$$P_W \cdot f = \frac{< w, f >}{\|w\|} \frac{w}{\|w\|} = \frac{< w, f > w}{< w, w >}. \tag{6.25}$$

By substituting eq. (6.25) into eq. (6.24), it has

$$< w, w > < c_0, c_1 > - < w, c_0 > < w, c_1 > = 0. \tag{6.26}$$

Together with solutions (6.20) and (6.26) yields

$$\begin{cases} \frac{7a}{8} - \frac{5b}{16} + \frac{7\sqrt{30}}{384} = 0 \\ \frac{11a^2}{1,536} - \frac{25ab}{1,536} + \frac{\sqrt{30}a}{12,288} - \frac{61b^2}{6,144} + \frac{5\sqrt{30}b}{24,576} + \frac{439}{589,824} = 0 \end{cases}. \tag{6.27}$$

By solving the equation, it yields

$$\begin{cases} a_1 = -0.17122977104588294194944 10627817 \\ a_2 = 0.0043608915192194306811886199929205 \\ b_1 = -0.15993853371712533794186926915495 \\ b_2 = 0.33171532146516130542389384261398 \end{cases}. \tag{6.28}$$

Take the first solution, substitute into eq. (6.19) and normalize

$$w = \begin{cases} (2x + 2\sqrt{30}ax + 4\sqrt{30}bx - 10x^2 - 4\sqrt{30}ax^2 - 16\sqrt{30}bx^2)/mww & x \in [0, 0.25] \\ (5/2 + 2\sqrt{30}b - 18x + 2\sqrt{30}ax - 12\sqrt{30}bx + 30x^2 - 4\sqrt{30}ax^2 + 16\sqrt{30}bx^2)/mww & x \in [0.25, 0.5] \\ (29/2 - 2\sqrt{30}a + 6\sqrt{30}b - 42x + 6\sqrt{30}ax - 20\sqrt{30}bx + 30x^2 - 4\sqrt{30}ax^2 + 16\sqrt{30}bx^2)/mww & x \in [0.5, 0.75] \\ (-8 - 2\sqrt{30}a - 12\sqrt{30}b + 18x + 6\sqrt{30}ax + 28\sqrt{30}bx - 10x^2 - 4\sqrt{30}ax^2 - 16\sqrt{30}bx^2)/mww & x \in [0.75, 1.0] \end{cases}, \tag{6.29}$$

$$mww = \sqrt{a^2 + b^2 + 7/96}. \tag{6.30}$$

According to the Gram–Schmidt regularization method, $\tilde{\tilde{\phi}}_1$ is obtained

$$\tilde{\tilde{\phi}}_1 = \tilde{\phi}_1 - <\tilde{\phi}_1, w(x+1)>w(x+1) - <\tilde{\phi}_1, w(x)>w(x). \tag{6.31}$$

After the normalization, $\tilde{\tilde{\tilde{\phi}}}_1$ is expressed as

$$\tilde{\tilde{\phi}}_1 = \begin{cases} (\tilde{\phi}_1(x) - a1w(x+1))/mm & x \in [-1.0, 0.0] \\ \tilde{\phi}_1(x) - b1w(x))/mm & x \in [0.0, 1.0] \end{cases}, \tag{6.32}$$

$$a1 = b1 = \frac{\sqrt{30}a}{96\sqrt{a^2 + b^2 + 7/96}} - \frac{1}{256\sqrt{a^2 + b^2 + 7/96}} + \frac{5\sqrt{30}b}{192\sqrt{a^2 + b^2 + 7/96}}, \tag{6.33}$$

$$mm = \left\| \tilde{\phi}_1 - <\tilde{\phi}_1, w(x+1) > w(x+1) - <\tilde{\phi}_1, w(x) > w(x) \right\|_2. \tag{6.34}$$

Define the three scale functions

$$\tilde{\tilde{\phi}}_2 = \tilde{\phi}_2 \quad \tilde{\tilde{\phi}}_3 = \tilde{\phi}_3 \quad \tilde{\tilde{\phi}}_4 = w. \tag{6.35}$$

Four scale functions after the secondary tangling are shown in Fig. 6.5. (For convenience, the superscript on the scale function is omitted.) Once the orthogonal scale function is obtained, the wavelet function is constructed by the scale function. Define V_0 as the space spanned by the scale function $\{\phi_1, \phi_2, \phi_3, \phi_4\}$. The wavelet function $\{\phi_1, \phi_2, \phi_3, \phi_4\}$ in the wavelet space W_0 should be orthogonal to any functions in V_0, that is, all the translation functions of $\{\phi_1, \phi_2, \phi_3, \phi_4\}$ and ψ are orthogonal. First, a symmetric wavelet ψ_1 is constructed with a support interval of [−1, 1]. Define ψ_1 as

$$\psi_1(x) = \phi_{1,0}^1(x) - <\phi_{1,0}^1(x), \phi_1(x)> \phi_1(x). \tag{6.36}$$

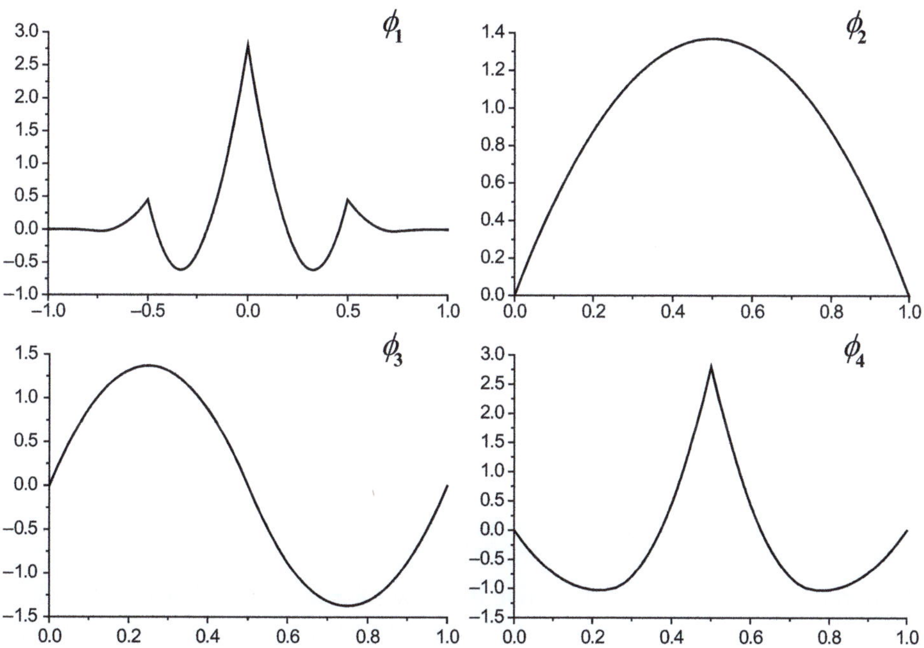

Fig. 6.5: Scale function obtained after secondary winding.

Normalize eq. (6.36) to get the first wavelet function. Define the second wavelet ψ_2 as 5 and the support interval is also [−1,1]

$$\psi_2(x) = \begin{cases} <\phi_1(x), \phi_{1,0}^1(x)> \phi_{1,0}^1(x) - \phi_1(x) & x \in [-1,0] \\ \phi_1(x) - <\phi_1(x), \phi_{1,0}^1(x)> \phi_{1,0}^1(x) & x \in [0,1] \end{cases}. \tag{6.37}$$

Since $\psi_2(x)$ is a skew-symmetric function, it guarantees to be orthogonal to $\phi_1(x)$ and $\psi_1(x)$, whilst the remaining two wavelet functions $\psi_3(x)$ and $\psi_4(x)$ have the support interval [0,1]. Derived from the two-scale equation, $\psi_3(x)$ and $\psi_4(x)$ are linearly combined by the scaling function forming the space V_1. Hence, the definitions $\psi_3(x)$ and $\psi_4(x)$ are as follows:

$$\begin{aligned} \psi_t(x) &= c_1\phi_{1,1}^1(x) + c_2\phi_{1,1}^2(x) + c_3\phi_{1,1}^3(x) + c_4\phi_{1,1}^4(x) \\ &\quad + c_5\phi_{1,0}^2(x) + c_6\phi_{1,0}^3(x) + c_7\phi_{1,0}^4(x), \quad t = 3,4. \end{aligned} \tag{6.38}$$

Equation (6.38) contains seven unknowns, while the following five constraint conditions can be obtained from the orthogonality of wavelet,

$$<\psi_t(x), \phi_i(x)> = 0, \quad i = 1,2,3,4, \tag{6.39}$$

$$<\psi_t(x), \phi_1(x-1)> = 0. \tag{6.40}$$

For the closure of the equation, two additional conditions are added as follows:

$$c_4 = 1, c_7 = -1, \tag{6.41}$$

$$c_4 = 1, c_7 = 1. \tag{6.42}$$

The skew-symmetric wavelet $\psi_3(x)$ is generated by eq. (6.41), while the symmetric wavelet $\psi_4(x)$ is generated by eq. (6.42). After the normalization, the orthogonal multiwavelet function is obtained (see Fig. 6.6).

6.2.3 Boundary-adaptive scaling function and wavelet

The purpose of this section is to derive first-order kernel and the second-order kernel of the Volterra series [24–28, 41] with wavelet-scale functions as the basis functions. Considering that the Volterra kernel has time history, it decays to zero after a certain time T_1. Therefore, in order to approximate the Volterra kernel function, the scale function and the wavelet function boundary need to be modified so that their support domains are on $[0, T_1]$. Considering the scale j (where the scale is for the first-order kernel, not the input/output sampling scale), the boundary-adaptive approximation space and wavelet space are defined as

$$\begin{aligned} V_j[0, T_1] := \text{span}\Big\{ &\phi_{j,0}^1 \chi_{[0,T_1]}, \phi_{j,k}^1\Big|_{k=1}^{2^j T_1 - 1}, \phi_{j,k}^2\Big|_{k=0}^{2^j T_1 - 1}, \phi_{j,k}^3\Big|_{k=0}^{2^j T_1 - 1}, \\ &\phi_{j,k}^4\Big|_{k=0}^{2^j T_1 - 1}, \phi_{j,2^j T}^1 \chi_{[0,T_1]} \Big\}, \end{aligned} \tag{6.43}$$

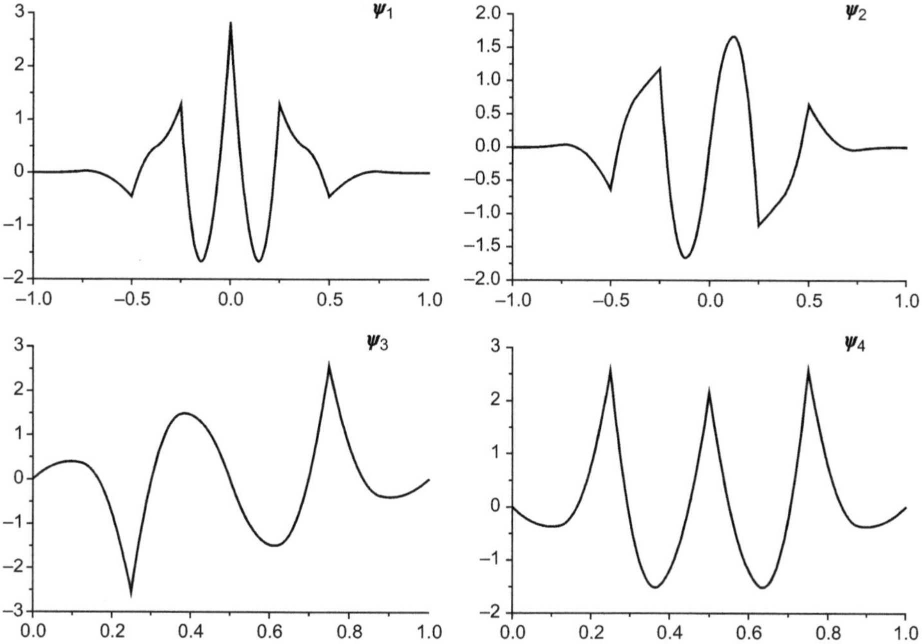

Fig. 6.6: Wavelet function.

$$W_j[0,\ T_1] := \text{span}\Big\{\psi_{j,0}^1 \chi_{[0,\ T_1]},\ \psi_{j,k}^1\Big|_{k=1}^{2^j T_1 - 1},\ \psi_{j,k}^2\Big|_{k=0}^{2^j T_1 - 1},\ \psi_{j,k}^3\Big|_{k=0}^{2^j T_1 - 1},$$
$$\psi_{j,k}^4\Big|_{k=0}^{2^j T_1 - 1},\ \psi_{j,2^j T_1}^1 \chi_{[0,\ T_1]}\Big\}, \tag{6.44}$$

where $\chi_{[0,\ T_1]}$ represents an eigenfunction $[0, T_1]$.

6.2.4 Orthogonal multiwavelet multiresolution analysis

Expand the function f with a multiscale function on a scale j. The domain of the function f is $[0,\ T_1]$

$$f_j(x) = \sum_{s=1}^{4} \sum_{k \in Z} \alpha_{j,k}^s \phi_{j,k}^s(x). \tag{6.45}$$

Equivalent two-scale equation is

$$f_j(x) = \sum_{s=1}^{4} \sum_{k \in Z} \alpha_{j-1,k}^s \phi_{j-1,k}^s(x) + \sum_{s=1}^{4} \sum_{k \in Z} \beta_{j-1,k}^s \psi_{j-1,k}^s(x). \tag{6.46}$$

Recursive to eq. (6.46), a multiscale description of the function f is obtained as

$$f_j(x) = \sum_{s=1}^{4}\sum_{k\in Z}\alpha^s_{j_0,k}\phi^s_{j_0,k}(x) + \sum_{l=j_0}^{j-1}\sum_{s=1}^{4}\sum_{k\in Z}\beta^s_{i,k}\psi^s_{l,k}(x). \tag{6.47}$$

Together with eq. (6.43), eq. (6.45) contains a total of $N_1 = 2^{2+j}T_1 + 1$ scale functions. Convert the function into a vector form

$$\begin{aligned} f_j(x) = {} & \begin{bmatrix}\alpha^1_{j,0} & \alpha^2_{j,0} & \alpha^3_{j,0} & \alpha^4_{j,0}\end{bmatrix}\begin{bmatrix}\phi^1_{j,0}\chi_{[0,T_1]}\\ \phi^2_{j,0}\\ \phi^3_{j,0}\\ \phi^4_{j,0}\end{bmatrix} \\ & + \begin{bmatrix}\alpha^1_{j,1} & \alpha^2_{j,1} & \alpha^3_{j,1} & \alpha^4_{j,1}\end{bmatrix}\begin{bmatrix}\phi^1_{j,1}\\ \phi^2_{j,1}\\ \phi^3_{j,1}\\ \phi^4_{j,1}\end{bmatrix} + \dots \\ & + \begin{bmatrix}\alpha^1_{j,(2^jT_1-1)} & \alpha^2_{j,(2^jT_1-1)} & \alpha^3_{j,(2^jT_1-1)} & \alpha^4_{j,(2^jT_1-1)}\end{bmatrix}\begin{bmatrix}\phi^1_{j,(2^jT_1-1)}\\ \phi^2_{j,(2^jT_1-1)}\\ \phi^3_{j,(2^jT_1-1)}\\ \phi^4_{j,(2^jT_1-1)}\end{bmatrix} \\ & + \begin{bmatrix}\alpha^1_{j,2^jT_1} & 0 & 0 & 0\end{bmatrix}\begin{bmatrix}\phi^1_{j,2^jT_1}\chi_{[0,T_1]}\\ 0\\ 0\\ 0\end{bmatrix}. \end{aligned} \tag{6.48}$$

Define

$$\underline{\boldsymbol{\alpha}}^T_{j,k} = \begin{cases}\begin{bmatrix}\alpha^1_{j,k} & 0 & 0 & 0\end{bmatrix} & k = 2^jT_1\\ \begin{bmatrix}\alpha^1_{j,k} & \alpha^2_{j,k} & \alpha^3_{j,k} & \alpha^4_{j,k}\end{bmatrix} & \text{else}\end{cases}, \tag{6.49}$$

$$\underline{\mathbf{\Phi}}^T = \begin{cases} \left[\phi^1_{j,k}\chi_{[0,T_1]} \quad \phi^2_{j,k} \quad \phi^3_{j,k} \quad \phi^4_{j,k}\right] & k=0 \\ \left[\phi^1_{j,k}\chi_{[0,T_1]} \quad 0 \quad 0 \quad 0\right] & k=2^jT_1, \\ \left[\phi^1_{j,k} \quad \phi^2_{j,k} \quad \phi^3_{j,k} \quad \phi^4_{j,k}\right] & k=\text{else} \end{cases} \tag{6.50}$$

$$\overline{\mathbf{\Phi}}^T = \left[\phi^1_{j,k} \quad \phi^2_{j,k} \quad \phi^3_{j,k} \quad \phi^4_{j,k}\right] \qquad k\in Z. \tag{6.51}$$

Similarly, the wavelet coefficients and wavelet functions corresponding to eqs. (6.49)–(6.51) are listed:

$$\underline{\mathbf{\beta}}^T_{j,k} = \begin{cases} \left[\beta^1_{j,k} \quad 0 \quad \beta^3_{j,k} \quad \beta^4_{j,k}\right] & k=0 \\ \left[\beta^1_{j,k} \quad 0 \quad 0 \quad 0\right] & k=2^jT_1, \\ \left[\beta^1_{j,k} \quad \beta^2_{j,k} \quad \beta^3_{j,k} \quad \beta^4_{j,k}\right] & \text{else} \end{cases} \tag{6.52}$$

$$\underline{\mathbf{\Psi}}^T = \begin{cases} \left[\psi^1_{j,k}\chi_{[0,T_1]} \quad 0 \quad \psi^3_{j,k} \quad \psi^4_{j,k}\right] & k=0 \\ \left[\psi^1_{j,k}\chi_{[0,T_1]} \quad 0 \quad 0 \quad 0\right] & k=2^jT_1\,, \\ \left[\psi^1_{j,k} \quad \psi^2_{j,k} \quad \psi^3_{j,k} \quad \psi^4_{j,k}\right] & k=\text{else} \end{cases} \tag{6.53}$$

$$\overline{\Psi}^T = \left[\psi^1_{j,k} \quad \psi^2_{j,k} \quad \psi^3_{j,k} \quad \psi^4_{j,k}\right] \quad k\in Z. \tag{6.54}$$

Rewrite eqs. (6.45) and (6.46) as the following vector form:

$$f_j(x) = \sum_{k\in Z} \underline{\mathbf{\alpha}}^T_{j,k}\underline{\mathbf{\Phi}}_{j,k}(x), \tag{6.55}$$

$$f_j(x) = \sum_{k\in Z} \underline{\mathbf{\alpha}}^T_{j-1,k}\underline{\mathbf{\Phi}}_{j-1,k}(x) + \sum_{k\in Z} \underline{\mathbf{\beta}}^T_{j-1,k}\underline{\mathbf{\Psi}}_{j-1,k}(x). \tag{6.56}$$

In order to derive the recursive form of multiscale coefficient expression, a two-scale equation in vector form is derived by substituting eq. (6.56) into eq. (6.55)

$$\sum_{k\in Z} \underline{\mathbf{\alpha}}^T_{j,k}\underline{\mathbf{\Phi}}_{j,k}(x) = \sum_{k\in Z} \underline{\mathbf{\alpha}}^T_{j-1,k}\underline{\mathbf{\Phi}}_{j-1,k}(x) + \sum_{k\in Z} \underline{\mathbf{\beta}}^T_{j-1,k}\underline{\mathbf{\Psi}}_{j-1,k}(x). \tag{6.57}$$

Equation (6.57) multiplies $\bar{\Phi}^T_{j-1,m}$ on both sides and integrates in the domain R

$$\begin{aligned} \sum_{k\in Z} \underline{\mathbf{\alpha}}^T_{j,k} \int_R \underline{\mathbf{\Phi}}_{j,k}(x)\bar{\mathbf{\Phi}}^T_{j-1,m}(x)\,dx = {} & \sum_{k\in Z} \underline{\mathbf{\alpha}}^T_{j-1,k} \int_R \underline{\mathbf{\Phi}}_{j-1,k}(x)\bar{\mathbf{\Phi}}^T_{j-1,m}(x)\,dx \\ & + \sum_{k\in Z} \underline{\mathbf{\beta}}^T_{j-1,k} \int_R \underline{\mathbf{\Psi}}_{j-1,k}(x)\bar{\mathbf{\Phi}}^T_{j-1,m}(x)\,dx \end{aligned} \tag{6.58}$$

According to the orthogonality of orthogonal multiwavelets, eq. (6.58) is simplified to

$$\sum_{k\in Z} \underline{\boldsymbol{\alpha}}_{j,k}^{T} \int_{R} \underline{\boldsymbol{\Phi}}_{j,k}(x)\, \bar{\boldsymbol{\Phi}}_{j-1,m}^{T}(x)\, dx = \sum_{k\in Z} \underline{\boldsymbol{\alpha}}_{j-1,k}^{T} \mathbf{J}_s \delta_{k,m} = \underline{\boldsymbol{\alpha}}_{j-1,m}^{T} \mathbf{J}_s, \tag{6.59}$$

where matrix $\mathbf{J}_s$ is

$$\begin{cases} \mathbf{J}_s = \mathbf{J}_{s,1} = \begin{bmatrix} 1/2 & 0_{1\times 3} \\ 0_{3\times 1} & \mathrm{I}_{3\times 3} \end{bmatrix} & m = 0 \\ \mathbf{J}_s = \mathbf{J}_{s,2} = I_{4\times 4} & m \in \left[1,\ 2^j T_1 - 1\right]. \\ \mathbf{J}_s = \mathbf{J}_{s,3} = \begin{bmatrix} 1/2 & 0_{1\times 3} \\ 0_{3\times 1} & \mathrm{I}_{3\times 3} \end{bmatrix} & m = 2^j T \end{cases} \tag{6.60}$$

According to the two-scale relationship, it yields

$$\bar{\boldsymbol{\Phi}}_{j-1,m}^{T} = \sum_{p} \bar{\boldsymbol{\Phi}}_{j,2m+p}^{T} \left[a_p\right]^{T}, \tag{6.61}$$

$$\bar{\boldsymbol{\Psi}}_{j-1,m}^{T} = \sum_{p} \bar{\boldsymbol{\Phi}}_{j,2m+p}^{T} \left[b_p\right]^{T}, \tag{6.62}$$

where $[a_p]$ and $[b_p]$ are wavelet filter matrices. Substitute eq. (6.61) into eq. (6.59), yielding

$$\underline{\boldsymbol{\alpha}}_{j-1,m}^{T} \mathbf{J}_s = \sum_{k\in Z} \underline{\boldsymbol{\alpha}}_{j,k}^{T} \int_{R} \underline{\boldsymbol{\Phi}}_{j,k}(x) \left(\sum_{p} \bar{\boldsymbol{\Phi}}_{j,2m+p}^{T}(x) \left[a_p\right]^{T} \right) dx. \tag{6.63}$$

Again, using multiwavelet orthogonality, it yields

$$\underline{\boldsymbol{\alpha}}_{j-1,m}^{T} \mathbf{J}_s^{m} = \sum_{k\in Z} \underline{\boldsymbol{\alpha}}_{j,k}^{T} \mathbf{J}_s^{k} \delta_{k,2m+p} \left[a_p\right]^{T} = \sum_{k\in Z} \underline{\boldsymbol{\alpha}}_{j,k}^{T} \mathbf{J}_s^{k} \left[a_{k-2m}\right]^{T}. \tag{6.64}$$

where the superscripts m and k indicate that the specific form of $\mathbf{J}_s$ is determined by the values of m and k, respectively. Transposition on both sides of the earlier equation can obtain multiwavelet multiresolution scale decomposition equation

$$\mathbf{J}_s^{m} \underline{\boldsymbol{\alpha}}_{j-1,m} = \sum_{k\in Z} \left[a_{k-2m}\right] \mathbf{J}_s^{k} \underline{\boldsymbol{\alpha}}_{j,k}. \tag{6.65}$$

Equation (6.57) is multiplied by $\bar{\boldsymbol{\Psi}}_{j-1,m}^{T}$ on both sides and integrate on R in the domain, yielding

$$\begin{aligned} \sum_{k\in Z} \underline{\boldsymbol{\alpha}}_{j,k}^{T} \int_{R} \underline{\Phi}_{j,k}(x)\, \boldsymbol{\Psi}_{j-1,m}^{T}(x)\, dx &= \sum_{k\in Z} \underline{\boldsymbol{\alpha}}_{j-1,k}^{T} \int_{R} \underline{\boldsymbol{\Phi}}_{j-1,k}(x)\, \boldsymbol{\Psi}_{j-1,m}^{T}(x)\, dx \\ &+ \sum_{k\in Z} \underline{\boldsymbol{\beta}}_{j-1,k}^{T} \int_{R} \underline{\boldsymbol{\Psi}}_{j-1,k}(x)\, \boldsymbol{\Psi}_{j-1,m}^{T}(x)\, dx \end{aligned} \tag{6.66}$$

Using multiple wavelet orthogonality, eq. (6.66) becomes

$$\sum_{k\in Z} \underline{\boldsymbol{\alpha}}_{j,k}^{T} \int_{R} \underline{\boldsymbol{\Phi}}_{j,k}(x)\, \bar{\boldsymbol{\Psi}}_{j-1,m}^{T}(x)\, dx = \sum_{k\in Z} \underline{\boldsymbol{\beta}}_{j-1,k}^{T} \mathbf{J}_{w}\, \delta_{k,m} = \underline{\boldsymbol{\beta}}_{j-1,m}^{T} \mathbf{J}_{w}, \tag{6.67}$$

where the expression of $\mathbf{J}_w$ is

$$\begin{cases} \mathbf{J}_w = \mathbf{J}_{w,1} = \begin{bmatrix} \begin{matrix} 1/2 & 0 \\ 0 & 0 \end{matrix} & \mathbf{0}_{2\times 2} \\ \mathbf{0}_{2\times 2} & \mathbf{I}_{2\times 2} \end{bmatrix} & m = 0 \\ \mathbf{J}_w = \mathbf{J}_{w,2} = \mathbf{I}_{4\times 4} & m \in \left[1,\ 2^j T_1 - 1\right]. \\ \mathbf{J}_w = \mathbf{J}_{w,3} = \begin{bmatrix} 1/2 & \mathbf{0}_{1\times 3} \\ \mathbf{0}_{3\times 1} & \mathbf{I}_{3\times 3} \end{bmatrix} & m = 2^j T \end{cases} \tag{6.68}$$

Substitute eq. (6.62) into eq. (6.67), yielding

$$\underline{\boldsymbol{\beta}}_{j-1,m}^{T} \mathbf{J}_w = \sum_{k\in Z} \underline{\boldsymbol{\alpha}}_{j,k}^{T} \int_{R} \underline{\boldsymbol{\Phi}}_{j,k}(x) \left(\bar{\boldsymbol{\Phi}}_{j,2m+p}^{T}(x) \left[b_p\right]^{T}\right) dx. \tag{6.69}$$

Similar to eq. (6.65), multiwavelet multiresolution scale decomposition equation is obtained

$$\mathbf{J}_w^m \underline{\boldsymbol{\beta}}_{j-1,m} = \sum_{k\in Z} \left[b_{k-2m}\right] \mathbf{J}_w^k \underline{\boldsymbol{\alpha}}_{j,k}. \tag{6.70}$$

Similarly, the multiwavelet reduction equation can be obtained as

$$\mathbf{J}_s^m \underline{\boldsymbol{\alpha}}_{j,m} = \sum_{k\in Z} \left\{ \mathbf{J}_s^m \left[a_{m-2k}\right]^T \underline{\boldsymbol{\alpha}}_{j-1,k} + \mathbf{J}_w^m \left[b_{m-2k}\right]^T \underline{\boldsymbol{\beta}}_{j-1,k} \right\}. \tag{6.71}$$

If the boundary-adaptive scaling function and the wavelet function are not used, the matrix $\mathbf{J}_{4\times 4} = \mathbf{I}_{4\times 4}$ is an identity matrix. Hence, the absence of boundary-adaptive scaling function is a special case of boundary adaptation.

6.2.5 Volterra kernel approximation

Consider a weakly nonlinear system with multiple inputs and multiple outputs and the second-order continuous Volterra series is expressed as [3, 9, 20]

$$y(t) = [y^1(t)\ \ y^2(t)\ \ \dots\ \ y^{mo}(t)]^T \tag{6.72}$$

$$
\begin{aligned}
y^l(t) &= y_1^l(t) + y_2^l(t) \\
&= \sum_{i=1}^{mi} \int_0^t h_1^i(\xi) u^i(t-\xi) d\xi \\
&+ \sum_{i_1, i_2 = 1}^{mi} \int_0^t \int_0^t h_2^{i_1 i_2}(\xi, \eta) u^{i_1}(t-\xi)\, u^{i_2}(t-\eta)\, d\xi d\eta,
\end{aligned}
\tag{6.73}
$$

where h_1^i denotes the first-order kernel corresponding to the ith input signal; $h_2^{i_1 i_2}$ is the second-order kernel corresponding to the ith and jth input signals. It is the autocorrelation second-order kernel in a symmetric matrix form as $i_1 = i_2$, and $h_2^{i_1 i_2}(\xi, \eta) = h_2^{i_1 i_2}(\eta, \xi)$. If $i_1 \neq i_2$, the equation is a cross-core with respect to inputs i_1 and i_2, and $h_2^{i_1 i_2}(\xi, \eta) = h_2^{i_2 i_1}(\eta, \xi)$ satisfies. Without loss of generality, all of the following derivations are based on a multi-input single-output system, and if no multiple-input single-output systems are considered, a multiple-input multiple-output system is obtained.

1. First-order Volterra kernel approximation

Assuming that the sampling frequency of the input signal is 2^j Hz (i.e., the input/output signal sampling scale is j), there are a total of N sampled data. Define the zero-order input signal as

$$u_j^i(t) = \sum_{k=0}^{N-1} u_{j,k}^i \chi_{j,k}(t), \tag{6.74}$$

where eigenfunction $\chi_{j,k}$ is defined as

$$\chi_{j,k}(t) = 2^{j/2} \chi(2^j t - k) = \begin{cases} 2^{j/2} & t \in [2^{-j}k, 2^{-j}(k+1)] \\ 0 & \text{else} \end{cases}. \tag{6.75}$$

χ is an eigenfunction that defines the domain on [0,1]. The input signal coefficient $u_{j,k}^i$ is equal to the scale sample of the input signal

$$u_{j,k}^i = 2^{-j/2} u^i(2^{-j}k) \quad k = 0, 1, \ldots, N-1. \tag{6.76}$$

N outputs can be obtained from N input samples, and then the first-order discrete output is

$$y_1(t_n) = \sum_{i=1}^{mi} \int_0^{t_n} h_1^i(\xi) u^i(t_n - \xi) d\xi, \quad n = 1, \ldots, N, \tag{6.77}$$

where $t_n = 2^{-j}n$. The zero-order discrete input signal is

$$u_j^i(t_n - \xi) = \sum_{k=0}^{n-1} u_{j,n-k-1}^i \chi_{j,k}(\xi). \tag{6.78}$$

The first-order Volterra kernel is expanded by the boundary-adaptive scaling function on the scale j_1:

$$h_{1,j_1}^i(\xi) = \sum_{s=1}^{4} \sum_{p} \alpha_{j_1,p}^{(i)s} \phi_{j_1,p}^s(x). \tag{6.79}$$

It should be noted that the specific expression of scale function $\phi_{j1,p}^s(x)$ in eq. (6.79) is the boundary-adaptive scaling function in eq. (6.43). Substitute eq. (6.78) and eq. (6.79) into eq. (6.77) and discrete first-order Volterra series output is derived

$$y_{1,j}(t_n) = \sum_{i=1}^{mi} \sum_{k=0}^{n_1-1} \sum_{p} \sum_{s=1}^{4} u_{j,n-k-1}^i \alpha_{j_1,p}^{(i)s} \int_0^{T_1} \phi_{j1,p}^s(\xi) \chi_{j,k}(\xi) d\xi, \tag{6.80}$$

where T_1 is the first-order kernel memory length and N_1 is defined as

$$n_1 = \begin{cases} n & n < 2^j T_1 \\ 2^j T_1 & n \geq 2^j T_1 \end{cases}. \tag{6.81}$$

After appropriate adjustment of the summation order in eq. (6.80), it yields

$$y_{1,j}(t_n) = \sum_{i=1}^{mi} \sum_{p} \sum_{s=1}^{4} \alpha_{j_1,p}^{(i)s} \sum_{k=0}^{n_1-1} u_{j,n-k-1}^i \int_0^{T_1} \phi_{j1,p}^s(\xi) \chi_{j,k}(\xi) d\xi. \tag{6.82}$$

Rewrite eq. (6.82) in matrix form

$$y_{1,j}(t_n) = \underline{\mathbf{U}}_1 \underline{\boldsymbol{\alpha}}_1 = [\mathbf{U}_1 \quad \cdots \quad \mathbf{U}_{mi}][\boldsymbol{\alpha}_1^{\mathbf{T}} \cdots \boldsymbol{\alpha}_{\mathbf{mi}}^{\mathbf{T}}]^T, \tag{6.83}$$

where α_i and $\mathbf{U_i}(n, m)$ is

$$\alpha_i = \mathbf{vec}\left(\sum_{p} \sum_{s=1}^{4} \alpha_{j_1,p}^{(i)s} \right) \quad i = 1, \ldots, mi, \tag{6.84}$$

$$\mathbf{U}_i(n, m) = \sum_{k=0}^{n_1-1} u_{j,n-k-1}^i \left(\int_0^{T_1} \phi_{j1,p}^s(\xi) \chi_{j,k}(\xi) d\xi \right) \quad \begin{cases} i = 1, \ldots, mi \\ n = 1, \ldots, N. \\ m = m(s, p) \end{cases} \tag{6.85}$$

It is noted that in eq. (6.84), **vec** is vectorizated. In eq. (6.85), m is the position of $\alpha_{j_1,p}^{(i)s}$ in $\boldsymbol{\alpha}_i$, which is a function of the scale function type s and the translation integer

p. The integral in eq. (6.85) can be calculated by the following three-point Gauss–Legendre numerical integration method

$$\begin{aligned}\int_0^{T_1} \phi_{j_1,0}^1(\xi)\chi_{[0,T_1]}\chi_{j,k}(\xi)d\xi &= 2^{j/2}\int_{2^{-j}k}^{2^{-j}(k+1)} \phi_{j_1,0}^1(\xi)\chi_{[0,T_1]}d\xi \\ &= 2^{(-1-j/2)}\int_{-1}^{1} \phi_{j_1,0}^1(2^{-j}k+2^{-j-1}+2^{-j-1}t)\chi_{[0,T_1]}dt \\ &\approx 2^{(-1-j/2)}[A_1\phi_{j_1,0}^1(2^{-j}k+2^{-j-1}+2^{-j-1}x_1)\chi_{[0,T_1]} \\ &\quad + A_2\phi_{j_1,0}^1(2^{-j}k+2^{-j-1}+2^{-j-1}x_2)\chi_{[0,T_1]} \\ &\quad + A_3\phi_{j_1,0}^1(2^{-j}k+2^{-j-1}+2^{-j-1}x_3)\chi_{[0,T_1]}\end{aligned}, \quad (6.86)$$

where the interpolation point and the interpolation coefficient are given as follows, respectively,

$$\begin{cases} x_1 = 0.7745966692 & x_2 = -0.7745966692 & x_3 = 0.0 \\ A_1 = 0.5555555556 & A_2 = 0.5555555556 & A_3 = 0.8888888889 \end{cases}. \quad (6.87)$$

Using the discrete wavelet transform, the relationship between wavelet coefficients and scale coefficients is obtained:

$$\boldsymbol{\beta}_i = [\mathbf{T}]\boldsymbol{\alpha}_\mathbf{i} \quad i = 1, \ldots, mi, \quad (6.88)$$

where $[\mathbf{T}]$ is a wavelet transform matrix.

2. Second-order Volterra kernel approximation

The second-order output of the multi-input single-output system is obtained by the same discrete scale as the input/output

$$y_2(t_n) = \sum_{i_1,i_2=1}^{mi}\int_0^{t_n}\int_0^{t_n} h_2^{i_1 i_2}(\xi,\eta)u^{i_1}(t_n-\xi)u^{i_2}(t_n-\eta)d\xi d\eta. \quad (6.89)$$

The input signal is approximated by a zero-order input,

$$u_j^{i_1}(t_n-\xi)u_j^{i_2}(t_n-\eta) = \sum_{k=0}^{n-1}\sum_{m=0}^{n-1} u_{j,n-k-1}^{i_1}u_{j,n-m-1}^{i_2}\chi_{j,(k,m)}(\xi,\eta). \quad (6.90)$$

The second-order kernel is expanded on a single-scale j_2 using a two-dimensional (2D) tensor product scale function

$$\begin{aligned} h_{2,j_2}^{i_1,i_2}(\xi,\eta) &= \sum_{p,q}\sum_{s,v=1}^{4} \alpha_{j_2,(p,q)}^{(i_1,i_2)(s,v)} \Phi_{j_2,(p,q)}^{(s,v)}(\xi,\eta) \\ &= \sum_{p,q}\sum_{s,v=1}^{4} \alpha_{j_2,(p,q)}^{(i_1,i_2)(s,v)} \phi_{j_2,p}^{s}(\xi)\phi_{j_2,q}^{v}(\eta) \end{aligned}, \tag{6.91}$$

Assume the memory length of second-order kernel as T_2, then the support region of the second-order kernel is $[0,T_2]\times[0,T_2]$. The total number of scale functions is $N_2=(2^{2+j_2}T_2+1)^2$ in eq. (6.91). Same as the single-scale expansion for first-order kernel, the scaling function is boundary-adaptive scaling function. By substituting eqs. (6.90) and (6.91) into eq. (6.89), discrete second-order output is obtained:

$$\begin{aligned} y_{2,j}(t_n) = &\sum_{i_1 i_2=1}^{mi}\sum_{k,m=0}^{n_2-1}\sum_{p,q}\sum_{s,v=1}^{4} u_{j,n-k-1}^{i_1} u_{j,n-m-1}^{i_2} \alpha_{j_2,(p,q)}^{(i_1,i_2)(s,v)} \\ &\int_0^{T_2}\int_0^{T_2} \Phi_{j_2,(p,q)}^{(s,v)}(\xi,\eta)\chi_{j,(k,m)}(\xi,\eta)d\xi d\eta \end{aligned}, \tag{6.92}$$

where the expression of n_2 is similar to n_1

$$n_2=\begin{cases} n & n\le 2^j T_2 \\ 2^j T_2 & n\ge 2^j T_2 \end{cases}. \tag{6.93}$$

Rewrite eq. (6.92) in a matrix form

$$y_{2,j}=\underline{\mathbf{U}}_2\underline{\alpha}_2=[\mathbf{U}_1\ \ \mathbf{U}_2\ \ \cdots\ \ \mathbf{U}_{mi(mi-1)}\ \ \mathbf{U}_{mi.mi}][\boldsymbol{\alpha}_1^T\ \ \boldsymbol{\alpha}_2^T\ \ \cdots\ \ \boldsymbol{\alpha}_{mi(mi-1)}^T\ \ \boldsymbol{\alpha}_{mi.mi}^T]^T, \tag{6.94}$$

where

$$\boldsymbol{\alpha}_i=\mathbf{vec}\left(\sum_{p,q}\sum_{s=1}^{4}\alpha_{j_2,(p,q)}^{(i_1,i_2)(s,v)}\right) \qquad i=1,\ldots,mi\cdot mi, \tag{6.95}$$

$$\begin{aligned} \mathbf{U}_i(n,o) = &\sum_{k,m=0}^{n_2-1} u_{j,n-k-1}^{i_1} u_{j,n-m-1}^{i_2} \int_0^{T_2}\int_0^{T_2} \Phi_{j_2,(p,q)}^{(s,v)}(\xi,\eta)\chi_{j,(k,m)}(\xi,\eta)d\xi d\eta \\ &\begin{cases} i=1,\ldots,mi \\ n=1,\ldots,N \\ o=o(s,v,p,q) \end{cases}. \end{aligned} \tag{6.96}$$

In eq. (6.96), o is the position of $\alpha_{j_1,p}^{(i_1,i_2)(s,v)}$ in $\boldsymbol{\alpha}_i$. The quadratic integral is calculated by

$$\begin{aligned}
&\int_0^{T_2}\int_0^{T_2} \Phi_{j_2,(p,q)}^{(s,v)}(\xi,\eta)\chi_{j,(k,m)}(\xi,\eta)d\xi d\eta \\
&= 2^j \int_{2^{-j}k}^{2^{-j}(k+1)} \int_{2^{-j}m}^{2^{-j}(m+1)} \phi_{j_2,p}^{S}(\xi)\,\phi_{j_2,q}^{v}(\eta)d\xi d\eta \\
&= 2^j \int_{2^{-j}k}^{2^{-j}(k+1)} \phi_{j_2,p}^{S}(\xi)d\xi \int_{2^{-j}m}^{2^{-j}(m+1)} \phi_{j_2,q}^{v}(\eta)d\eta \\
&= 2^j JF_1 \cdot JF_2
\end{aligned} \tag{6.97}$$

JF_1 is also calculated using the Gauss–Legendre numerical integration method. After the derivation of the first- and second-order outputs, the discrete output of the system is

$$\mathbf{y}_j = [\underline{\mathbf{U}}_1\mathbf{T}^{-1} \quad \underline{\mathbf{U}}_2]\begin{bmatrix}\underline{\boldsymbol{\beta}}_1 \\ \underline{\boldsymbol{\alpha}}_2\end{bmatrix}. \tag{6.98}$$

It is worth noting that the discrete wavelet transform is not performed on the 2D scale coefficient because the multiwavelet method has expensive computation cost in the 2D discrete wavelet transform. The coarser scale j_2 is generally chosen to reduce the computational time.

6.2.6 Adaptive QR decomposition recursive least squares algorithm

The Volterra series eq. (6.98) in the wavelet domain is derived from the previous section, which can be considered as a pseudo-linear Finite impulse response (FIR) adaptive filter, where $[\underline{\mathbf{U}}_1\mathbf{T}^{-1} \; \underline{\mathbf{U}}_2]$ is the input vector, $[\underline{\boldsymbol{\beta}}_1 \; \underline{\boldsymbol{\alpha}}_2]^T$ the adaptive parameter and output $\mathbf{y}_j$ for the desired signal. The main idea of the adaptive filter [19, 36] is to automatically adjust the filter coefficients to the optimal state according to the difference between the output signal and the actual observed signal. The flowchart is shown in Fig. 6.7.

Adaptive filtering algorithms [36] can be divided into two categories: least mean square (LMS) method and recursive least squares (RLS) method. The LMS method uses the instantaneous value of the data error correlation matrix to estimate the error

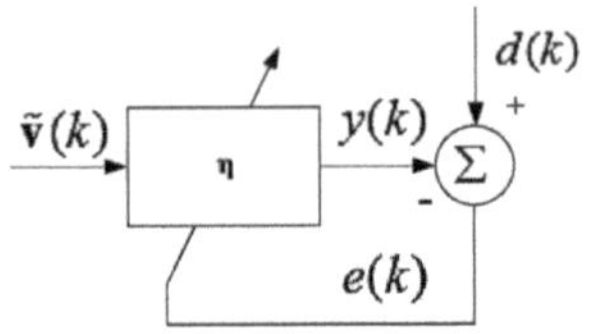

Fig. 6.7: Diagram of an adaptive filter.

gradient vector, thereby adaptively to adjust the weighting vector size. Although it is simple to implement, the convergence speed is relatively slow, and it is very sensitive to the change of the eigenvalue expansion degree (the ratio of the maximum eigenvalue to the minimum eigenvalue) of autocorrelation matrix of the input data.

The RLS algorithm uses the recursive formula of Kalman filter to derive the adaptive filter weight vector update equation. It has faster convergence speed than LMS algorithm and is insensitive to eigenvalue expansion. However, the inverse matrix of the time-averaged autocorrelation matrix loses nonnegative characterization due to the accumulated numerical error, causing the algorithm to diverge rapidly. In order to improve the stability of the RLS algorithm, the QR decomposition is introduced to directly decompose the input matrix, which effectively reduces the conditional number of the autocorrelation matrix and improves the stability of the algorithm. The details of the QRD-RLS algorithm are

1) When $k = 0$, the upper diagonal matrix $\mathbf{U}(0) = \delta \begin{bmatrix} & & 1 \\ & \cdot^{\cdot^{\cdot}} & 1 \\ 1 & & \end{bmatrix}$ is initialized, where δ is a small amount.
2) For each $k = 1, 2, \ldots,$

$$\begin{bmatrix} \mathbf{0}^T \\ \mathbf{U}(k) \end{bmatrix} = \mathbf{Q}_\theta(k) \begin{bmatrix} \bar{\mathbf{v}}^T(k) \\ \lambda^{1/2}\mathbf{U}(k-1) \end{bmatrix}, \tag{6.99}$$

$$\mathbf{Q}_{\theta i}(k) = \begin{bmatrix} \cos\theta_i(k) & 0 & -\sin\theta_i(k) & 0 \\ 0 & I_i & 0 & 0 \\ \sin\theta_i(k) & 0 & \cos\theta_i(k) & 0 \\ 0 & 0 & 0 & I_{N-i} \end{bmatrix}. \tag{6.100}$$

The matrix $\mathbf{Q}_\theta$ is a unitary matrix obtained by continuous Givens transform. Matrix $\mathbf{Q}_\theta$ makes the new input signal vector rotating onto the upper triangular matrix of the main diagonal.
3) Perform continuous Givens transformation on the desired output to obtain a priori error e_{q1}

$$\begin{bmatrix} e_{q1} & (k) \\ \mathbf{d}_{q2} & (k) \end{bmatrix} = \mathbf{Q}_\theta(k) \begin{bmatrix} d(k) \\ \lambda^{1/2}\mathbf{d}_{q2}(k-1) \end{bmatrix}. \tag{6.101}$$

4) Iterate the filter coefficients backward

$$\mathbf{U}(k)\eta(k) = d_{q2}(k). \tag{6.102}$$

5) Calculate the posterior error

$$\varepsilon(k) = e_{q1}(k)\,\gamma(k), \quad \gamma(k) = \prod_{i=0}^{N} \cos\theta_i(k), \tag{6.103}$$

where γ is the rotation factor, defined as the ratio of a posteriori error to a priori error. λ is a forgetting factor which is not greater than unity. It should be noted that the results show that for the linear Volterra series, the pseudo-inverse method for the Volterra kernel is more accurate than the QRD-RLS algorithm. Therefore, the QRD-RLS algorithm is only used to identify the nonlinear Volterra series [9].

6.2.7 Numerical examples

1. Flutter analysis of vertical tail with control surface based on the first-order Volterra series

In this study, linear Volterra series is used to simulate the unsteady aerodynamics, and the coupled structural dynamics equation is used to analyze the flutter with the control surface [40]. Figure 6.8 shows the aerodynamic mesh of the control surface. The Mach number is $M_\infty = 0.6$; angle of attack is $\alpha = 0.0°$. The spatial discretization adopts the Roe scheme, and the time advancing technique uses the LU-SGS-τTs method. The grid is divided into seven blocks, and the total number of grids is about 180,000.

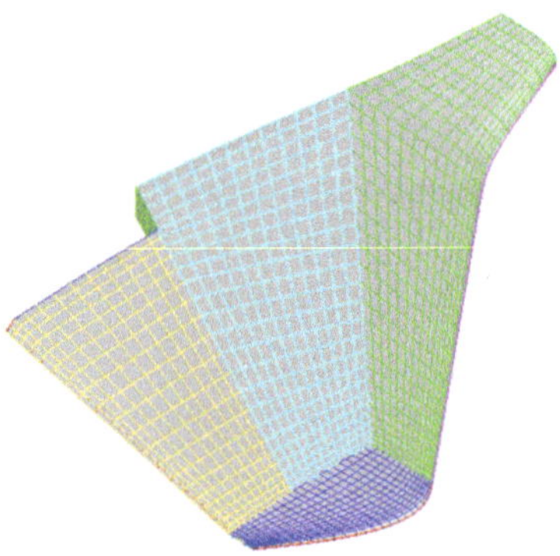

Fig. 6.8: Aerodynamic grid for control surface.

Figure 6.9 is a pressure coefficient contour of the control surface in a steady state. For further flutter analysis, the finite element modal of control surface and the vertical tail structure are interpolated to the aerodynamic mesh using infinite-plate splines interpolation method. In order to obtain reliable identification results, a multifrequency sinusoidal signal is used (as shown in Fig. 6.10). The first, third and fifth-order modals use

Fig. 6.9: Surface pressure coefficient distribution at $M_\infty = 0.6$.

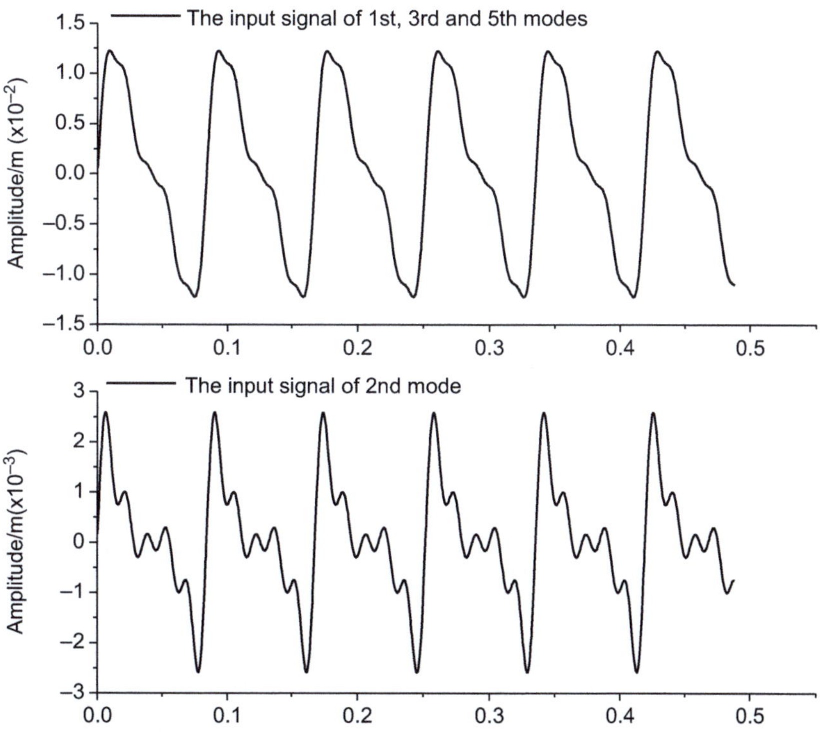

Fig. 6.10: Identification of the input signal.

the same input signal, whilst the second-order mode adopts another type of signal. It is because the second-order mode is mainly based on the rigid deflection of the control surface, and considering the second-order modal-generalized displacement amplitude is of the order of 0.001 m, the magnitude of the second-order modal identification signal is slightly smaller than the others.

In the computation, the input/output signal sampling scale is j_s=12, and the first-order kernel discrete scale is $j_1 = 11$. The first-order kernel memory time $T_1 = 100 \times 2^{j_1}$ s, while the input signal action time is $T_I = 2,000 \times 2^{j_s}$ s. The dynamic pressure is $Q_{\text{ref}} =$ 215,000 Pa. The results are shown in Fig. 6.11. The results are in good agreement with full model, with an error of 0.5% at this time.

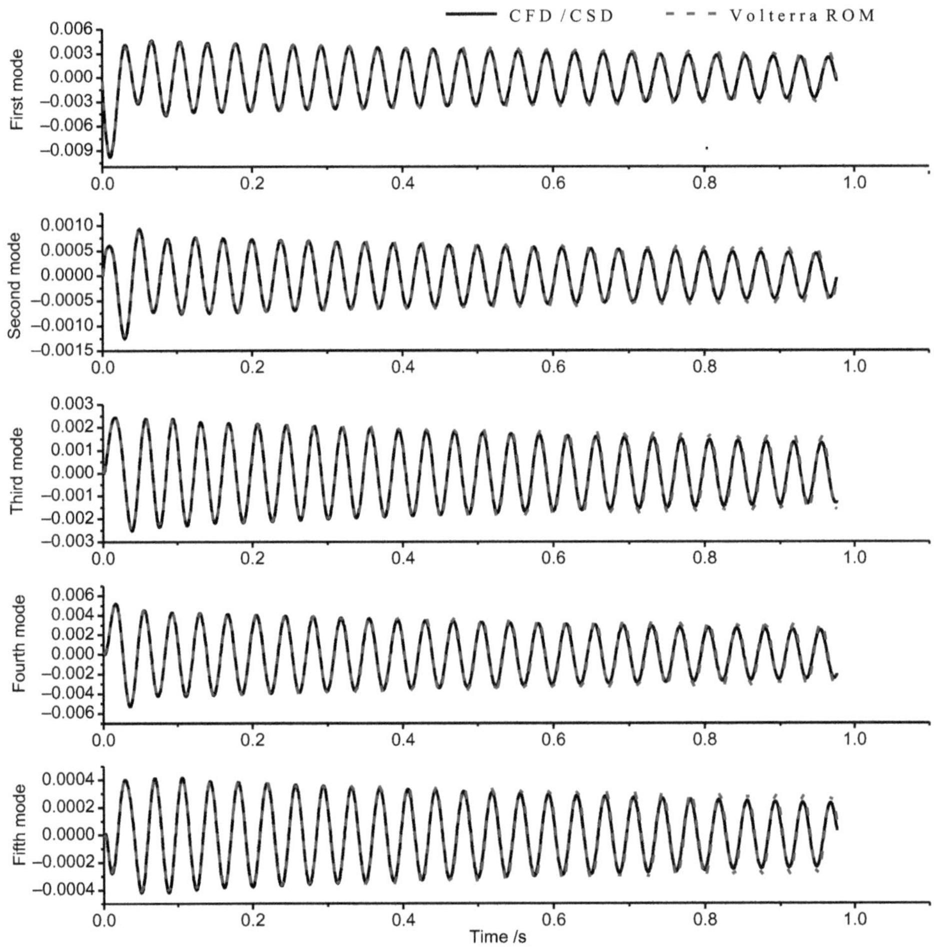

Fig. 6.11: Comparison of generalized displacements from the first to the fifth order ($Q_{\text{ref}} = 215,000$ Pa).

By comparing the two cases in this example, the unsteady aerodynamic ROM based on the Volterra series is of high accuracy and agrees well with CFD/CSD coupling results under small perturbations. The dynamic pressure error of the ROM and the

CFD/CSD coupling are both within 1%, which can meet the requirements of aeroelastic analysis.

However, as time advances, the displacement error is gradually accumulating during the mutual influence in the aeroelastic computation, causing increasing difference between the structural response of ROM and the CFD/CSD coupling result.

2. Two degrees of freedom nonlinear forced oscillation of airfoil based on multi-input Volterra series method

For the single degree-of-freedom motion, to construct a second-order Volterra series reduction model only needs the autocorrelation first-order kernel and second-order kernel [8, 42]. When multiple degrees of freedom case is considered, as the nonlinear system no longer satisfies the principle of linear superposition, it is necessary to introduce a cross-kernel to consider the coupling between multiple degrees of freedom. In this section, the fourth-order generalized aerodynamic force is identified with the consideration of the first and second-order modal motion of the AGARD445.6 wing. The Mach number is $M_\infty = 0.96$, initial angle of attack is set $\alpha = 4.0°$. The equation of motion is

$$\begin{cases} \xi_1 = 0.01\sin(40\pi t) \\ \xi_2 = 0.01\sin(32\pi t) \end{cases}. \tag{6.104}$$

The sampling scale is $j_S = 12$. The first-order kernel expansion scale is $j_1 = 9$, while the second-order kernel expansion scale is $j_2 = 7$, and the cross-kernel expansion scale is $j_c = 7$. Assume that the sampling time of first-order kernel is $T_1 = 20 \times 2^{-j_1}$ s, for the second-order kernel is $T_2 = 1.5 \times 2^{-j_2}$ s and time for the cross-kernel is the same as the second-order kernel.

The cross-kernel can be identified by multiinput Volterra series identification with synthesis input signals in each channel. The two different pseudo-random sequences are used with a Kaiser low-pass filter, in which passband frequency is $\omega_p = 3 \times 40\pi$ rad, stopband frequency $\omega_s = 4 \times 40\pi$ rad. The filtered input signal is shown in Fig. 6.12. The QRD-RLS algorithm is used to identify the Volterra series, in which the coefficient is set $\delta = 0.0008$ in the initial upper diagonal matrix, and the forgetting factor $\lambda = 0.995$. After iterating 4,000 steps, the adaptive filter coefficients are obtained, that is, the coefficients of Volterra kernel scaling function in the wavelet domain. Figure 6.13 compares the actual output and the desired output after iteration. It is found that the error between actual output and the expected output agree reasonably except the disparity at the beginning time. In order to verify the model of the multiinput Volterra series, the input convolution of Volterra kernel in eq. (6.104) is used to compare with the CFD/CSD coupling result shown in Fig. 6.14. The convolution of the first-order kernel is corresponding to the first-order components in the response, that is, the components of $f \approx 16$ Hz and $f \approx 20$ Hz, while the convolution of the autocorrelation second-order kernel is related to the quadratic

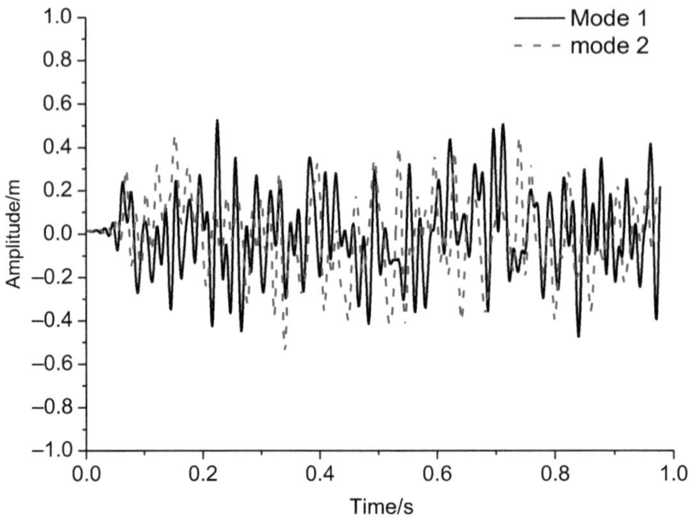

Fig. 6.12: Identification of the input signal.

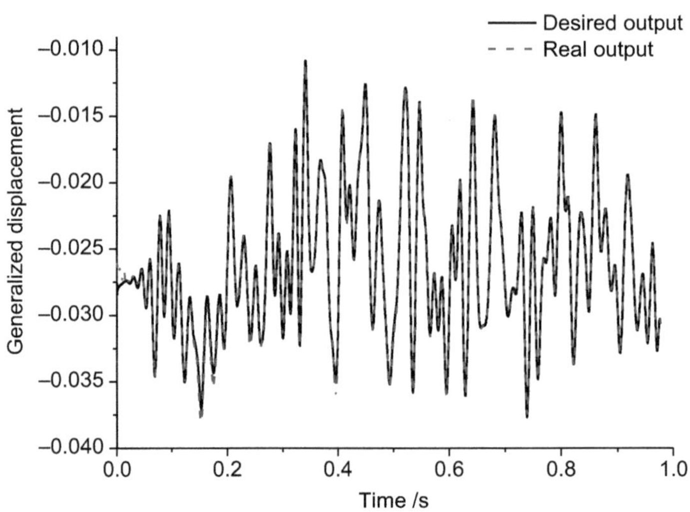

Fig. 6.13: Comparison between actual output and expected output.

autocorrelation component in the response, that is, the components of $f \approx 32$ Hz and $f \approx 40$ Hz. The convolution of cross-kernel is for the quadratic cross-correlation components in response with $f \approx 20 - 16 = 4$ Hz and $f \approx 20 + 16 = 36$ Hz.

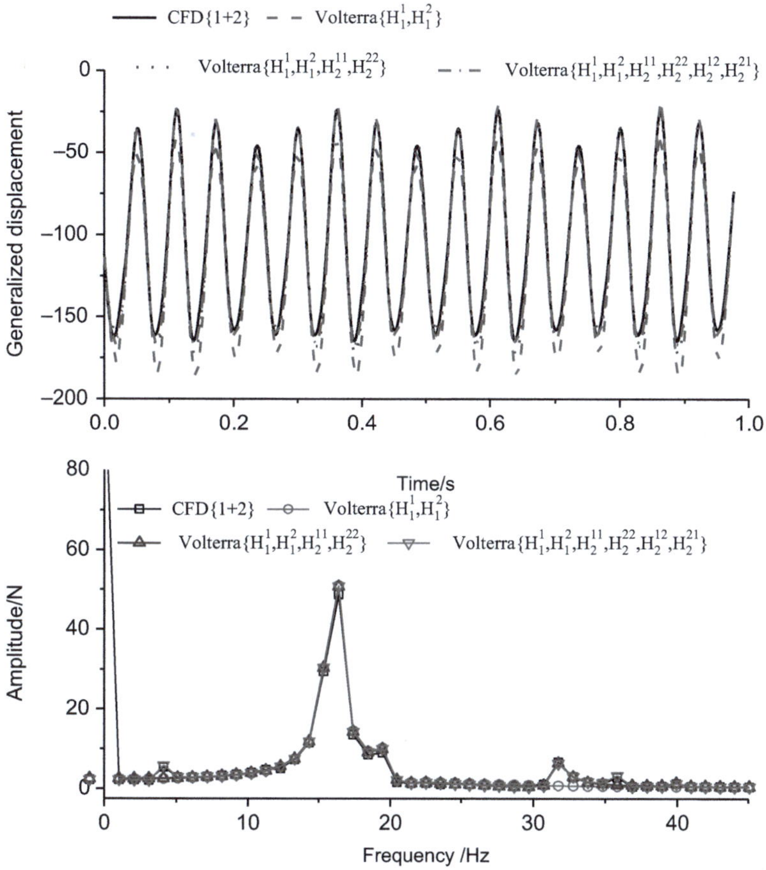

Fig. 6.14: Comparison of second-order Volterra series model and CFD results.

6.3 ROMs for unsteady aerodynamic forces based on state space

Although the Volterra series can well describe the response under any input, the expression of the series is not easy to use. This is because the stability of the system cannot be judged by the eigenvalue of the system. On the other hand, it is difficult for the design of the control system in series form.

Therefore, a ROM of the state space would be an option. In this section, two unsteady aerodynamic ROMs based on state space are discussed, the pulse/eigensystem realization algorithm (ERA) method and the single-composite input (SCI)/ ERA method [20, 31]. The aeroelastic equation is obtained by coupling with structural dynamics equation. The efficiency and accuracy of the two models are compared in the case of flutter analysis of the vertical wing with control surface.

6.3.1 Pulse/ERA method

The ERA algorithm is a time domain identification method for multi-input multi-output system. It only needs a free response data in a short time period as the identification data, with less computational time and strong ability for low frequency, dense frequency and refrequency identification. Moreover, it can achieve the minimum realization of the system for the convenience of control design. The main idea of the ERA algorithm is to construct a Hankel matrix using impulse response for a minimum implementation of the system by singular value decomposition, and then transform into an eigenvalue normal form.

In the case of small perturbations, unsteady aerodynamic forces can be represented by the following linear time-invariant discrete state space model

$$\begin{cases} \mathbf{X}_A(n+1) = \mathbf{A}_A\mathbf{X}_A(n) + \mathbf{B}_A\xi(n) \\ \mathbf{F}_A(n) = \mathbf{C}_A\mathbf{X}_A(n) + \mathbf{D}_A\xi(n) \end{cases}, \tag{6.105}$$

where $\boldsymbol{\xi}(n)$ is the generalized displacement of the structure, $\mathbf{X}_A(n)$ is the aerodynamic state variable, and $\mathbf{F}_A(\mathrm{n})$ is the generalized aerodynamic coefficient. $\mathbf{A}_A$, $\mathbf{B}_A$, $\mathbf{C}_A$, $\mathbf{D}_A$ denotes the system matrix, the input matrix, the output matrix and the feedforward matrix, respectively. Then the system's impulse response output is

$$\mathbf{F}_A(n) = \begin{cases} \mathbf{D}_A & n=0 \\ \mathbf{C}_A\mathbf{A}_A^{n-1}\mathbf{B}_A & n \ge 1 \end{cases}. \tag{6.106}$$

For a linear system, impulse response of the zero initial state system is equal to the first-order kernel of linear Volterra series $\mathbf{h}_1(n)$. Let $\mathbf{h}_1(n)$ be an $N_o \times N_i$-order matrix and construct a Hankel matrix as follows:

$$\mathbf{H}(n-1) = \begin{bmatrix} \mathbf{h}_1(n) & \mathbf{h}_1(n+1) & \mathbf{h}_1(n+2) & \cdots & \mathbf{h}_1(n+M-1) \\ \mathbf{h}_1(n+1) & \mathbf{h}_1(n+2) & \mathbf{h}_1(n+3) & \cdots & \mathbf{h}_1(n+M) \\ \mathbf{h}_1(n+2) & \mathbf{h}_1(n+3) & \mathbf{h}_1(n+4) & \cdots & \mathbf{h}_1(n+M+1) \\ \vdots & \vdots & \vdots & \vdots & \vdots \\ \mathbf{h}_1(n+K-1) & \mathbf{h}_1(n+K) & \mathbf{h}_1(n+K+1) & \cdots & \mathbf{h}_1(n+K+M+1) \end{bmatrix}, \tag{6.107}$$

where K and M are chosen as the appropriate integer value for the minimum implementation. The dimension of matrix $\mathbf{H}(0)$ is $KN_O \times MN_i$ and $MN_i < KN_o$. Singular value decomposition of $\mathbf{H}(0)$ is

$$\mathbf{H}(0) = \mathbf{U}\boldsymbol{\Sigma}\mathbf{V}^T. \tag{6.108}$$

The first $MN_i \times MN_i$ order singular value matrix Σ and the first MN_i-column matrix $\mathbf{U}$ are reserved. Let

$$\begin{cases} \mathbf{E}_{N_O}^T = \begin{bmatrix} \mathbf{I}_{N_O} & \mathbf{0}_{N_O} & \cdots & \mathbf{0}_{N_O} \end{bmatrix}_{N_O \times KN_O} \\ \mathbf{E}_{N_i}^T = \begin{bmatrix} \mathbf{I}_{N_i} & \mathbf{0}_{N_i} & \cdots & \mathbf{0}_{N_i} \end{bmatrix}_{N_i \times MN_i} \end{cases}, \tag{6.109}$$

then the matrix in eq. (6.105) is

$$\begin{cases} \mathbf{A}_A = \Sigma^{-1/2}\mathbf{U}^T\mathbf{H}(1)\mathbf{V}\Sigma^{-1/2} \\ \mathbf{B}_A = \Sigma^{1/2}\mathbf{V}^T\mathbf{E}_{N_i} \\ \mathbf{C}_A = \mathbf{E}_{N_O}^T\mathbf{U}\Sigma^{1/2} \\ \mathbf{D}_A = \mathbf{h}_1(0) \end{cases} \tag{6.110}$$

6.3.2 SCI/ERA method

In the pulse/ERA algorithm described in Section 6.3.1, the structural impulse response required for identification needs to be calculated separately. Thus, the time for model construction is proportional to that of the order of structural modal, that is, the more coupled modes, the longer the construction consumes. In order to overcome the shortcomings of the pulse/ERA method, Silva proposed using a pulse algorithm to extract modal impulse responses from a single input response. Then, Kim proposed the SCI/ERA method, which can be used for the extraction of structure impulse response from one input signal. This method is explicitly independent on structure impulse response. Figure 6.15 shows the flowchart of the pulse/ERA method and the SCI/ERA method. Considering a small perturbation case, state space model for the unsteady aerodynamic force is represented by eq. (6.105) and constructs the following single-signal composite input:

$$\begin{cases} \mathbf{b}_{\mathrm{SCI}}^n = \sum_{i=1}^{N_i} \mathbf{b}_i r_i^n \\ \mathbf{d}_{\mathrm{SCI}}^n = \sum_{i=1}^{N_i} \mathbf{d}_i r_i^n \end{cases}, \tag{6.111}$$

where $\mathbf{b}_i$ is the ith column vector of $\mathbf{B}$; $\mathbf{d}_i$ is the ith column vector of $\mathbf{D}$; r_i^n is arbitrary value. With eq. (6.111), the output of eq. (6.105) is $\mathbf{y}^n, n = 0,1,\ldots,M-1$. Define the reduced dimension of the state variable measured as

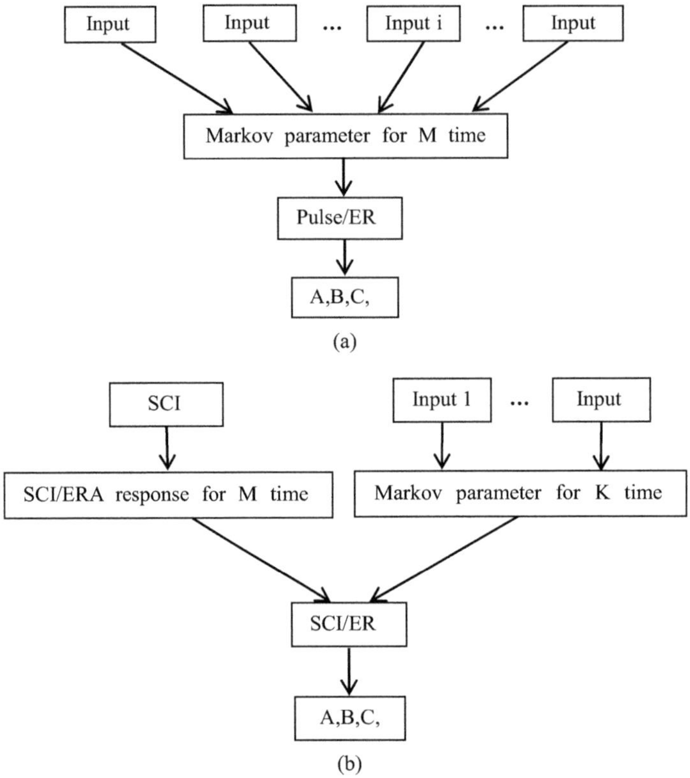

Fig. 6.15: Flowchart of pulse/ERA and SCI/ERA methods: (a) impulse/ERA and (b) SCI/ERA.

$$\begin{aligned}\mathbf{y}_{ck}^{n} &= \mathbf{C}\mathbf{A}^{k}\mathbf{x}^{n} \\ &= \mathbf{y}^{n+k} - \sum_{i=1}^{N_i} \mathbf{y}_i^0 r_i^{n+k} - \sum_{i=1}^{N_i} \mathbf{y}_i^1 r_i^{n+k-1} - \cdots - \sum_{i=1}^{N_i} \mathbf{y}_i^k r_i^n . \\ &= \mathbf{y}^{n+k} - \sum_{j=1}^{k+1} \sum_{i=1}^{N_i} \mathbf{y}_i^{j-1} r_i^{n+k+1-j} \quad (k=0,1,2,\ \ldots, K+1)\end{aligned} \tag{6.112}$$

Similar to the pulse/ERA method by constructing a Hankel matrix with impulse response, the measurement of reduced dimension for the state variable constructs the following Hankel matrices $\mathbf{H}_{c0}$ and $\mathbf{H}_{c1}$:

$$\mathbf{H}_{c0} = \begin{bmatrix} \mathbf{y}_{c0}^1 & \mathbf{y}_{c0}^2 & \cdots & \mathbf{y}_{c0}^{M-1} \\ \mathbf{y}_{c1}^1 & \mathbf{y}_{c1}^2 & \cdots & \mathbf{y}_{c1}^{M-1} \\ \cdots & \cdots & \cdots & \cdots \\ \mathbf{y}_{cK}^1 & \mathbf{y}_{cK}^2 & \cdots & \mathbf{y}_{cK}^{M-1} \end{bmatrix}, \tag{6.113}$$

$$\mathbf{H}_{c1} = \begin{bmatrix} \mathbf{y}_{c1}^1 & \mathbf{y}_{c1}^2 & \cdots & \mathbf{y}_{c1}^{M-1} \\ \mathbf{y}_{c2}^1 & \mathbf{y}_{c2}^2 & \cdots & \mathbf{y}_{c2}^{M-1} \\ \cdots & \cdots & \cdots & \cdots \\ \mathbf{y}_{cK+1}^1 & \mathbf{y}_{cK+1}^2 & \cdots & \mathbf{y}_{cK+1}^{M-1} \end{bmatrix}. \tag{6.114}$$

Singular value decomposition of $\mathbf{H}_{c0}$ is expressed as follows:

$$\mathbf{H}_{c0} = \mathbf{U}\Sigma\mathbf{V}^T = [\,\mathbf{U}_1 \quad \mathbf{U}_2\,] \begin{bmatrix} \Sigma_1 & 0 \\ 0 & 0 \end{bmatrix} \begin{bmatrix} \mathbf{V}_1^T \\ \mathbf{V}_2^T \end{bmatrix}. \tag{6.115}$$

Then the state matrix, the input matrix, the output matrix and the feedforward matrix of the ROMs are

$$\begin{cases} \mathbf{D} = \left[\mathbf{I}_{N_O \times N_O} \quad 0 \quad \cdots \quad 0\right]_{N_O \times (K+2)N_O} \mathbf{Y}_c^0 \\ \mathbf{C} = \left[\mathbf{I}_{N_O \times N_O} \quad 0 \quad \cdots \quad 0\right]_{N_O \times (K+1)N_O} \mathbf{U}_1 \Sigma_1^{1/2} \\ \mathbf{B} = \Sigma_1^{-1/2} \mathbf{U}_1^T \mathbf{Y}_c^1 \\ \mathbf{A} = \Sigma_1^{-1/2} \mathbf{U}_1^T \mathbf{H}_{c1} \mathbf{V}_1 \Sigma_1^{-1/2} \end{cases}, \tag{6.116}$$

where $\mathbf{Y}_c^0$ and $\mathbf{Y}_c^1$ are initial state responses of the system under unit pulse signal of the ith-order structure modal:

$$\mathbf{Y}_c^n = \begin{bmatrix} \mathbf{y}_1^n & \mathbf{y}_2^n & \cdots & \mathbf{y}_{N_i}^n \\ \mathbf{y}_1^{n+1} & \mathbf{y}_2^{n+1} & \cdots & \mathbf{y}_{N_i}^{n+1} \\ \cdots & \cdots & \cdots & \cdots \\ \mathbf{y}_1^{n+K} & \mathbf{y}_2^{n+K} & \cdots & \mathbf{y}_{N_i}^{n+K} \end{bmatrix} \quad (n = 0, 1), \tag{6.117}$$

$$\begin{cases} \mathbf{y}_i^n = \mathbf{d}_i, n = 0 \\ \mathbf{y}_i^n = \mathbf{C}\mathbf{A}^{n-1}\mathbf{b}_i, n \neq 0 \end{cases} \quad i = 1, 2, \ldots, N_i. \tag{6.118}$$

In order to compare the computational efficiency of the SCI/ERA method and the pulse/ERA method, the dimension of $\mathbf{H}_{c0}$ in the pulse/ERA method is set $KN_o \times MN_i$, and the method needs to calculate at least $(1+K+M)N_i$ times for systems response. In the SCI/ERA method, on the other hand, the total number of responses required to be calculated is $(1+K+M) + (K+1)N_i$ times (including the M-step SCI response and the first K+1 step impulse response). Two methods have difference on computational step by $\Delta_{\text{step}} = MN_i-(1+K+M)$ and the number of $\mathbf{H}_{c0}$ elements by $\Delta_{\text{elm}} = K(M-1)N_oN_i$. It is found that both Δ_{step} and Δ_{elm} are proportional to the number of structural modals with scaling factors M and $K(M-1)N_o$. The matrices K, M and N_i are generally on the order of 10^2 for engineering applications. Therefore, the SCI/ERA method saves considerable computational costs.

6.3.3 Numerical examples

In this section, the pulse/ERA method is used to construct the state space model for the flutter analysis of the tail with the control surface [42]. The Hankel matrix is constructed by first-order linear Volterra kernel. The sampling parameters are $K = 90$ and $M = 50$, respectively. Based on the model from Singular Value Decomposition (SVD), the aeroelastic response at dynamic pressure $Q_{ref} = 215{,}000$ Pa is compared with the CFD/CSD coupling result (see Fig. 6.16). It is found that even after more than 20 time cycles, the structural response still agrees well. The results indicate that the ROM can accurately represent the dynamics of the full-order system. After obtaining the state space model, we can observe the characteristics of the system through the response in

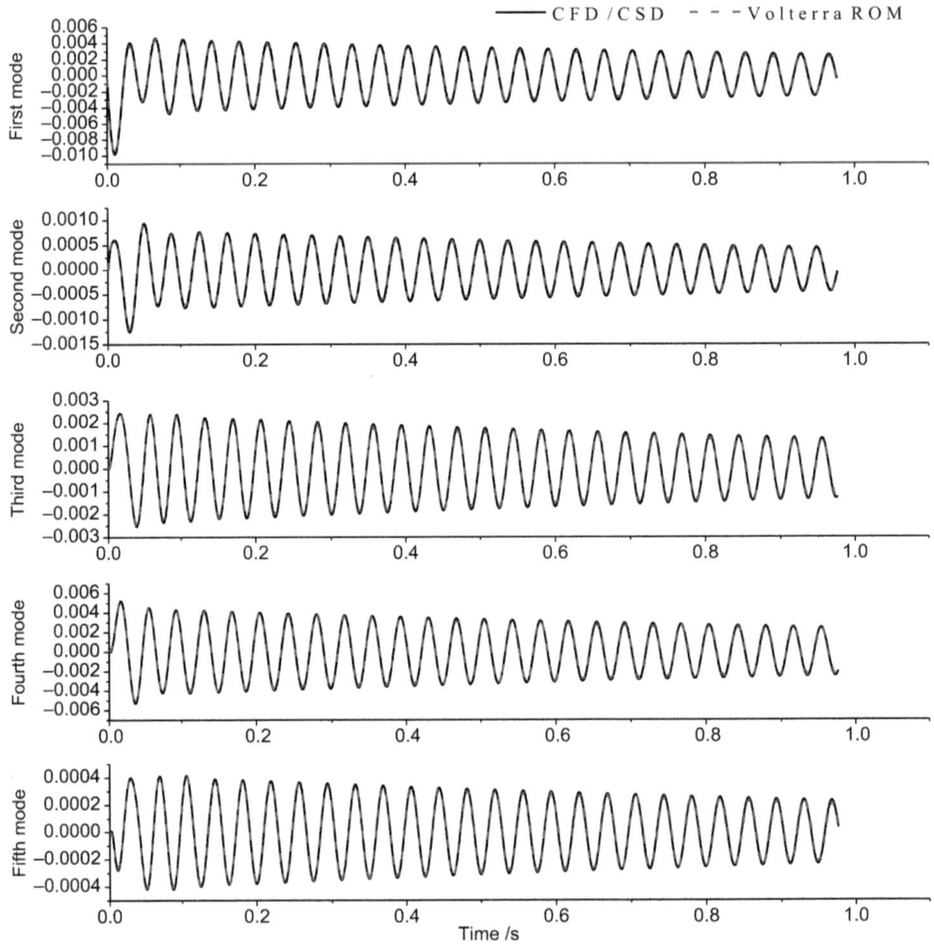

Fig. 6.16: First- to fifth-order generalized displacement ($Q_{ref} = 215{,}000$ Pa).

the time domain. Furthermore, the model can be used to judge the stability of the system through its eigenvalues. Figure 6.17 shows the root locus map of the aeroelastic system with respect to the increase of dynamic pressure. It can be found that the system changes from stable to unstable with the increase of dynamic pressure, and the third-order mode of the structure is the cause for system instability. By comparing Figs. 6.11 and 6.17, it can be seen that ROM using state space is more accurate than the model based on Volterra series because the calculation of the generalized aerodynamics at a time step in Volterra series convolution operations uses the current and

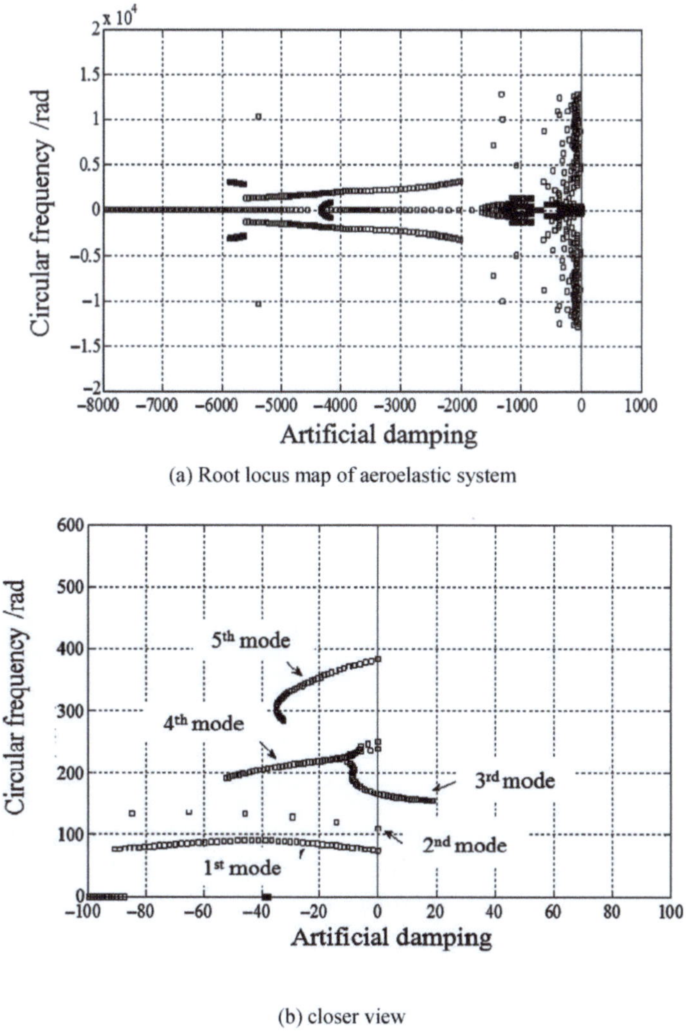

Fig. 6.17: Root locus diagram of aeroelastic system: (a) root locus map of aeroelastic system and (b) closer view.

previous N_1 time step which includes the previous N_1 step error. In the state space model, the generalized aerodynamic force of the current step only needs the displacement of the previous step, and accordingly the error accumulation is only related to the error of the previous step. That is the reason that the ROM constructed by the state space model has higher precision.

6.4 Reduced-order model of aeroelastic system based on POD method

POD method is a modal-based reduced-order method by extracting a set of orthogonal basis from a series of sampled data (or called snapshots) and mapping the original system to the low-order subspace spanned by the basis to obtain a low-order model of the system [1, 6, 10, 21, 29, 30]. Appropriate selection of the subspace dimension can guarantee the low-order model to accurately reflect the dynamic features of the original system. This method has been applied to many fields such as image processing, signal analysis and fluid mechanics. Lumley [23] first introduced the POD method in turbulence research. Kenneth and Jeffrey [15] applied POD to the ROMs for aeroelasticity of the airfoil in the time domain and frequency domain Euler equations, respectively. Subsequently, the POD method was extended to flow control problems and aeroelastic analysis of three-dimensional (3D) aerodynamic configuration with complex geometry. Pettit and Beran [5] used POD to establish nonlinear ROMs of the Euler equations and studied the nonlinear flutter characteristics of transonic panels.

According to the construction method of POD model, this section is divided into two parts. The first part is based on the traditional POD method to establish ROMs for a small perturbation linearized flow field, which is used for aeroelastic problem. The other part is based on POD-Galerkin method to establish a ROM for nonlinear unsteady compressible flow.

6.4.1 POD-based snapshot method

Define $Q = \{\mathbf{q}^m \in H; m = 1, \ldots, M\}$ as the set of snapshots in the Hilbert space, and snapshot $\mathbf{q}^m$ is the solution of the full-order equation at time t_m, where $\mathbf{q}_i^m = [q_{i1}^m, \cdots q_{inv}^m]^T$ on grid element Ω_i. For 2D problems, $nv = 4$, and for 3D problem, $nv = 5$. Define the inner product in the Hilbert square-integrable space as

$$(\mathbf{q}^i, \mathbf{q}^j) = \int_\Omega \sum_{k=1}^{nv} q_k^i q_k^j d\Omega. \tag{6.119}$$

Define $\|\mathbf{q}\|^2 = (\mathbf{q}, \mathbf{q})$. Snapshot $\mathbf{q}^{\mathrm{m}}$ can be expressed as the sum of the average value and perturbation, that is,

$$\mathbf{q}^m = \bar{\mathbf{q}} + \tilde{\mathbf{q}}^m, \tag{6.120}$$

where $\bar{\mathbf{q}} = <\mathbf{q}^{\mathrm{m}}>$ and $< >$ represents the time average value.

The goal of POD method is to find a low-dimensional orthogonal subspace $\Psi = \mathrm{span}\{\boldsymbol{\varphi}^1, \cdots, \boldsymbol{\varphi}^r\}$ to represent a high-dimensional variable $\tilde{\mathbf{q}}^m$ in the full-order space

$$\tilde{\mathbf{q}}^m{}_{\mathrm{POD}} = \sum_{j=1}^{r} a_j^m \boldsymbol{\varphi}^j, \tag{6.121}$$

and further satisfies

$$\begin{cases} min < \| \tilde{\mathbf{q}}^m - \sum\limits_{j=1}^{r} (\tilde{\mathbf{q}}^m, \boldsymbol{\varphi}^j)\boldsymbol{\varphi}^j \|^2 > \\ (\boldsymbol{\varphi}^i, \boldsymbol{\varphi}^j) = \delta_{ij} \end{cases}, \tag{6.122}$$

where $\boldsymbol{\varphi}$ is the POD base (or POD mode). Sivorch proposed a method of snapshots to solve the earlier problem. According to the method of snapshots, the POD base can be represented by a linear combination of snapshots

$$\boldsymbol{\varphi}^j = \sum_{m=1}^{M} d_j^m \tilde{\mathbf{q}}^m. \tag{6.123}$$

Solving the coefficient vector $\mathbf{d}_j$ is equivalent to solving the following eigenvalue problem

$$\mathbf{R}\mathbf{d}_j = \lambda_j \mathbf{d}_j, \tag{6.124}$$

where the autocorrelation matrix is $\mathbf{R}_{i,j} = \left(\tilde{\mathbf{q}}^i, \tilde{\mathbf{q}}^j\right)$. Then, the POD base is expressed as

$$\boldsymbol{\varphi}^j = \frac{1}{\sqrt{\lambda_j}} \sum_{m=1}^{M} d_j^m \tilde{\mathbf{q}}^m. \tag{6.125}$$

Each POD base $\boldsymbol{\varphi}^j$ corresponds to a real eigenvalue λ_j. The physical meaning of λ_j representation is the contribution of vector $\boldsymbol{\varphi}^j$ to the snapshot matrix. The larger its value, the greater the contribution of $\boldsymbol{\varphi}^j$. The eigenvalues λ_j are arranged from large to small, $\lambda_1 > \lambda_2 > \cdots > \lambda_m$. The first r eigenvectors are substituted into eq. (6.121) to obtain the energy optimal approximation of the full-order system. It should be pointed out that for linearized equations with small perturbation, the inner product can be regarded as the inner product of the discrete vector, that is, the element volume of grid in eq. (6.119) is 1.

6.4.2 Reduced-order modeling of aeroelastic system based on POD method

For aeroelastic problems with small perturbation, a linearized model can be used to approximate the nonlinear flow field model. The linearized model is then reduced by applying a direct POD mapping method [19, 36, 38].

1. Linearization of aeroelastic system

The aeroelastic system is the coupling system between fluid and structure. The discrete governing equations for fluids and structures can be approximated by finite volume method and finite element method, respectively, as follows:

$$(\mathbf{A}(\mathbf{u})\mathbf{Q})_t + \mathbf{F}(\mathbf{Q}, \mathbf{u}, \mathbf{v}) = 0, \tag{6.126}$$

$$\mathbf{M}\mathbf{v}_t + \mathbf{C}\mathbf{v} + \mathbf{K}\mathbf{u} = \mathbf{f}^{\mathrm{ext}}(\mathbf{u}, \mathbf{v}, \mathbf{Q}), \tag{6.127}$$

where **u** and **v** represent the displacement and velocity vectors of the structure, respectively. **A** is the volume matrix of the fluid grid element; **Q** is the fluid conservation variable; **F** the nonlinear numerical flux function. **M, C** and K are the structural mass matrix, the damping matrix and the stiffness matrix, respectively. $\mathbf{f}^{\mathrm{ext}}$ is the equivalent load acting on the structure. The subscript t is the partial derivative with respect to time.

Here only ROM for aerodynamic force is considered. Hence, the POD method is used to approximate eq. (6.126) only. The reduced-order modeling for eq. (6.127) is based on modal superposition method. For the aeroelastic system, the equilibrium state of aeroelastic system is generally selected as the reference point for linearization of the equation. Takes the Euler equation as an example and performs a first-order Taylor expansion on the governing equation at equilibrium point

$$\begin{cases} (\mathbf{A}(\mathbf{u})\mathbf{Q})_t = (\mathbf{A}(\mathbf{u}_0)\mathbf{Q}_0)_t + \left(\left. \dfrac{\partial(\mathbf{A}\mathbf{Q})}{\partial \mathbf{u}} \right|_{t=t_0} \right) (\mathbf{u} - \mathbf{u}_0) + \mathbf{A}(\mathbf{u}_0)(\mathbf{Q} - \mathbf{Q}_0)_t + \cdots \\ \mathbf{F}(\mathbf{Q}, \mathbf{u}, \mathbf{v}) = \mathbf{F}(\mathbf{Q}_0, \mathbf{u}_0, \mathbf{v}_0) + \dfrac{\partial \mathbf{F}}{\partial \mathbf{Q}}(\mathbf{Q} - \mathbf{Q}_0) + \dfrac{\partial \mathbf{F}}{\partial \mathbf{u}}(\mathbf{u} - \mathbf{u}_0) + \dfrac{\partial \mathbf{F}}{\partial \mathbf{v}}(\mathbf{v} - \mathbf{v}_0) + \cdots \end{cases}. \tag{6.128}$$

Substitute eq. (6.128) into eq. (6.126):

$$\left(\frac{\partial \mathbf{A}}{\partial \mathbf{u}} \mathbf{Q}_0 \right) \tilde{\mathbf{v}} + \mathbf{A}_0 \tilde{\mathbf{Q}}_t + \frac{\partial \mathbf{F}}{\partial \mathbf{Q}} \tilde{\mathbf{Q}} + \frac{\partial \mathbf{F}}{\partial \mathbf{u}} \tilde{\mathbf{u}} + \frac{\partial \mathbf{F}}{\partial \mathbf{v}} \tilde{\mathbf{v}} = 0. \tag{6.129}$$

After merging similar items, it becomes

$$\mathbf{A}_0\tilde{\mathbf{Q}}_t + \frac{\partial \mathbf{F}}{\partial \mathbf{Q}}\tilde{\mathbf{Q}} + \frac{\partial \mathbf{F}}{\partial \mathbf{u}}\tilde{\mathbf{u}} + \left[\left(\frac{\partial \mathbf{A}}{\partial \mathbf{u}}\mathbf{Q}_0\right) + \frac{\partial \mathbf{F}}{\partial \mathbf{v}}\right]\tilde{\mathbf{v}} = 0. \tag{6.130}$$

The superscript ~ indicates the perturbation value. By considering the initial value $\mathbf{Q}_0$, $\mathbf{Q} = \mathbf{Q}_0 + \tilde{\mathbf{Q}}$. For the purpose of convenience, the superscript ~ is omitted. Then, eq. (6.130) can be written as follows:

$$\begin{cases} \mathbf{A}_0\mathbf{Q}_t + \mathbf{H}\mathbf{Q} + (\mathbf{E}+\mathbf{C})\mathbf{v} + \mathbf{G}\mathbf{u} = 0 \\ \mathbf{f}^{\mathrm{ext}} = 0.5\rho V^2 \mathbf{P}\mathbf{Q} \end{cases}, \tag{6.131}$$

where $\mathbf{P}$ is the derivative of the dimensionless external force with respect to the fluid conservation variable. The expression of the matrix in eq. (6.131) is

$$\begin{cases} \mathbf{H} = \dfrac{\partial \mathbf{F}}{\partial \mathbf{u}}(\mathbf{Q}_0, \mathbf{u}_0, \mathbf{v}_0) \\ \mathbf{G} = \dfrac{\partial \mathbf{F}}{\partial \mathbf{u}}(\mathbf{Q}_0, \mathbf{u}_0, \mathbf{v}_0) \\ \mathbf{E} = \dfrac{\partial \mathbf{A}}{\partial \mathbf{u}}\mathbf{Q}_0 \\ \mathbf{C} = \dfrac{\partial \mathbf{F}}{\partial \mathbf{v}}(\mathbf{Q}_0, \mathbf{u}_0, \mathbf{v}_0) \\ \mathbf{P} = \dfrac{2}{\rho_\infty V_\infty^2}\dfrac{\partial \mathbf{f}^{\mathrm{ext}}}{\partial \mathbf{Q}}(\mathbf{Q}_0, \mathbf{u}_0, \mathbf{v}_0) \end{cases} \tag{6.132}$$

The variables $\mathbf{Q}$, $\mathbf{Q}_t$, $\mathbf{u}$ and $\mathbf{v}$ represent the perturbation quantities. $0.5\rho_\infty {V_\infty}^2$ represent the free stream dynamic pressure. Assume the CFD system have n grids, then for the 2D and 3D Euler equations, the order of eq. (6.131) is $J = 4 \times n$ and $J = 5 \times n$, respectively.

2. Aeroelastic reduced-order modeling

The linearized fluid dynamics (6.131) can be written in the form of a linear system as follows:

$$\begin{cases} \dot{\mathbf{Q}} = \mathbf{A}\mathbf{Q} + \mathbf{B}\mathbf{y} \\ \mathbf{f}^{\mathrm{ext}} = 0.5\rho V^2 \mathbf{P}\mathbf{Q} \end{cases} \tag{6.133}$$

where $\mathbf{A} = -\mathbf{A}_0^{-1}\mathbf{H}$, $\mathbf{B} = -\mathbf{A}_0^{-1}(\mathbf{E}+\mathbf{C}\mathbf{G})$, $\mathbf{y} = [\mathbf{v}; \mathbf{u}]^T$. Solve eq. (6.133) in the time domain to get snapshots of the system. Since the explicit time advance scheme for CFD solution is limited by the stability condition, it is suggested that the implicit method or the dual time explicit method is used to solve the equation.

1) Implicit method

Use first-order backward difference and center difference for eq. (6.133) and yield

$$\frac{\mathbf{Q}^{n+1} - \mathbf{Q}^{n}}{\Delta t} = \mathbf{A}\frac{\mathbf{Q}^{n+1} + \mathbf{Q}^{n}}{2} + \mathbf{B}\mathbf{y}^{n+1/2}, \tag{6.134}$$

where Δt is the physical time step. Reorganize eq. (6.134)

$$\left(\mathbf{I} - \frac{\mathbf{A}\Delta t}{2}\right)\mathbf{Q}^{\mathrm{n}+1} = \left(\mathbf{I} + \frac{\mathbf{A}\Delta t}{2}\right)\mathbf{Q}^{n} + \Delta t\,\mathbf{B}\mathbf{y}^{\mathrm{n}+1/2}. \tag{6.135}$$

Since the matrix $(\mathbf{I} - 0.5\,\mathbf{A}\Delta t)$ is constant during the solution process, LU decomposition can be performed on the matrix in advance.

2) Explicit method

The solution of an implicit equation such as eq. (6.135) requires inversion of the coefficient matrix. For the case of large-scale grid, the dimension of the coefficient matrix is very large (usually above $10^6 \times 10^6$), which makes it difficult to implement LU decomposition. In order to deal with such difficulty, the dual-time explicit advancing method can be adopted:

$$\frac{\mathbf{Q}^{\mathrm{m}+1} - \mathbf{Q}^{\mathrm{m}}}{\Delta \tau} = \frac{\mathbf{Q}^{\mathrm{n}} - \mathbf{Q}^{\mathrm{m}}}{\Delta t} + \mathbf{A}\mathbf{Q}^{\mathrm{m}} + \mathbf{B}\mathbf{y}^{\mathrm{n}}, \tag{6.136}$$

where $\Delta\tau$ denotes the local pseudo-time step. Δt denotes the physical time step. The superscript m denotes the mth iteration for the pseudo-time step.

Equation (6.136) is solved by the classic four-step Runge–Kutta method. It should be noted that the coefficient matrix $\mathbf{A}$ is of particularly high dimensionality. A block coefficient matrix can be used instead of a single matrix $\mathbf{A}$ to solve large-scale problem to save the massive memory cost of computer.

A pulse function of displacement and velocity on each order modal in eq. (6.135) or eq. (6.136) is used as the excitation to obtain the corresponding unit impulse response as a snapshot. The method explained in Section 6.3 is used to obtain the corresponding POD base matrix Ψ_r. Then, the ROM of the fluid system is obtained by mapping eq. (6.133) onto Ψ_r

$$\begin{cases} \dot{\mathbf{Q}}_r = \boldsymbol{\Psi}_r^T \mathbf{A}\boldsymbol{\Psi}_r \mathbf{Q}_r + \boldsymbol{\Psi}_r^T \mathbf{B}\mathbf{y} \\ \mathbf{f}^{\mathrm{ext}} = 0.5\rho V^2 \mathbf{P}\boldsymbol{\Psi}_r \mathbf{Q}_r \end{cases}. \tag{6.137}$$

The order of eq. (6.137) is r, which is much smaller than the full-order system. The structural dynamics equation (6.127) is rewritten as a first-order ordinary differential equation:

$$\begin{bmatrix} \dot{\mathbf{u}} \\ \dot{\mathbf{v}} \end{bmatrix} = \begin{bmatrix} 0 & \mathbf{I} \\ -\mathbf{M}^{-1}\mathbf{K} & -\mathbf{M}^{-1}\mathbf{C} \end{bmatrix} \begin{bmatrix} \mathbf{u} \\ \mathbf{v} \end{bmatrix} + \begin{bmatrix} 0 \\ \mathbf{M}^{-1}\mathbf{f}^{\text{ext}} \end{bmatrix} \tag{6.138}$$

The aeroelastic ROM is obtained by combining eqs. (6.137) and (6.138):

$$\begin{bmatrix} \dot{\mathbf{Q}}_r \\ \dot{\mathbf{u}} \\ \dot{\mathbf{v}} \end{bmatrix} = \begin{bmatrix} -\mathbf{\Psi}_r^T \mathbf{A}_0^{-1} \mathbf{H} \mathbf{\Psi}_r & -\mathbf{\Psi}_r^T \mathbf{A}_0^{-1} \mathbf{G} & -\mathbf{\Psi}_r^T \mathbf{A}_0^{-1} (\mathbf{E} + \mathbf{C}) \\ \mathbf{0} & \mathbf{0} & \mathbf{I} \\ 0.5\rho V^2 \mathbf{M}^{-1} \mathbf{P} \mathbf{\Psi}_r & -\mathbf{M}^{-1}\mathbf{K} & -\mathbf{M}^{-1}\mathbf{C} \end{bmatrix} \begin{bmatrix} \mathbf{Q}_r \\ \mathbf{u} \\ \mathbf{V} \end{bmatrix}. \tag{6.139}$$

Equation (6.139) is more convenient and faster than the direct CFD/CSD coupling computation either solved using system eigenvalues or the time domain computation.

6.5 Reduced order of aeroelastic system based on BPOD method

BPOD is a combination with both the advantages of balanced truncation method and POD method [4, 7, 22, 39]. The main idea of the BPOD is to remeasure the controllability and observability of the system by dividing the system into strong subsystems and weak subsystems. The weak subsystems are eliminated due to less influence on the input and output of the system in the full-order model, which benefit this method with similar input and output characteristics with the full-scale model.

6.5.1 Balance truncated reduction theory

Consider the following linear time-invariant asymptotic stability system:

$$\begin{cases} \dot{\mathbf{Q}} = \mathbf{A}\mathbf{Q} + \mathbf{B}\mathbf{y} \\ \mathbf{f}^{\text{ext}} = 0.5\rho V^2 \mathbf{P}\mathbf{Q} \end{cases}. \tag{6.140}$$

The definition of the controllable Gramian matrix of the system in the time domain is

$$\begin{cases} \mathbf{W}_c = \int_0^\infty e^{\mathbf{A}t} \mathbf{B}\mathbf{B}^T e^{\mathbf{A}^T t} \, dt \\ \mathbf{W}_o = \int_0^\infty e^{\mathbf{A}^T t} \mathbf{P}^T \mathbf{P} e^{\mathbf{A}t} \, dt \end{cases}, \tag{6.141}$$

where both $\mathbf{W}_c$ and $\mathbf{W}_o$ are symmetric semidefinite matrices and satisfy the following Lyapunov equations

$$\begin{cases} \mathbf{A}\mathbf{W}_c + \mathbf{W}_c\mathbf{A}^T + \mathbf{B}\mathbf{B}^T = 0 \\ \mathbf{A}^T\mathbf{W}_o + \mathbf{W}_o\mathbf{A} + \mathbf{P}^T\mathbf{P} = 0 \end{cases}. \tag{6.142}$$

If the system is controllable and observable, both $\mathbf{W}_c$ and $\mathbf{W}_o$ should be nonsingular. The equilibrium of the system is to find a nonsingular transformation matrix $\mathbf{T}$ to transform the controllable and observable Gramian matrix into the same diagonal matrix:

$$\begin{cases} \hat{\mathbf{W}}_c = \mathbf{T}^{-1}\mathbf{W}_c\mathbf{T}^{-T} \\ \hat{\mathbf{W}}_o = \mathbf{T}^T\mathbf{W}_o\mathbf{T} \\ \hat{\mathbf{W}}_o = \hat{\mathbf{W}}_c = \Sigma \end{cases}, \tag{6.143}$$

where $\Sigma = diag\left(\sigma_1^2\sigma_2^2 \cdots \sigma_n^2\right)$; $\sigma_1 \geq \sigma_2 \geq \cdots \geq \sigma_n$ is the Hankel singular value. The magnitude of the Hankel singular value indicates the influence of state quantity on the input and output of the system. The larger the Hankel singular value, the greater the influence. Therefore, only the first r singular values of $\sum$ are retained to obtain the matrix $\sum_1 = \mathrm{diag}(\sigma_1^2\ \sigma_2^2 \cdots \sigma_r^2)$. The first r column corresponding to $\sum_1$ in the transform matrix T is retained as matrix $\mathbf{T}_r$. The first r row corresponding to $\sum_1$ in the reserved matrix T^{-1} is obtained as the matrix $\mathbf{S}_r$. Finally, eq. (6.140) can be reduced to the following r-order system

$$\begin{cases} \dot{\mathbf{Q}}_r = \mathbf{S}_r\mathbf{A}\mathbf{T}_r\mathbf{Q}_r + \mathbf{S}_r\mathbf{B}\mathbf{y} \\ \mathbf{f}^{ext} = 0.5\rho V^2\mathbf{P}\mathbf{T}_r\mathbf{Q}_r \end{cases} \tag{6.144}$$

Equation (6.144) is the resulting ROM, which is much smaller than the full-order system.

6.5.2 Construction of transformation matrix

The construction of transformation matrix $\mathbf{T}_r$ and $\mathbf{S}_r$ is crucial for BPOD. The steps of construction the transformation matrix are as follows:

1) Gramian matrices $\mathbf{W}_c$ and $\mathbf{W}_o$ of the system is obtained by eq. (6.141).
2) Perform Cholesky decomposition on the Gramian matrix $\mathbf{W}_c = \mathbf{X}\mathbf{X}^T$, $\mathbf{W}_o = \mathbf{Z}\mathbf{Z}^T$.
3) Perform singular value decomposition on $\mathbf{Z}^T\mathbf{X}$

$$\mathbf{Z}^T\mathbf{X} = \mathbf{U}\Sigma\mathbf{V}^T = [\mathbf{U}_1\ \mathbf{U}_2]\begin{bmatrix} \Sigma_1 & \\ & \Sigma_2 \end{bmatrix}\begin{bmatrix} \mathbf{V}_1^T \\ \mathbf{V}_2^T \end{bmatrix}. \tag{6.145}$$

4) Construct the transformation matrix

$$\begin{cases} \mathbf{T}_r = \mathbf{X}\mathbf{V}_1 \sum_1^{-1/2} \\ \mathbf{S}_r = \sum_1^{-1/2} \mathbf{U}_1^T \mathbf{Z}^T \end{cases} \tag{6.146}$$

6.5.3 The connection between POD snapshot and Gramian matrix

The key to constructing a transformation matrix is the derivation of the Gramian matrix from the analysis of the previous section. However, it is inconvenient to accurately solve the controllable Gramian matrix directly for the high-dimensional system by eq. (6.141) or (6.142). It can be found by observing the analytical expression (6.141) that the Gramian matrix is associated with the snapshots of the system, implying the Gramian matrix can be approximated by snapshots. In eq. (6.140), the first p column vectors of the input matrix B are $\mathbf{b_1}$, $\mathbf{b_2}$, ..., $\mathbf{b_p}$, and the input is $u = \left[u_1, u_2, \ldots, u_p\right]^T$. In the initial state, each input of system is excited by a unit pulse excitation $u_i(t) = \delta(t)$, then the response of eq. (6.140) is

$$\mathbf{x}_i(t) = e^{\mathbf{A}t}\mathbf{b}_i, \tag{6.147}$$

where $\mathbf{x}_i(t)$ is a snapshot of the system. Define $\bar{\mathbf{x}}(t) = [\mathbf{x}1(t)\ \mathbf{x}_2(t)\ \cdots\ \mathbf{x}_p(t)]$, satisfying $\bar{\mathbf{x}}(t) = e^{\mathbf{A}t}\mathbf{B}$. Then the controllable Gramian matrix of the system (6.141) can be expressed as

$$\mathbf{W}_c = \int_0^\infty \left(\mathbf{x}_1(t)\mathbf{x}_1(t)^T + \cdots + \mathbf{x}_p(t)\mathbf{x}_p(t)^T\right) dt. \tag{6.148}$$

Integrating eq. (6.148), yield

$$\mathbf{W}_c = \mathbf{X}\mathbf{X}^T, \tag{6.149}$$

$$\mathbf{X} = \left[\bar{\mathbf{x}}(t_1)\sqrt{\delta_1}\ \ \bar{\mathbf{x}}(t_2)\sqrt{\delta_2}\ \cdots\ \bar{\mathbf{x}}(t_m)\sqrt{\delta_m}\right], \tag{6.150}$$

where δ_i denotes the integral weighted value at time t_i, and its selection is related to the integral method. Herein the complex Simpson integral formula is used. By using eqs. (6.149) and (6.150), it is convenient to approximate the system's controllable Gramian matrix with time-domain snapshots, and the approximation of the observable Gramian matrix is similar to that of the controllable Gramian matrix. The time domain impulse response of the dual system can be taken as a system snapshot. The impulse response of the dual system is

$$\bar{\mathbf{z}}(t) = e^{\mathbf{A}^T t}\mathbf{P}^T. \tag{6.151}$$

By solving the impulse response of the system under time $t_1, t_2, \cdots, t_n$, the observable Gramian matrix can be expressed as

$$\mathbf{W}_o = \mathbf{Z}\mathbf{Z}^T \tag{6.152}$$

$$\mathbf{Z} = \left[\bar{\mathbf{z}}(t_1)\sqrt{\delta_1} \quad \bar{\mathbf{z}}(t_2)\sqrt{\delta_2} \quad \cdots \quad \bar{\mathbf{z}}(t_n)\sqrt{\delta_n}\right]. \tag{6.153}$$

Equations (6.152) and (6.153) indicate an observable Gramian matrix that can be approximated by the snapshots of the dual system. Compared with the POD method, the BPOD method not only considers the greatest impact on the system input but also considers the greatest impact on the system output, which is not missed in the POD method.

Similar to (6.139), the aeroelastic ROM based on BPOD method is

$$\begin{bmatrix} \dot{\mathbf{Q}}_r \\ \dot{\mathbf{u}} \\ \dot{\mathbf{v}} \end{bmatrix} = \begin{bmatrix} -\mathbf{S}_r\mathbf{A}_0^{-1}\mathbf{H}\mathbf{T}_r & -\mathbf{S}_r\mathbf{A}_0^{-1}\mathbf{G} & -\mathbf{S}_r\mathbf{A}_0^{-1}(\mathbf{E}+\mathbf{C}) \\ 0 & 0 & \mathbf{I} \\ \mathbf{M}^{-1}\mathbf{C}\mathbf{T}_r & -\mathbf{M}^{-1}\mathbf{K}+\mathbf{M}^{-1}\mathbf{D} & -\mathbf{M}^{-1}\mathbf{C}+\mathbf{M}^{-1}\mathbf{D} \end{bmatrix} \begin{bmatrix} \mathbf{Q}_r \\ \mathbf{u} \\ \mathbf{v} \end{bmatrix} \tag{6.154}$$

6.5.4 Numerical examples

1. Aeroelastic analysis of two-dimensional airfoil based on time domain POD method

The key part for the aeroelastic ROM based on the POD method is the linearization of the flow governing equation. In this section, an aeroelastic problem of 2D airfoil is studied and compared with the direct CFD/CSD coupling method to validate the linearization of the governing equation. The airfoil is NACA0012 symmetrical airfoil. The O-type Euler mesh (61×21) is used to discretize the computational domain. AUSM+-up scheme is used for the spatial discretization, and LU-SGS-τTs scheme is for the time advancement. The motion equation of 2D airfoil is given as follows:

$$\begin{cases} m\ddot{h} + S_{h\alpha}\ddot{\alpha} + K_h h = -L \\ S_{h\alpha}\ddot{h} + I_\alpha\ddot{\alpha} + K_\alpha\alpha = M_{ea} \end{cases}, \tag{6.155}$$

where h is the plunging displacement of airfoil, downward is positive; α is the pitch angle; m is the mass of airfoil; $S_{h\alpha}$ is the static moment of airfoil; I_α is the moment of inertia; K_h and K_α are the plunging and pitching stiffness, respectively; L and M_{ea} are the lift of the airfoil and the pitching moment, respectively. Figures 6.18 and 6.19 are schematics and aerodynamic grids of the computational domain. The structural parameters and incoming flow parameters are selected as

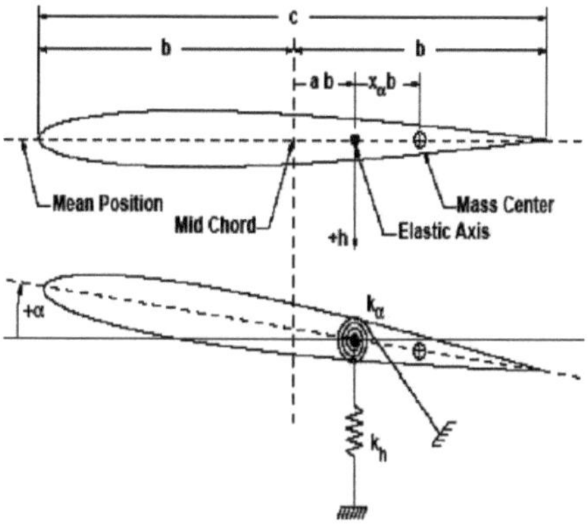

Fig. 6.18: Schematic diagram of two-dimensional airfoil structure.

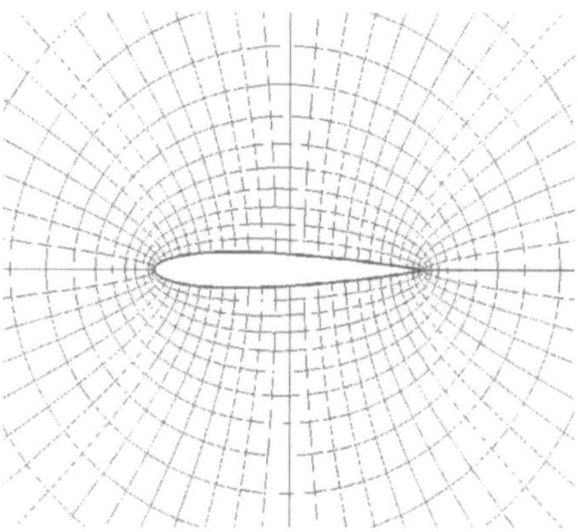

Fig. 6.19: O-type Euler grid.

$$\begin{cases} m = 794.49kg & S_{h\alpha} = -39.73kg \cdot m \\ I_\alpha = 411.28kg \cdot m & K_h = 6.08 \times 10^5 N/m \ , \\ K_\alpha = 4.05 \times 10^5 N/m & b = 0.5m \quad a = -0.5 \end{cases} \tag{6.156}$$

$$M_\infty = 0.5 \ \ \rho_\infty = 1.225 \ kg/m^3 \ \ p_\infty = 101,325 \ Pa. \tag{6.157}$$

The initial condition is $\dot{h} = 0.01 \ m/s$. Comparing the linearized model with the direct CFD/CSD coupling results (as shown in Fig. 6.20), it can be seen that except for the slight difference in the moment coefficients, the displacements and lift coefficient agree well with the full-order model. The results indicate that in the case of small perturbations, the linearized flow field equation can be used as a replacement for the nonlinear CFD system.

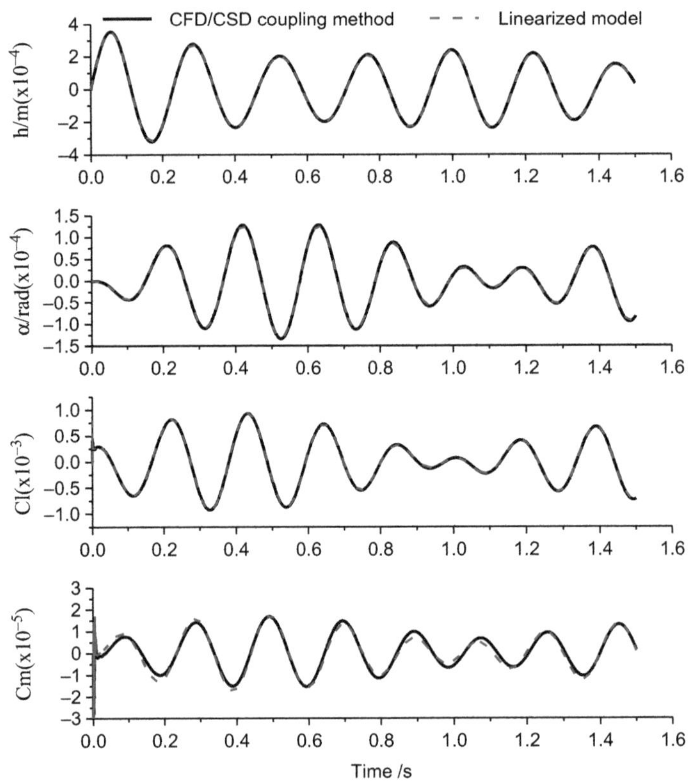

Fig. 6.20: Comparison between linearized model and direct CFD/CSD coupling calculation.

2. Supersonic flutter analysis of control surface based on time domain POD method

The flutter analysis of a supersonic control surface is performed using the POD method. The Mach number is chosen $M_\infty = 2.0$, angle of attack $\alpha = 0.0°$. Figures 6.21 and 6.22 show the computational grid and steady-state surface pressure contour of the control surface. Figure 6.23 shows the finite element model of the wing structure, and Tab. 6.2 lists the fourth-order structure modal frequency and type calculated by finite element software. Because the structural stiffness is relatively large and the displacement of the response is small, the TFI method is used for better grid orthogonality for flow field calculation.

First, the aeroelastic response of the wing is calculated by the direct CFD/CSD coupling method under the dynamic pressure $Q_{\text{ref}} = 222{,}306$ Pa. Then, the Euler equation is linearized and the response of the linearization equation is calculated under the same conditions. It can be seen from Fig. 6.23 that the full-order linearization model agrees well with the direct CFD/CSD coupling response, indicating that the equation is linearized correctly. A pulse excitation is applied to each order displacement and velocity. The unit impulse response of the linearized model is

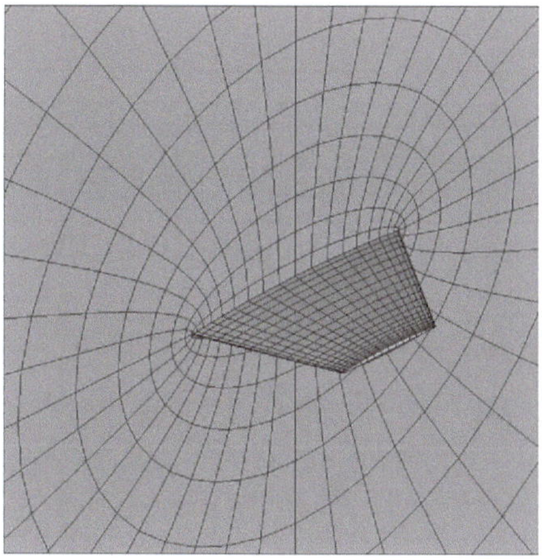
Fig. 6.21: Aerodynamic grid of flow field.

Fig. 6.22: Steady pressure contour.

Fig. 6.23: Structure finite element model.

Tab. 6.2: Finite element model of control surface.

Order	1	2	3	4
Frequency (Hz)	32.356	52.97	300.87	430.78
Modal Shape	Shaft Bending	Shaft Torsion	Surface Bending	Surface Torsion

calculated to obtain the snapshots of the system, and each input takes 50 snapshots to form the total snapshot matrix. Finally, the POD basis is obtained by solving eq. (6.124). Figure 6.24 compares the accuracy of the linearized model, the 50th-order and the 150th-order POD-based ROM. It is be found that the 150th-order POD-based ROM reasonably agrees with the full-order linear system response.

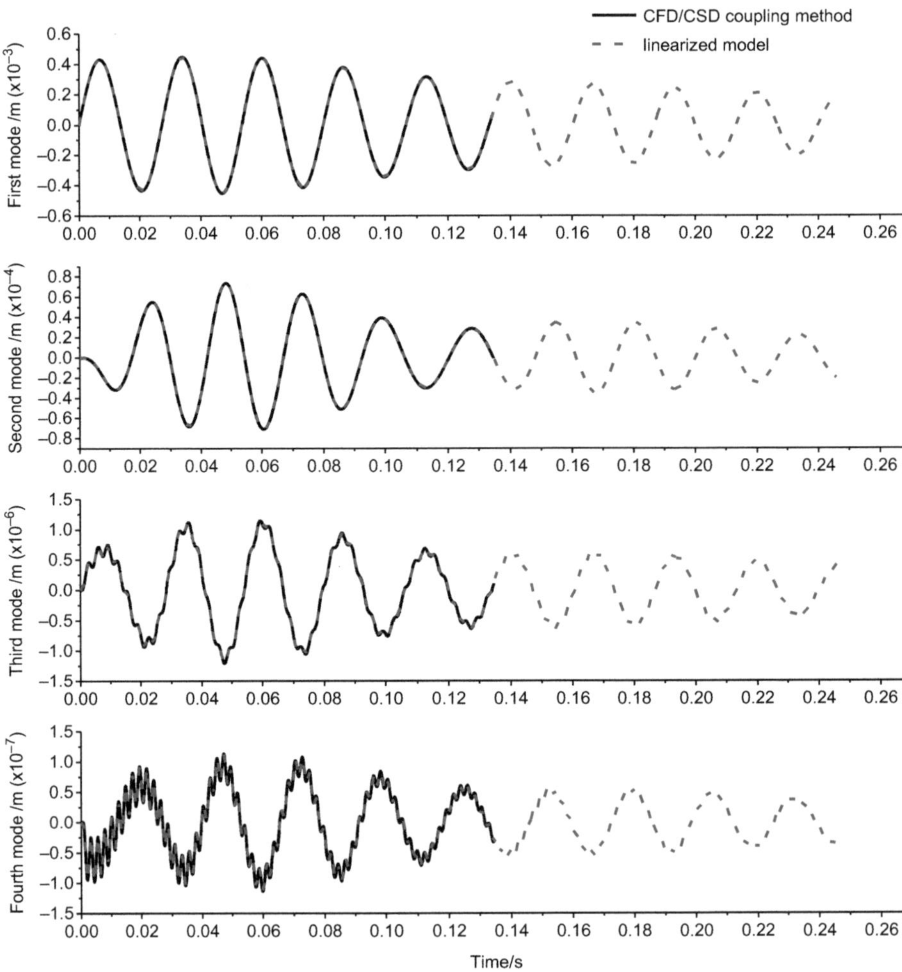

Fig. 6.24: Comparison of direct CFD/CSD coupling response and linearized model response.

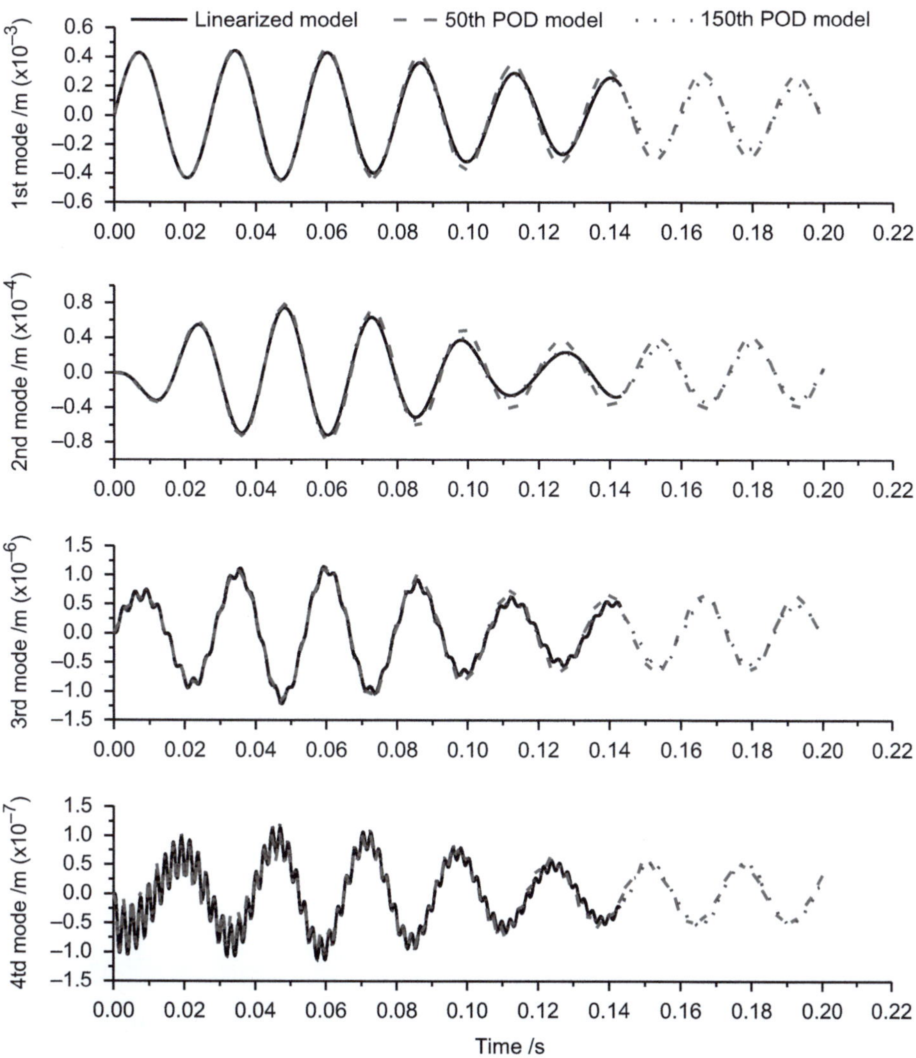

Fig. 6.25: Comparison of linearized model and POD reduced-order model ($Q_{\mathrm{ref}} = 222{,}306$ Pa).

6.6 Nonlinear aerodynamic reduction model based on POD-Galerkin projection

The ROMs for unsteady aerodynamic force based on POD-Galerkin projection is proposed to preserve the nonlinear characteristics of flow field, which is different from the POD method with the assumption of small perturbation [17, 21, 37]. This projection method does not need to linearize the flow governing equation, instead directly project the governing equation onto the subspace spanned by the POD basis to form a low-order ordinary differential equation. Most of the research on this method focused on the field of incompressible flow, such as vortex street, back step flow and cavity driven flow. For compressible flows with structural perturbations, the difficulty in using the POD-Galerkin projection method lies in

1) how to modify the compressible fluid governing equation to ensure the resulting low-order model is a quadratic nonlinear equation
2) stability improvement for the ROM

This section transforms the compressible Euler equation with explicit convection term to facilitate the construction of the Galerkin system. Furthermore, a correction method is present to improve the stability of the ROM.

6.6.1 Modification of flow governing equation

When dealing with compressible flows, POD-Galerkin projection method requires the modification of the governing equations. This is because the governing equation in conservation variables is not a quadratic linear form, and the resulting equation by directly applying Galerkin projection method is an implicit equation.

Define $\mathbf{q} = (\vartheta, \mathbf{u}, p)^T$ as modified main variables. Then, compressible Euler equation can be written as a Cartesian coordinate system

$$\begin{cases} \dfrac{\partial \vartheta}{\partial t} + (\mathbf{u} - \mathrm{s}) \cdot \nabla \vartheta = \vartheta div\mathbf{u} \\ \dfrac{\partial \mathbf{u}}{\partial t} + (\mathbf{u} - \mathrm{s}) \cdot \nabla \mathbf{u} = -\vartheta \nabla p \\ \dfrac{\partial p}{\partial t} + (\mathbf{u} - \mathrm{s}) \cdot \nabla p = -\gamma p div\mathbf{u} \end{cases}, \tag{6.158}$$

where $\vartheta = 1/\rho$, $\mathbf{s}$ is the velocity of the grid. Define the following dimensionless variables:

$$\begin{cases} \vartheta^* = \vartheta/\vartheta_\infty, \quad u^* = u/c_\infty, \quad v^* = v/c_\infty \\ p^* = p/(\rho_\infty c_\infty^2), \quad t^* = t/(L/c_\infty) \\ x^* = x/L, \quad y^* = y/L \end{cases}. \tag{6.159}$$

Substitute eq. (6.159) into eq. (6.158), and the dimensionless governing equation is obtained (for the sake of convenience, omit the dimensionless superscript). It should be noted that eq. (6.158) is only used as the governing equations for

Galerkin projection. The solution of the full-order flow field still uses the integral conservation equation and the details of the computational method are introduced in Chapter 2.

6.6.2 Grid velocity definition

In order to construct an unsteady aerodynamic force ROM caused by structural perturbations, it is required to define the grid velocity resulting from structural perturbations, that is, **s** in eq. (6.158). This section will focus on the case of rigid body motion of structure because the volume of the structural grid changes at each time step in the case of elastic deformation, which does not meet the definition of the inner product in Section 6.3. For small perturbation aeroelastic problems such as flutter, the change of mesh volume can be negligible. However, such assumption will not hold for the case of large structural perturbations.

For further illustration of the mesh velocity, choose a fixed point in the space as the origin for fixed coordinate system, and the motion coordinate system is fixed on the computational grid. Then the position of any point **x** in the fixed coordinate system can be expressed as

$$\mathbf{x}_A = \mathbf{x}_{A,0} + \mathbf{R}\mathbf{x}_E, \tag{6.160}$$

where the subscript A denotes a fixed coordinate system, and the subscript E is a motion coordinate system. **R** is the transformation matrix. For a 2D problem, the transformation matrix **R** is

$$\mathbf{R} = \begin{bmatrix} \cos\alpha & -\sin\alpha \\ \sin\alpha & \cos\alpha \end{bmatrix}, \tag{6.161}$$

where α is the angle between the two coordinate systems. The derivation of both sides of eq. (6.160) is

$$\mathbf{s} = \frac{d\mathbf{x}_A}{dt} = \frac{d\mathbf{x}_{A,0}}{dt} + \frac{dR}{dt}\mathbf{x}_E + R\frac{d\mathbf{x}_E}{dt}. \tag{6.162}$$

The first term on the right-hand side of eq. (6.162) represents the velocity of the origin of the motion coordinate system; the second term represents the rotational speed of **x**; the third term represents the deformation velocity of **x**. For the problem studied herein, only rigid motion is considered so that the third term is zero.

6.6.3 POD-Galerkin projection

In order to obtain the ROM of eq. (6.158), eq. (6.158) is rewritten as the following general quadratic form:

$$\dot{\mathbf{q}} = \mathbf{Q}^C(\mathbf{q}, \mathbf{q}) + \mathbf{T}(\mathbf{q}, \mathbf{s}), \tag{6.163}$$

where the convection term $\mathbf{Q}^C$ and the source term $\mathbf{T}$ are defined, respectively,

$$\mathbf{Q}^C = \begin{bmatrix} -\mathbf{u} \cdot \nabla \vartheta + \vartheta \,\mathrm{div}\mathbf{u} \\ -\mathbf{u} \cdot \nabla \mathbf{u} - \vartheta \nabla p \\ -\mathbf{u} \cdot \nabla p - \gamma \,\mathrm{div}\mathbf{u} \end{bmatrix}, \quad \mathbf{T} = \begin{bmatrix} \mathbf{S} \cdot \nabla \vartheta \\ \mathbf{S} \cdot \nabla \mathbf{u} \\ \mathbf{S} \cdot \nabla p \end{bmatrix}. \tag{6.164}$$

Substitute eq. (6.120) into eq. (6.163) and from the inner product of $\boldsymbol{\varphi}^i$ on both sides, the following Galerkin project ion system is obtained:

$$\begin{aligned} \dot{a}_i(t) = K_i + \sum_{j=1}^{r} L_{ij} a_j(t) + \sum_{j=1}^{r} \sum_{k=1}^{r} Q_{ijk} a_j(t)\, a_k(t) \\ - \dot{\alpha} \sin \alpha SK1_i + \dot{\alpha} \cos \alpha\, SK2_i - \dot{\alpha} \sin \alpha \sum_{j=1}^{r} SL1_{ij} a_j(t) + \dot{\alpha} \cos \alpha \sum_{j=1}^{r} SL2_{ij} a_j(t), \end{aligned} \tag{6.165}$$

where K_i, L_{ij} and Q_{ijk} are autonomous coefficients. $SK1_i$, $SK2_i$, $SL1_{ij}$ and $SL2_{ij}$ are non-autonomous coefficients. The expressions of the coefficients are as follows:

$$K_i = (\mathbf{Q}^C (\bar{\mathbf{q}}, \bar{\mathbf{q}}), \boldsymbol{\varphi}^j, \tag{6.166}$$

$$L_{ij} = (\mathbf{Q}^C (\bar{\mathbf{q}}, \boldsymbol{\varphi}^i) + \mathbf{Q}^C(\boldsymbol{\varphi}^i, \bar{\mathbf{q}}), \boldsymbol{\varphi}^i), \tag{6.167}$$

$$Q_{ijk} = (\mathbf{Q}^C(\boldsymbol{\varphi}^j, \boldsymbol{\varphi}^k), \boldsymbol{\varphi}^i), \tag{6.168}$$

$$SK1_i = (\mathbf{T}(\bar{\mathbf{q}}, \mathbf{X}), \boldsymbol{\varphi}^i), \tag{6.169}$$

$$SK2_i = (\mathbf{T}(\bar{\mathbf{q}}, \mathbf{X}1), \boldsymbol{\varphi}^i), \tag{6.170}$$

$$SL1_{ij} = (\mathbf{T}(\boldsymbol{\varphi}^j, \mathbf{X}), \boldsymbol{\varphi}^i), \tag{6.171}$$

$$SL2_{ij} = (\mathbf{T}(\boldsymbol{\varphi}^j, \mathbf{X}1), \boldsymbol{\varphi}^i). \tag{6.172}$$

In 2D problems, **x1** is

$$\mathbf{x1} = \begin{bmatrix} 0 & -1 \\ 1 & 0 \end{bmatrix} \mathbf{x}. \tag{6.173}$$

In order to calculate formulas (6.166)–(6.172), the derivative of the variable at the center point of the grid needs to be calculated. Herein the Green's formula is used to calculate the derivative at the grid point

$$(\nabla\Psi)_i = \frac{1}{V(\Omega_i)}\int_{\Omega_i}\nabla\Psi d\,\Omega = \frac{1}{V(\Omega_i)}\sum_{j=1}^{n_f}\int_{\partial\Omega_{i,j}}\Psi\mathbf{n}d\partial\Omega, \tag{6.174}$$

where n_f is the number of boundaries of the grid cell, and $n_f = 4$ for the 2D problem. The value of the variable at the grid interface is calculated using the central difference scheme. After deriving the POD-Galerkin ROM, the solution of the equation becomes a Cauchy problem with the initial condition

$$a_i^0 = (\tilde{\mathbf{q}}^0, \boldsymbol{\varphi}^i). \tag{6.175}$$

6.6.4 Correction method

When the solution of eq. (6.165) is integrated using the fourth-order Runge–Kutta method, the resulting solution will be inaccurate and even unstable. This is because the ROM obtained by the POD-Galerkin projection method lacks dissipative term and tends to be unstable. Hence, a correction is required to stabilize the equation. Equation (6.165) can be rewritten in the following vector form:

$$\dot{\mathbf{a}}(t) = \mathbf{f}(\mathbf{y}, \mathbf{a}\,(t)). \tag{6.176}$$

The calibration methods can be categorized into three types by different definitions of error estimation: state correction method with dynamic constraints, state correction method and flow field correction method.

The first method is to obtain an optimal solution of Cauchy problem, generally using an iterative solution. However, as pointed out by Couplet that constrained optimization problems is often an ill-conditioned problem that there may exist multiple solutions and there is no guarantee of the convergence of the solution. Therefore, Couplet proposed two different error definitions by the integration and differentiation of the error definitions in the first method. The correction methods based on these two new error definitions are the state correction method and the flow field correction method. In this section, the flow field correction method is used, and the error is defined as follows

$$\mathbf{e}(\mathbf{y}, t) = \dot{\mathbf{a}}^P(t) - \mathbf{f}(\mathbf{y}, \mathbf{a}^P\,(t)), \tag{6.177}$$

where the superscript P denotes the POD basis coefficient obtained by directly mapping the snapshot to the POD basis. The purpose of flow field correction is to minimize the following equation:

$$\min <\|\mathbf{e}(\mathbf{y}, t)\|^2> \;=\; \min \frac{1}{N}\sum_{K=1}^{N}\sum_{j=1}^{r}(\dot{a}_i^P(t) - f_i(\mathbf{y}_i,\, \mathbf{a}^P\,(t_k)))^2, \tag{6.178}$$

where N is the total number of solutions in the sample time. It can be proved that the earlier minimum problem is equivalent to the least squares problem as follows:

$$\mathbf{A}_e \mathbf{y} = \mathbf{b}_e, \tag{6.179}$$

where matrix $\mathbf{A}_e$ and vector $\mathbf{b}_e$ are

$$\begin{cases} \mathbf{A}_e = <\mathbf{E}^T(t)\,\mathbf{E}(t)> \\ \mathbf{b}_e = -<\mathbf{E}^T\mathbf{e}(0,t)> \end{cases}. \tag{6.180}$$

The matrix $\mathbf{E}$ denotes the following affine matrix:

$$\mathbf{E}(t)\mathbf{y} = -\mathbf{f}(\mathbf{y}, \mathbf{a}^P(t)). \tag{6.181}$$

Equation (6.179) actually solves an inverse problem. Usually the inverse problem solves the ill-conditioned equations. Placzek and Cordier suggested using the Tikhonov normalization method to solve the problem. The idea of the method is to add additional constraints (modulus of the solution vector) to the equations, so that the problem is transformed into a minimization problem in the following equations:

$$\min \|A_e\mathbf{y} - \mathbf{b}_e\|_2^2 + \rho_T\|\mathbf{y} - \mathbf{y}_0\|_2^2, \tag{6.182}$$

where ρ_T represents the normalized parameter and can be obtained using the L-curve method.

6.6.5 Numerical examples

In order to illustrate the establishment of ROM for the unsteady compressible flow field based on the POD-Galerkin projection method, the unsteady flow case for AGARD CT5 wing in Chapter 2 is considered. The details of computational scheme and conditions are the same as those in Chapter 2. The snapshots for sampling starts from the steady state solution (herein sampling after 10 time period cycles), and the sampling time step is $\Delta t^* = 2\pi/\omega_\alpha/50$, and 100 snapshots (equivalent to two time period cycles) are sampled. The POD basis and eigenvalues are obtained by snapshot method.

Figure 6.26 shows the power spectrum of the first 50-order POD eigenvalues. It is seen that the first 10 eigenvalues constitute a total of 99.74% of the energy field, which is used for the representation of the full-order flow field. By projecting eq. (6.163) onto the subspace spanned by the first ten-order POD basis, eq. (6.165) is obtained and integrated in the time domain with time step $\Delta t^*/20$. Figure 6.27 presents the time history of the response of ROM. It is seen that the ROM obtained by direct projection is unstable, even as the number of POD bases required for the projection and the number of snapshots are increased. It is because of small numerical

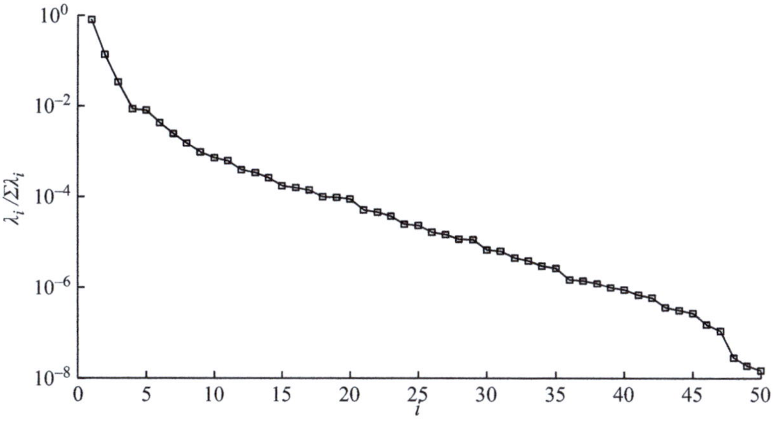

Fig. 6.26: Energy distribution of POD eigenvalues.

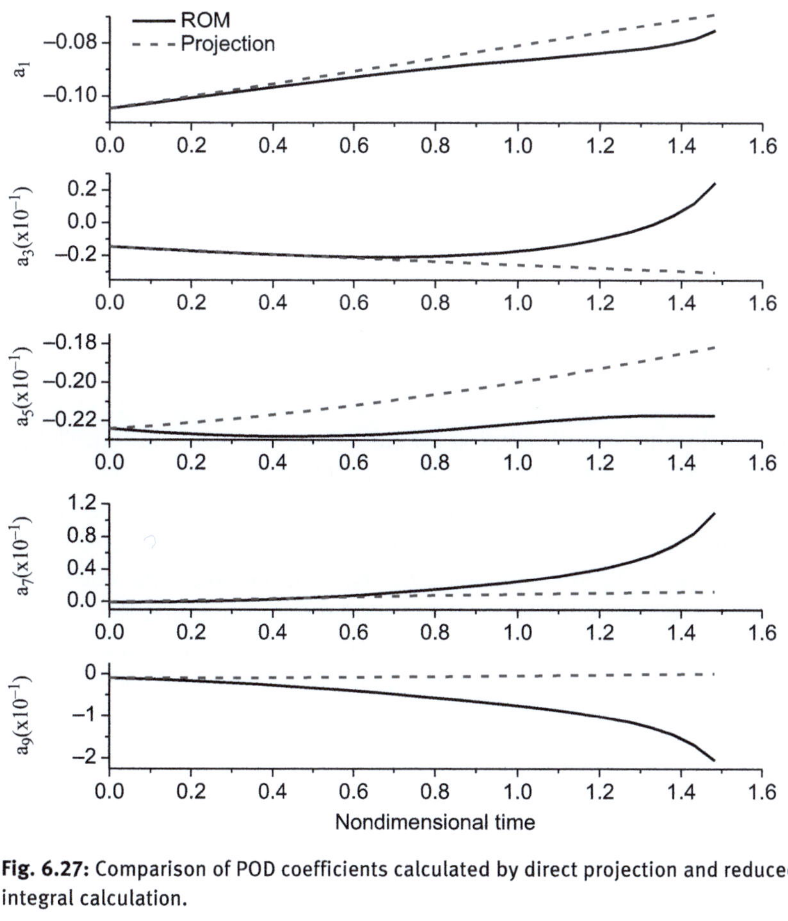

Fig. 6.27: Comparison of POD coefficients calculated by direct projection and reduced-order model integral calculation.

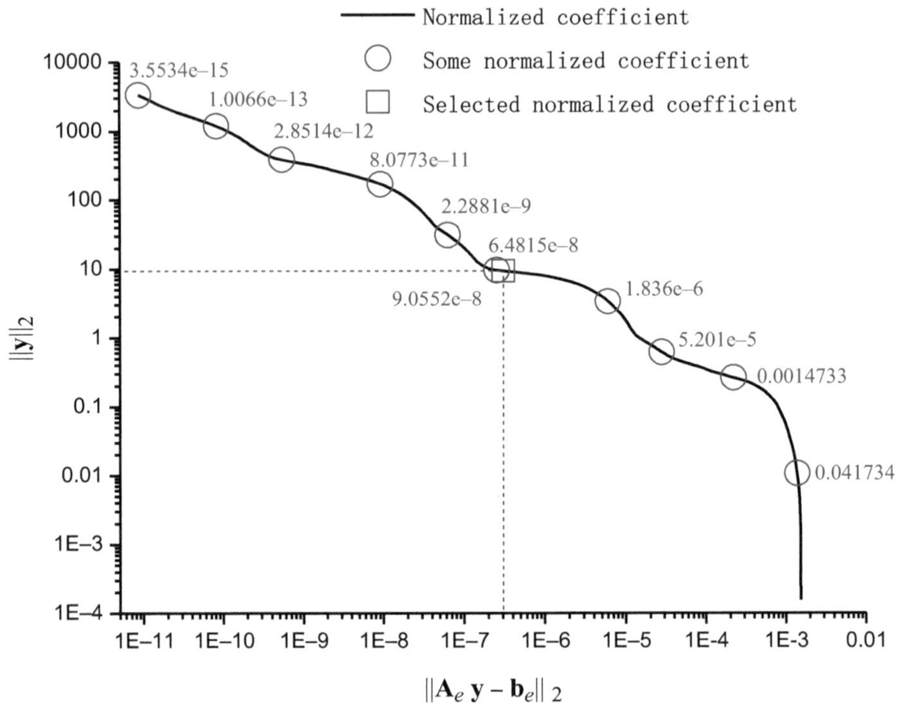

Fig. 6.28: Normalized coefficient *ρT*.

dissipation as mentioned earlier for ROM, while the Roe scheme used in the computation of the full-order Euler equations implicitly contains numerical dissipative terms to make the computation of the full-order equations stable.

In order to overcome this problem, the flow field correction method is used to introduce additional numerical dissipation for stability of the computation. The interpolation technique is used to obtain the POD-based mapping coefficients by direct mapping in the snapshot sampling period. $N = 3{,}961$ base mapping coefficients are used in the flow field correction method, and the L-curve method is used to calculate the Tikhonov normalized coefficients. As shown in Fig. 6.28, the coefficient ρ_T as the blue square is the equilibrium of the minimum residual and the solution norm of eq. (6.182). The correction coefficient y minimizes the residual of eq. (6.182) and reasonable approximation of the solution. Figure 6.29 shows the time history of response of the corrected ROM. It is found that the short-period response of the ROM (the first two periods of the first-order POD basis) is stable compared with Fig. 6.27 with improved stability and accuracy. Figure 6.30 shows the phase diagram of the POD-based projection coefficients. It is seen that the long-period response (the first-order POD-based 16 cycles) of ROM is slightly different from the full-order model, especially the last six-order POD basis projecting coefficient.

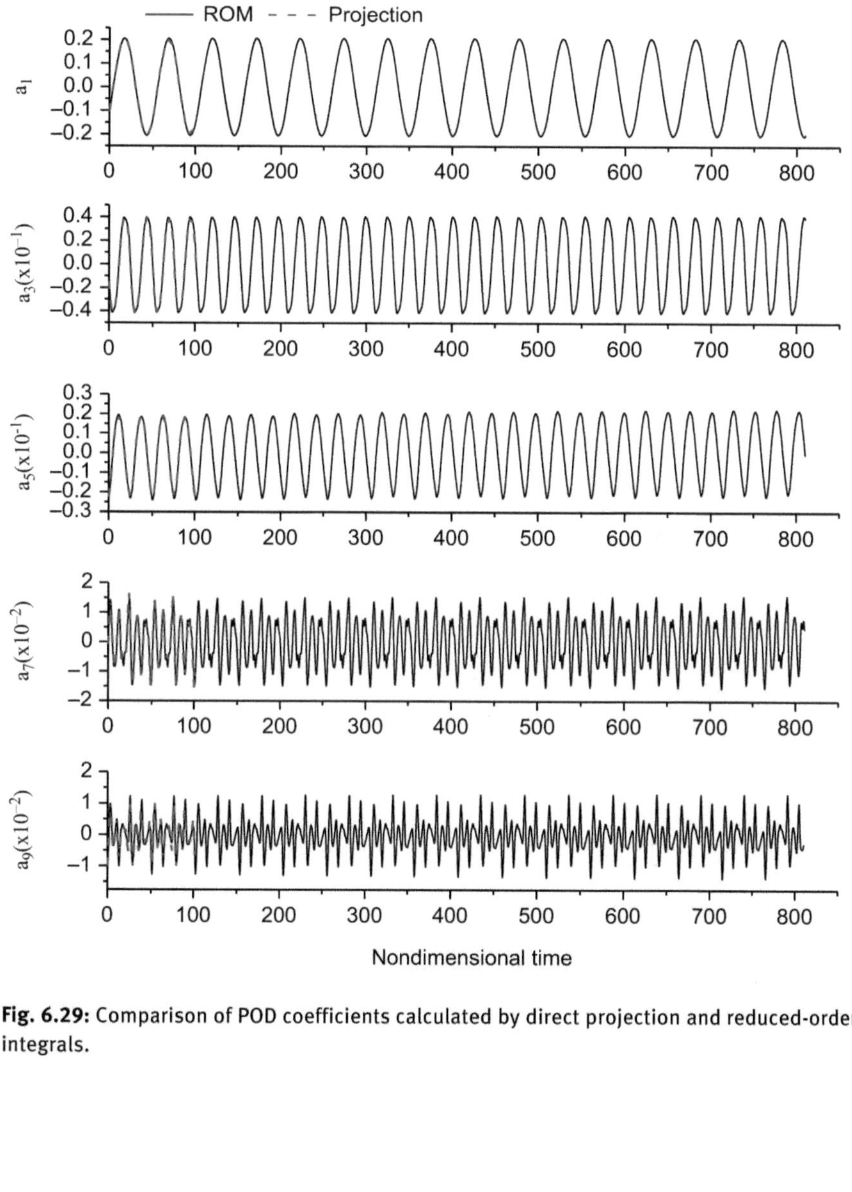

Fig. 6.29: Comparison of POD coefficients calculated by direct projection and reduced-order model integrals.

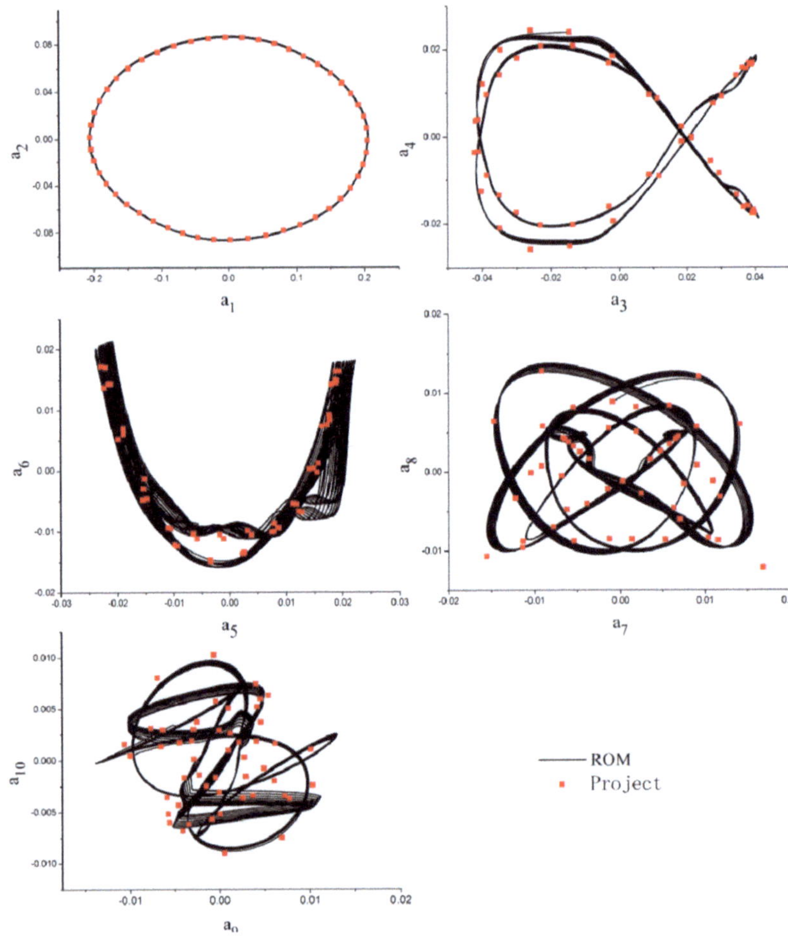

Fig. 6.30: Comparison of POD coefficients calculated by direct projection and reduced-order model integral (phase diagram).

Bibliography

[1] Algazi V, Sakrison D. On the optimality of the Karhunen-Loève expansion (Corresp.). IEEE Transactions on Information Theory. 1969, 15(2), 319–321.

[2] Allen MJ, Dibley RP Modeling aircraft wing loads from flight data using neural networks. 2003.

[3] An X. CFD/CSD coupled Solution for Aeroelasticity. Xi'an: Northwestern Polytechnical University; 2006. (in Chinese).

[4] An X. Research on Nonlinear Aeroelasticity Based on CFD/CSD Coupled Solution. Xi'an: Northwestern Polytechnical University; 2009.

[5] Beran PS, Lucia DJ, Pettit CL. Reduced-order modelling of limit-cycle oscillation for aeroelastic systems. Journal of Fluids and Structures. 2004, 19(5), 575–590.
[6] Cai XC, Ladeinde F, editors. A comparison of two POD methods for airfoil design optimization. 35th AIAA Fluid Dynamics Conference and Exhibit; 2005.
[7] Chen G, Li Y, Yan G, Xu M, Zeng X. A fast aeroelastic response prediction method based on proper orthogonal decomposition reduced order model. Journal of Astronautics. 2009, 30(5), 1765–1769. (in Chinese).
[8] Chen G, Sun J, Li Y-M. Adaptive reduced-order-model-based control-law design for active flutter suppression. Journal of Aircraft. 2012, 49(4), 973–980. (in Chinese).
[9] Chen G. Reduced Order Model for Unsteady Aerodynamics and its Application. Xi'an: Northwestern Polytechnical University; 2004. (in Chinese).
[10] Couplet M, Basdevant C, Sagaut P. Calibrated reduced-order POD-Galerkin system for fluid flow modelling. Journal of Computational Physics. 2005, 207(1), 192–220.
[11] Cowan TJ, Arena AS, Gupta KK. Accelerating computational fluid dynamics based aeroelastic predictions using system identification. Journal of Aircraft. 2001, 38(1), 81–87.
[12] Donovan GC, Geronimo JS, Hardin DP. Intertwining multiresolution analyses and the construction of piecewise-polynomial wavelets. SIAM Journal on Mathematical Analysis. 1996, 27(6), 1791–1815.
[13] Dowell E, Hall K, Thomas J, Florea R, Epureanu B, Heeg J, editors. Reduced Order Models In Unsteady Aerodynamics. 40th Structures. Structural Dynamics, and Materials Conference and Exhibit; 1999.
[14] Dowell EH, Curtiss HC, Scanlan RH, Sisto F. A Modern Course in Aeroelasticity. Springer; 1989.
[15] Hall KC, Thomas JP, Dowell EH. Proper orthogonal decomposition technique for transonic unsteady aerodynamic flows. AIAA Journal. 2000, 38(10), 1853–1862.
[16] Johnson MR, Denegri CM. Comparison of static and dynamic neural networks for limit cycle oscillation prediction. Journal of Aircraft. 2003, 40(1), 194–203.
[17] Kang W, Zhang J, Ren S, Lei P. Nonlinear Galerkin method for low-dimensional modeling of fluid dynamic system using POD modes. Communications in Nonlinear Science and Numerical Simulation. 2015, 22(1–3), 943–952.
[18] Khawar J, Zhigang W, Chao Y. Volterra kernel identification of MIMO aeroelastic system through multiresolution and multiwavelets. Computational Mechanics. 2012, 49(4), 431–458.
[19] Li M, Huang H, Xia J. Performance of adaptive equalization algorithm based on IQRD-RLS in DS-SS systems. Microcomputer & Its Applications. 2011, 15, (in Chinese).
[20] Li Y. Unsteady Aerodynamics and Aeroelastic Analysis Based on Volterra-Series 2007. (in Chinese)
[21] Lieu T, Farhat C, Lesoinne M, editors. POD-based aeroelastic analysis of a complete F-16 configuration: ROM adaptation and demonstration. 46th AIAA/ASME/ASCE/AHS/ASC Structures, Structural Dynamics and Materials Conference; 2005.
[22] Lucia DJ, Beran PS, Silva WA. Aeroelastic system development using proper orthogonal decomposition and volterra theory. Journal of Aircraft. 2005, 42(2), 509–518.
[23] Lumley JL Coherent structures in turbulence. Transition and turbulence: Elsevier, 1981. 215–242.
[24] Munteanu S, Rajadas J, Nam C, Chathopadhyay A, editors. An Efficient Approach for Solving Nonlinear Aeroelastic Phenomenon Using Reduced-Order Modeling. 45th AIAA/ASME/ASCE/AHS/ASC Structures, Structural Dynamics & Materials Conference; 2004.
[25] Munteanu S, Rajadas J, Nam C, Chattopadhyay A, editors. A volterra kernel reduced-order model approach for nonlinear aeroelastic analysis. 46th AIAA/ASME/ASCE/AHS/ASC Structures, Structural Dynamics and Materials Conference; 2005.

[26] Nowak RD, Van Veen BD. Random and pseudorandom inputs for Volterra filter identification. IEEE Transactions on Signal Processing. 1994, 42(8), 2124–2135.
[27] Prazenica RJ, Kurdila AJ. Multiwavelet constructions and Volterra kernel identification. Nonlinear Dynamics. 2006, 43(3), 277–310.
[28] Prazenica RJ Wavelet-based Volterra series representations of nonlinear dynamical systems. 2003.
[29] Preisendorfer R. Principal component analysis in meteorology and oceanography. Elsevier Science Publisher. 1988, 17, 425.
[30] Raveh DE. Reduced-order models for nonlinear unsteady aerodynamics. AIAA Journal. 2001, 39(8), 1417–1429.
[31] Shi Z. Research on CFD/CSD Coupled Interface Technology: Xi'an. Northwestern Polytechnical University; 2003. (in Chinese).
[32] Silva W, Raveh D, editors. Development of unsteady aerodynamic state-space models from CFD-based pulse responses. 19th AIAA Applied Aerodynamics Conference; 2001.
[33] Silva WA, Bartels RE. Development of reduced-order models for aeroelastic analysis and flutter prediction using the CFL3Dv6. 0 code. Journal of Fluids and Structures. 2004, 19(6), 729–745.
[34] Silva WA. Application of nonlinear systems theory to transonic unsteady aerodynamic responses. Journal of Aircraft. 1993, 30(5), 660–668.
[35] Silva WA. Simultaneous excitation of multiple-input/multiple-output CFD-based unsteady aerodynamic systems. Journal of Aircraft. 2008, 45(4), 1267–1274.
[36] Syed MA, Mathews VJ. QR-decomposition based algorithms for adaptive Volterra filtering. IEEE Transactions on Circuits and Systems I: Fundamental Theory and Applications. 1993, 40(6), 372–382.
[37] Tang D, Kholodar D, Juang J, Dowell EH System identification and POD method applied to unsteady aerodynamics. 2001.
[38] Tang L, Chen P, Liu D, Gao X, Shyy W, Utturkar Y, et al. editors. Proper orthogonal decomposition and response surface method for tps/rlv structural design and optimization: X-34 case study. 43rd AIAA Aerospace Sciences Meeting and Exhibit; 2005.
[39] Xu M, An X, Zeng X, Chen S. Model reduction using BPOD and its application to aeroelastic active control. Sciencepaper Online. 2008, (10), 17. (in Chinese).
[40] Yao W, Xu M. Aeroelasticity numerical analysis via Volterra series approach. Journal of Astronautics. 2008, 6, (in Chinese).
[41] Zeng X, X M. Volterra-series-based reduced-order model for unsteady aerodynamics. Structure and Environment Engineering. 34(5), 22–28, (in Chinese).
[42] Zeng X. Studies of Model Reduction Technique for Aeroelasticity. Xi'an: Northwestern Polytechnical University; 2007. (in Chinese).

Chapter 7
Aeroelastic software application based on CFD/CSD coupling method

7.1 Introduction

Time domain simulation based on computational fluid dynamics (CFD)/computational structure dynamics (CSD) coupling has become much popular in aeroelastic research community. However, the current CFD/CSD coupling technique for aeroelastic program still lacks generality because of complex implements of its unique features, like the complexity of the coupling interface between the fluid and structure solver, the dynamic mesh deformation technology in the computational fluid dynamics and the complex aerodynamic shape handling for engineering application. Combined with the experiences of wide-range of engineering applications and aeroelastic research of authors in years, a software for aeroelastic applications is developed, targeting to provide an alternative of complex aerodynamic shape handling and the dynamic mesh technique for full-scale aircraft or wings.

This chapter briefly describes the basic ideas of our software development and introduces the basic functions and application of the software. Finally, the CFD/CSD coupling software is used to analyze several typical aeroelastic problems.

7.2 Introduction to computational aeroelastic software

7.2.1 Software features

The software based on CFD/CSD coupling is a result of continuous effort of our research team for many years. The name of software is Multidisciplinary Flight Dynamics- Aeroelastic Dynamicity (MIFD-AED) [2, 32, 6, 7, 22]. The basic idea of the software is based on the coupling of CFD and computational structural dynamics to solve structural response and loads, and further analyze nonlinear aeroelastic responses with structural large deformation and large perturbation. The software is equipped with unique features as follows.

First of all, the user-defined function (UDF) technology is adopted for the CFD solver, which provides a standard interface for data exchange between multidisciplinary information.

Second, the multivariable nonlinear finite element technique is developed for structural solver. It avoids shear deadlock and volume deadlock, which makes it

https://doi.org/10.1515/9783110576689-007

possible to analyze in a grid with large aspect ratio. Furthermore, the constitutive relationship of the antiisotropic materials is introduced for the composite materials.

Third, besides the conformal dynamic mesh technology, the branch modal method by the split and combination of the modal as a whole is applied to the aero-thermal coupled load analysis of the aircraft.

Fourth, nonlinear acoustic solver is introduced into a multidisciplinary coupled computational framework to account for the effects of noise on the structure.

Fifth, since the conformal dynamic mesh technique adopts the projection technique based on finite element interpolation, the rigid motion or elastic deformation of the aerodynamic components can be considered separately or simultaneously. It lays a solid foundation for the aeroelastic analysis of rigid-flexible coupling problems by combination of the rigid body dynamics and CFD. Moreover, by introducing elastic body dynamics into the multidisciplinary coupled computational framework, the ballistic load analysis of the free-flying aircraft can be performed, without the requirement of the feature points.

Details of the software features are

(1) Aeroelastic analysis of aircraft with complex shape including flutter analysis, buffeting analysis, gust response, control surface reverse effect and limit cycle analysis.
(2) Analysis of dynamic load including transient inertial load analysis of elastic structures, transient stress analysis of elastic structures and unsteady aerodynamic load analysis with elastic structural coupling.
(3) Aerodynamic/aerodynamic noise/elastic analysis, that is, structural response analysis in a comprehensive load environment with the consideration of aerodynamic loads/aerodynamic noise loads.
(4) Aeroelastic analysis with the rigid body motion.

In addition, the software can be used on a single solver for the single-physical problem, in which aerodynamic and structural dynamics can be calculated individually. The software has a friendly user interface with excellent versatility and extension ability. It has been well tested and applied in engineering application, and the application functions of the software are developed and expanded further by our research group [12, 33, 9, 19, 11, 3, 26, 16].

7.2.2 Overall design of the software

Since the 1990s, CFD has been developed rapidly and be able to reliably solve the Euler and RANS equations [23]. Finite element method in structural dynamics and CFD techniques has laid a good technical foundation for the development of aeroelastic computation technology. From then on, aeroelastic analysis and loading analysis based on CFD/CSD coupling technique have been paid much attention.

While extensive researches have been carried out, a number of codes for CFD/CSD coupling simulation in time domain have been developed. Moreover, with the advancing computer hardware, the algorithms and techniques based on CFD/CSD coupling are gradually targeting from scientific research to engineering applications.

Since the 1990s, several foreign software companies and organizations used object-oriented technologies to develop Computer Aided Design/Computer Aided Engineering/Computer aided manufacturing (CAD/CAE/CAM) codes which have launched a large number of well-known commercial software for aeroelastic analysis, as shown in Tab. 7.1 [33, 9, 15, 5, 8, 4]:

Tab. 7.1: Typical aeroelastic software in the world.

Software	Developer	Algorithm for aeroelastic solution	GUI
FLEXSTAB	Boeing	Linear potential flow theory for unsteady aerodynamic force/linear beam theory, frequency domain coupling	N/A
DYLOFLEX	Boeing	Lift surface theory, slender body theory, vortex lattice method or quasi-steady theory + structure modal method + linearized small perturbation motion equation, frequency domain computation, introduction of dynamic analysis	N/A
Flds	MSC	Vortex lattice method + modal superposition method	Uses Patran as the user interface; it has preprocessing and postprocessing functions
ZAERO	Zona tech	Transonic small perturbation + linear modal + linear time-varying state space control model, frequency domain computation.	Has user interface for pre- and postprocessing
STARS	NASA Dryden	CFD method for solving aerodynamic force + structural modal analysis	No user interface, pre- and postprocessing functions depend on other software, input files need to be manually configured
ENSAERO	NASA Ames Center	Finite difference solution for Euler/ N-S equation + linear structure modal coupling solution, interface data transfer using bilinear interpolation method	The same as earlier
CFL3D V6	NASA Langley Center	CFD method + linear structure modal coupling solution	The same as earlier

Tab. 7.1 (continued)

Software	Developer	Algorithm for aeroelastic solution	GUI
Fluent	Ansys	Aeroelastic analysis using CFD solver coupling with CSD solver with UDF interface.	Has the CFD solver interface and the grid processing software, the structural coupling data and coupling interface require manual configuration by the user
CFX	AEA Technology	Aeroelastic analysis using CFD solver coupling with CSD solver with custom interface	The same as earlier

The aeroelastic software shown in Tab. 7.1 is mainly divided into four categories:

a) Classic method-based software without graphical user interface (GUI), such as FLEXSTA and DYLOFLEX. It is completely dependent on the user to specify the model data with limited modeling function. Because such software uses empirical method for the simplification of model, the reliability and accuracy of the software is reduced.
b) Classic solution method with GUI such as Flds and ZARERO.
c) Modern aeroelastic solvers based on CFD/CSD coupling, such as STARS, ENSAERO and CFL3DV6. The softwares mentioned earlier have no perfect GUI.
d) CFD software is coupling with the structure solver by using its user interface or the UDF for the aeroelastic computation, such as Fluent and CFX.

In light of the analysis and implementing the idea of foreign CFD/CSD software, the development scheme for aeroelastic software is presented based on the basic principle of CFD/CSD coupling (see Fig. 7.1). Figure 7.2 shows the design chart of the CFD/CSD coupling software.

In Fig. 7.2, CFD code is the foundation of the whole framework. Users can use secondary development interface of CFD software to set initial values, boundary conditions and dynamic mesh settings. Furthermore, CSD solver and other solvers (e.g., NHT solver for heat transfer) can couple with CFD solver through the interface for multiphysical simulation.

7.2.3 Software system framework

The system is divided into two parts: Database and DatabaseUser. Solver, FieldInterface as coupling interface and PostProcessor as postprocessing module are subclasses of DatabaseUser. When DatabaseUser uses the data from Database, there is a one-to-n relationship between them, as shown in Fig. 7.3.

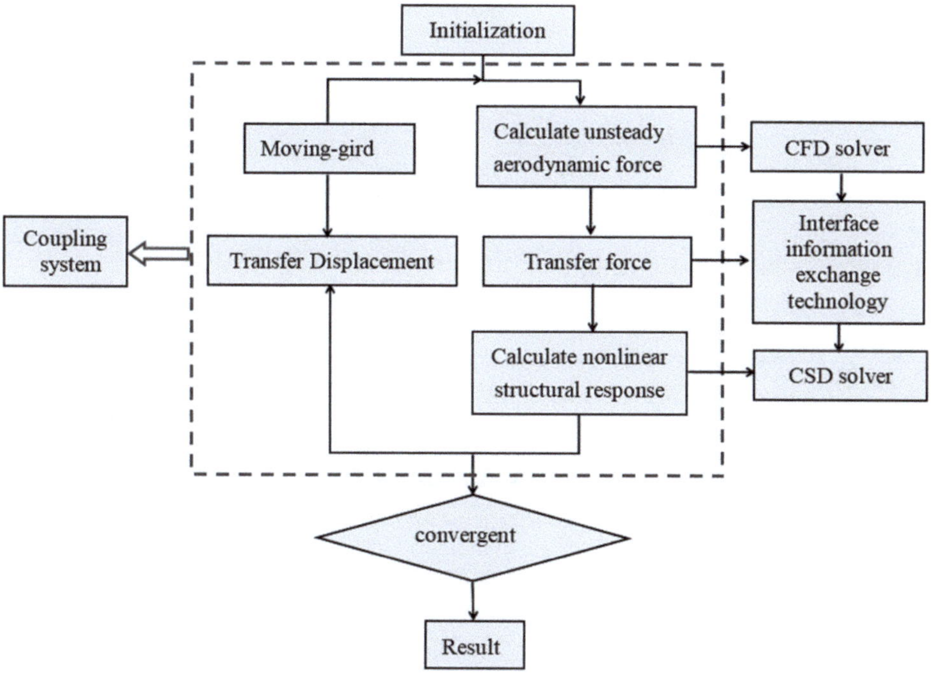

Fig. 7.1: CFD/CSD coupling principle for aeroelastic simulation.

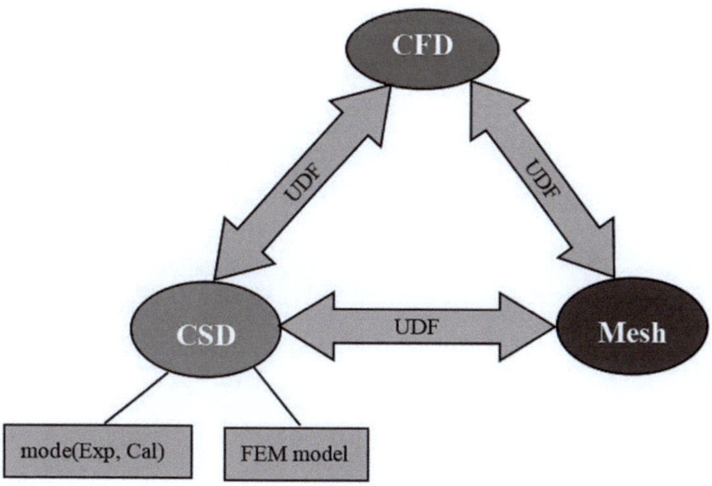

Fig. 7.2: The aeroelastic software design chart.

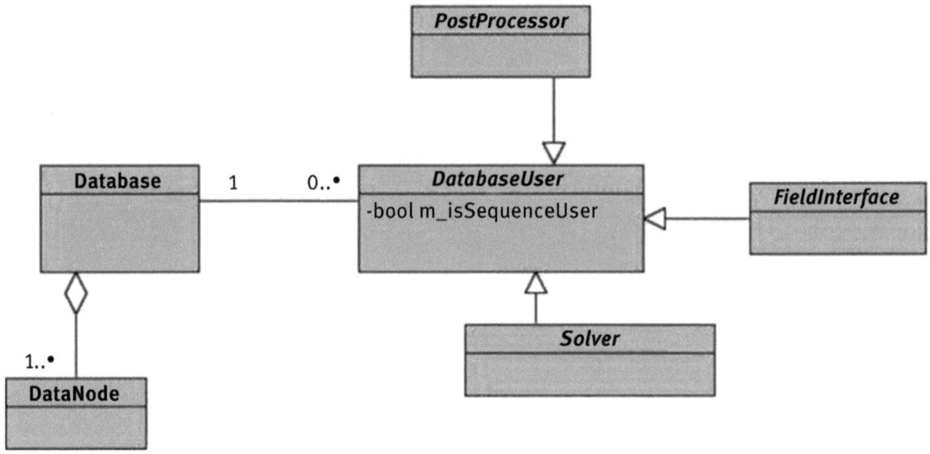

Fig. 7.3: Class diagram of system overall framework design.

1. Database design

Task is a container of data that determines whether a data user can manipulate data and provide an operational interface for data transfer. It is not responsible for the setup and modification of data. The setup of the data is implemented by the preprocessed interface module, and the modification of the data is the task of the solver.

Implementation is referred to the CGNS standard, based on the layers. It is implemented in a tree structure. A schematic diagram of the tree structure is given in Fig. 7.4.

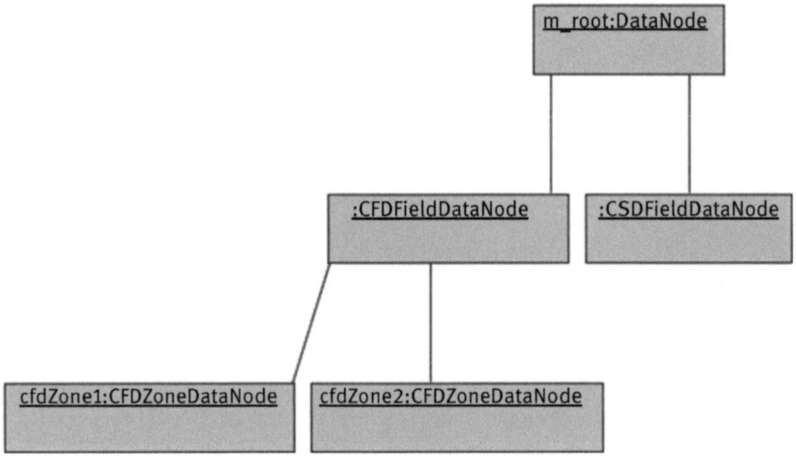

Fig. 7.4: Tree structure of the node under Database.

Code implementation:

```
class Database
{
    DataNode *m_root;
};

class DataNode
{
    DataNode *m_parent;
    std::vector<DataNode *> m_children;
};
```

2. Workflow of usage of Database by DatabaseUser

DatabaseUser should explicitly indicate to the Database which node to be used by the path. The node path is represented by an integer array. If available, the return value of requestUseNode(std::vector<int> const&,DatabaseUser*) is a pointer to the node, otherwise NULL. If it is NULL, DatabaseUser will wait in the queue of the node. When each DatabaseUser is finished to use, the end call signal of Database usedOver() is called, which is used to call the node or the child node to use the next DatabaseUser's wakeUser() in the queue to wake up the waiting user. The user processes in wakeUser() to restart the usage process (see Fig. 7.5).

The main purpose of the process is to coordinate the behavior between the solver and the postprocessing, because the process of the solver coupling is performed sequentially, and the refresh action of the postprocessing is random. In order to prevent conflicts between them, it is necessary to setup such a process (see Fig. 7.6).

The earlier sequence diagram is to coordinate the data stored in the Database with the structure of the private data saved in solver.

7.2.4 Data user and its subclasses

Figure 7.7 shows an abstract class for database user. Database users have three categories: solver, postprocessing and data transfer interfaces.

Postprocessing is achieved by Tecplot. The solver consists of CFD solver, CSD solver and coupled solver. Of course, each type of solver may be subdivided with the requirement of further functions and will not be described in detail here.

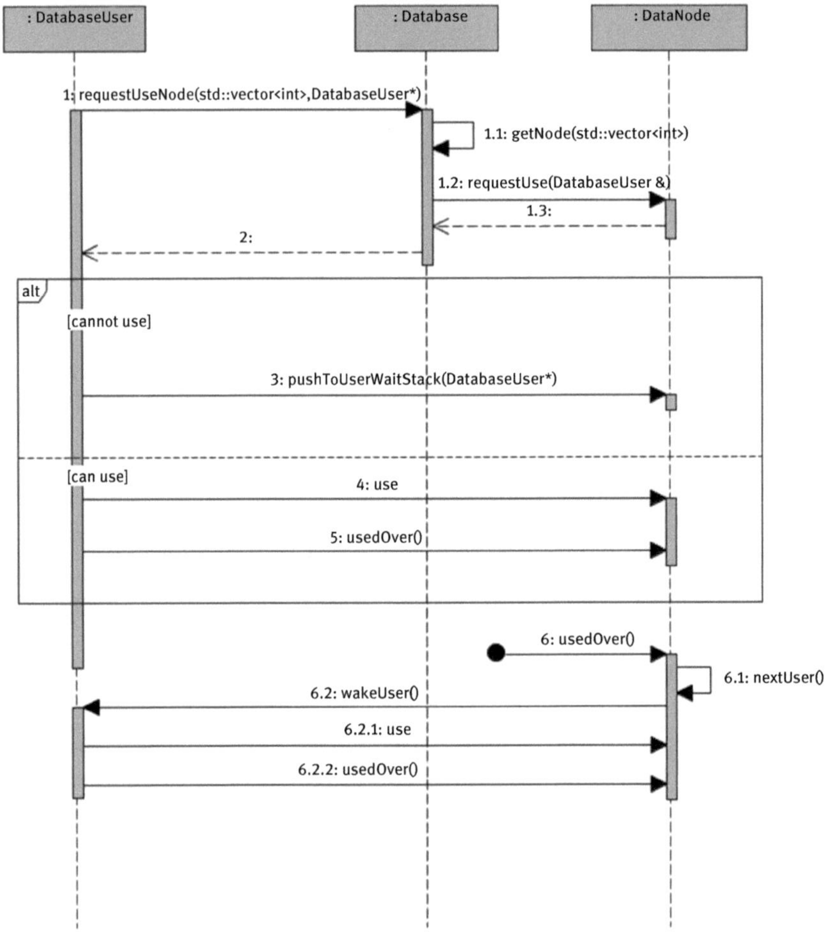

Fig. 7.5: DatabaseUser's workflow using Database.

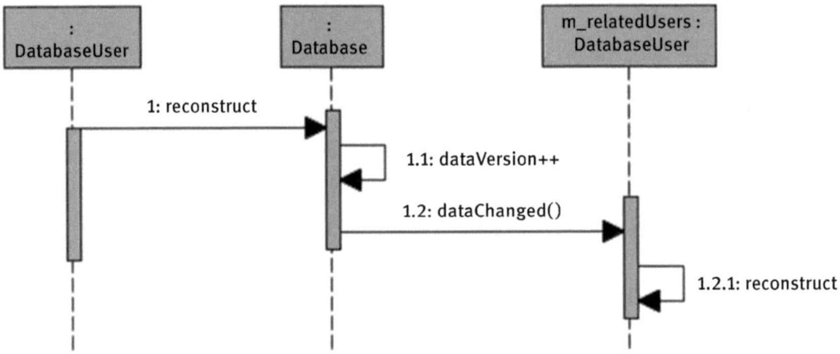

Fig. 7.6: Process after DatabaseUser changes the data structure of a node on Database.

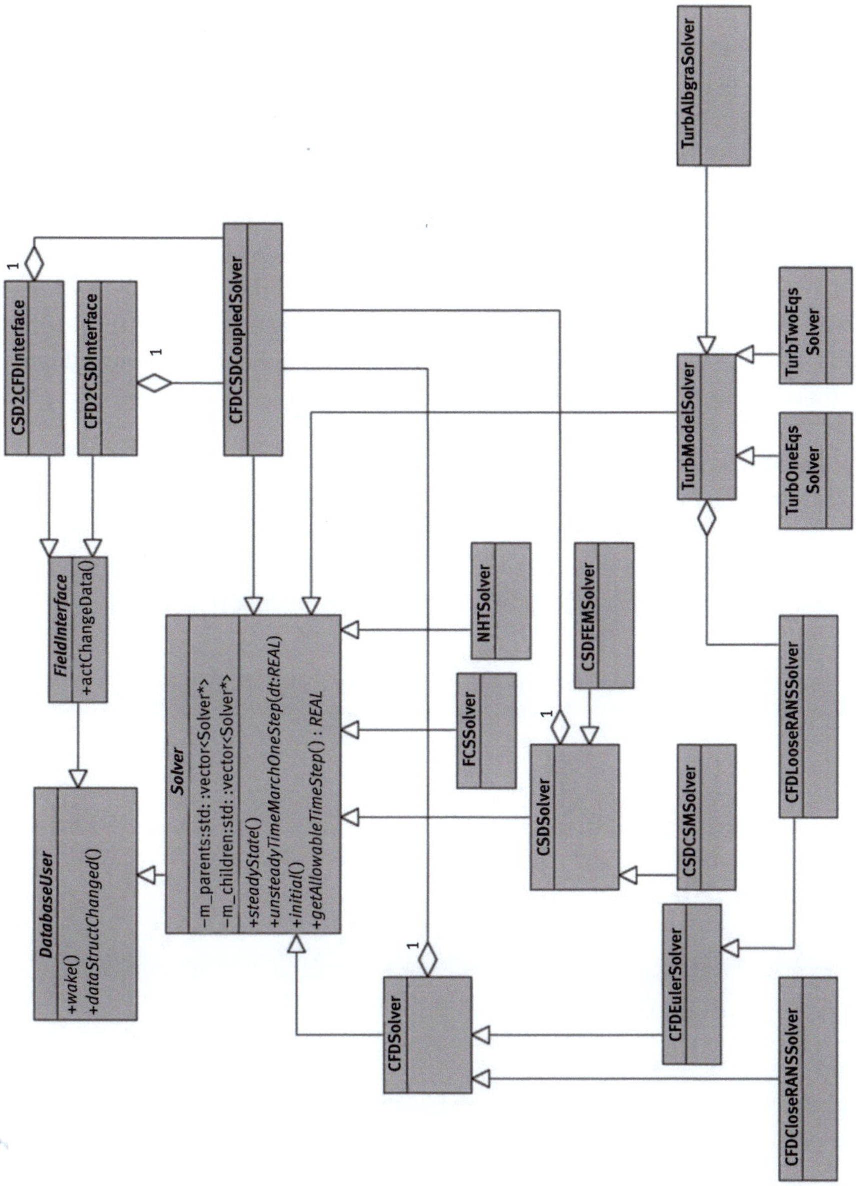

Fig. 7.7: Relationship between data user and subclasses.

1. Solver object

Solver is an abstract solver class, and all other classes are inherited from it.

Solver class needs to provide the following operations: steady solution, single-step time advancement, initialization operation and the requirement of the time step.

The member variables of the coupled solver class include a single-field solver and a coupling interface. There is no call relationship between the single-field solvers and neither between the coupled interfaces and between the single-field solver and the coupled interface. The coupling between them depends on the Database, that is, they change the data values in the Database, and other database users read the data in the latest Database and continue their own function. The coupled solver only calls the single-field solver that it owns and the operation of the coupled interface, without the operation on the Database.

2. Coupling interface

A coupling interface is responsible for the one-way coupling operation. For example, the data transfer from CFD to CSD is handled by a coupling interface CFD2CSDInterface, and the one from CSD to CFD is handled by CSD2CFDInterface.

The sequence diagram for CFD/CSD coupling is given later (Fig. 7.8).

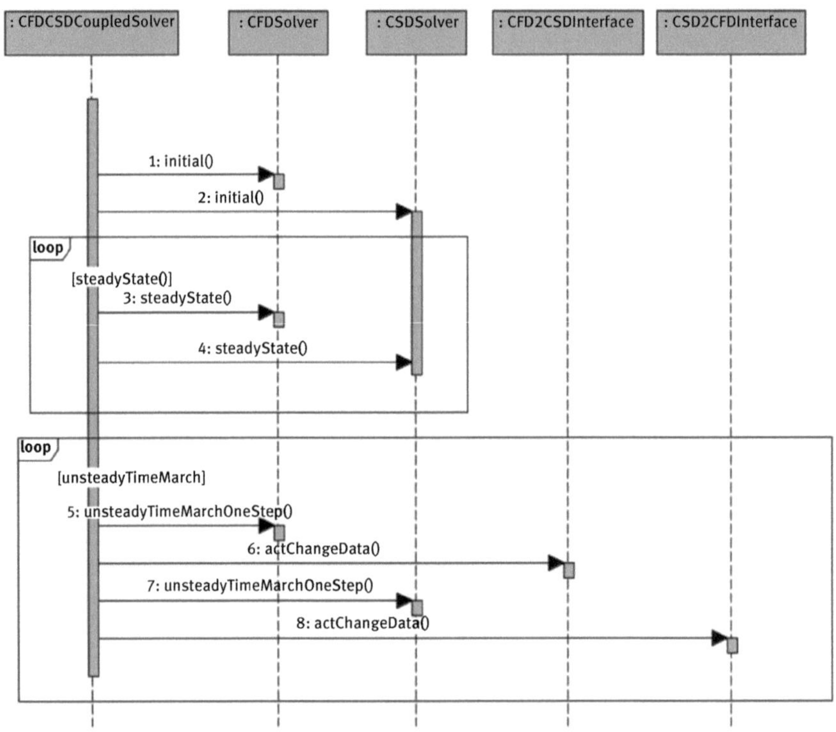

Fig. 7.8: Sequence diagram of CFD/CSD coupling.

7.3 Software usage

Input and output files

The input and output files and the specific functions of each file required for a typical CFD task are shown in Tab. 7.2.

Tab. 7.2: Input and output files in CFD task.

Input file	Grid file	Stores all coordinate information of the CFD model in PLOT3D format
	Boundary condition file	Stores the generic format of boundary condition in each face in Gridgen15 format
	Control file	Setup input and output requirements, gas parameters, solution methods (temporal/spatial scheme, etc.), residual requirements and so on
Output file	Residual file	Contains the equation residual value during calculation
	Force and moment files	Contains aerodynamic forces and torque of the aircraft
	Log file	Record program information during calculation
	Surface contour	Output wing, airfoil surface solution
	Flow field contour	Output flow variable in whole flow field
	Restart file	Save intermediate results for restart computation

Work process

Before solving, prepare the grid file, boundary condition file, and write the control file according to the task requirements. The grid file and boundary condition file can be generated by using the commercial grid generation software like ICEM or Pointwise. After reading the three input files, the computational result is written into the output file. The flowchart is shown in Fig. 7.9.

7.3.1 Preprocessing software

The interface of the preprocessing software is shown in Fig. 7.10. From top to bottom is the scalar bar, menu bar, toolbar and main window area, respectively. The main window is divided into four parts. The largest area in the upper right corner is the grid display and operation area, whilst the lower right is the output area of the information. The upper left is the display area of the CFD grid boundary information, whilst the lower left is the display area of the structure information.

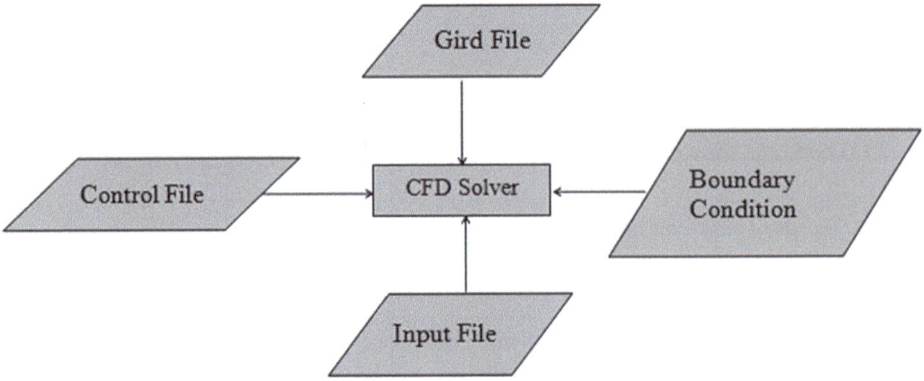

Fig. 7.9: Flowchart of CFD simulation task.

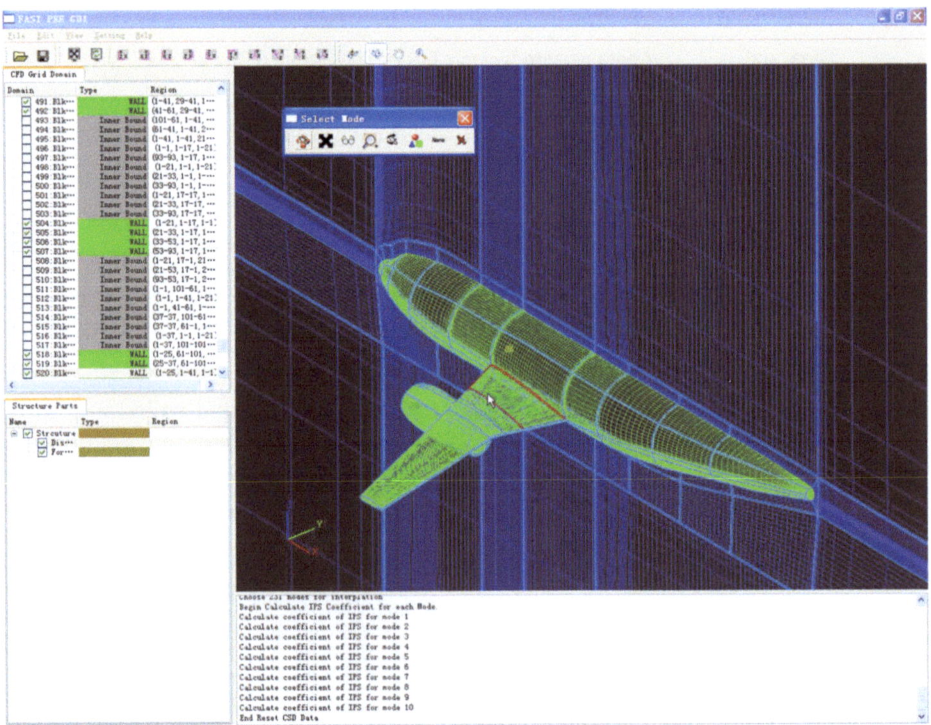

Fig. 7.10: Main interface of preprocessing software.

It is mainly used to specify the interpolation interface between the CFD grid and the finite element model and preview the interpolation effect. It also prepares input information such as CFD grid, boundary conditions, interpolation interface information, structural generalized mass matrix and generalized stiffness matrix for the solver. It is interactive GUI for user to specify the interpolation interface by mouse pick-up for the boundary of the CFD grid block on the screen and to specify which structure block to exchange data with. It is implemented by select function in OpenGL. The main interface of the preprocessing software is shown in Fig. 7.10.

1. CFD grid import and display settings

The preprocessing software has an interface with the commercial grid generation software such as ICEM and Gridgen/Pointwise. By importing the grid file and the boundary condition file (the import interface is shown in Fig. 7.11), the topology and boundary information of the multiblock grid can be established (Fig. 7.12) while the grid is displayed on the main interface, as shown in Fig. 7.10.

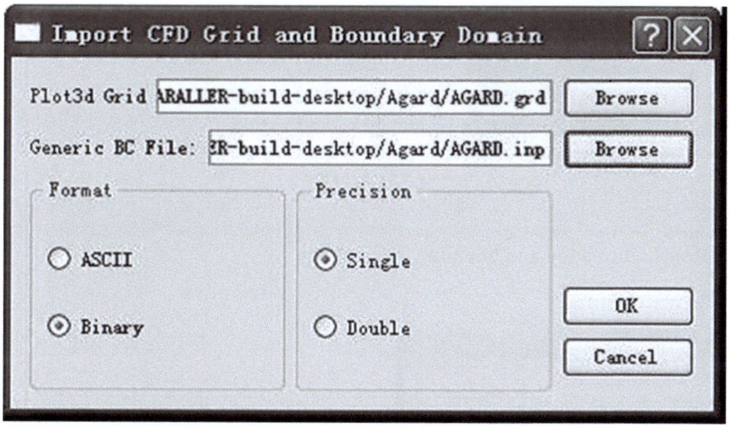

Fig. 7.11: Specify the imported CFD grid file and grid boundary information file.

2. Import of structure information

The preprocessing software is compatible with the file format of Nastran, which is widely used for structural analysis in aerospace engineering. By importing Nastran BDF files and F06 files, (import interface as shown in Fig. 7.13), the structure nodes and modal information can be obtained to automatically calculate the interface interpolation parameters. Users can freely select the interpolation direction via the interface and the modal for the coupling computation. These information can be displayed or modified through the interface shown in Fig. 7.14. The imported structural nodes are displayed on the main interface, as shown in Fig. 7.15.

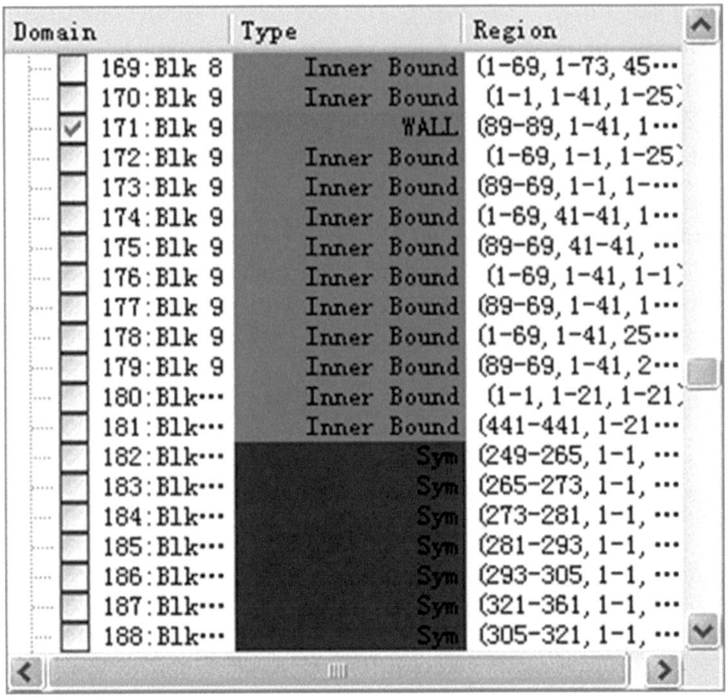

Fig. 7.12: CFD mesh topology information and boundary information.

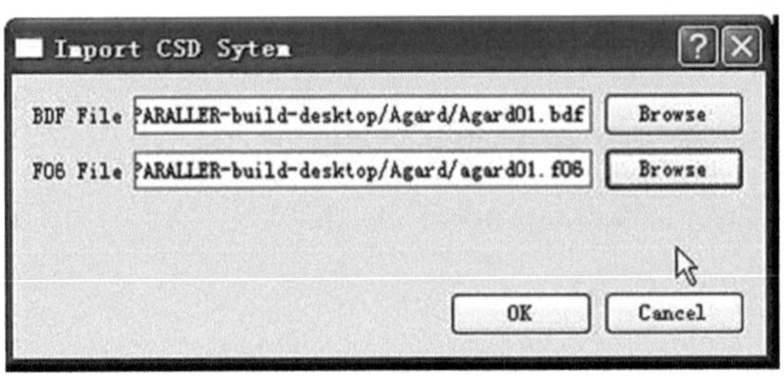

Fig. 7.13: Import of BDF file and F06 file.

3. Specification of interpolation interface

After the CFD grid and the structure finite element model are sequentially imported, right-click the structure item on the structure information display area (when a structural component is imported, a corresponding item will be created in this area).

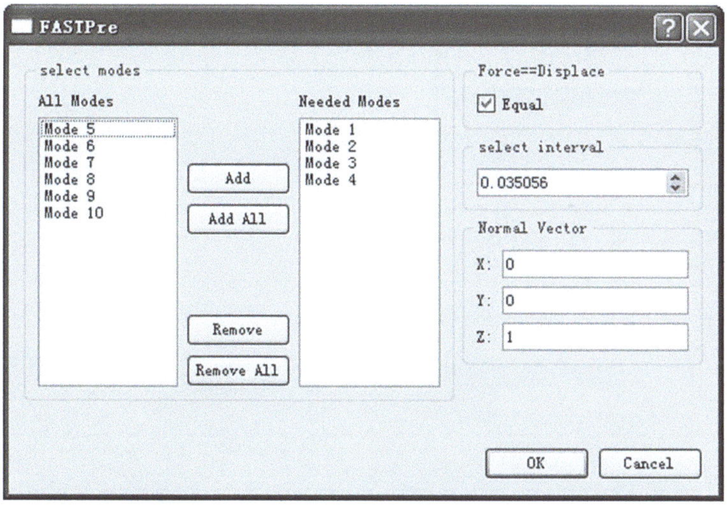

Fig. 7.14: Display and setting of structural parameters.

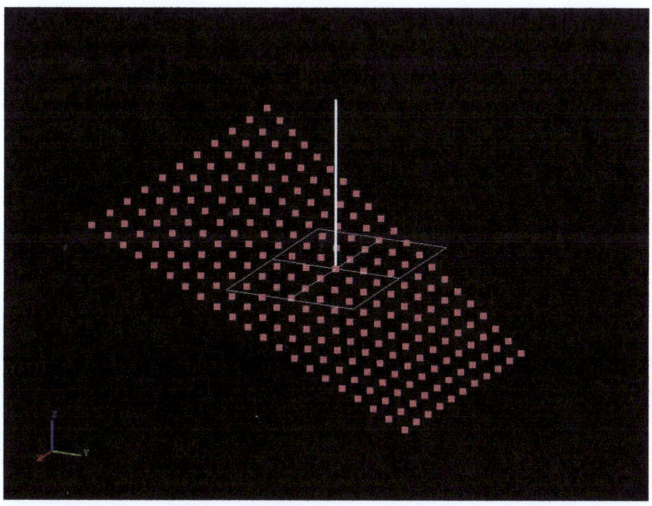

Fig. 7.15: Imported structure nodes.

Click the Associate button when the menu for selecting the interpolation interface pops up, as shown in Fig. 7.16. After the earlier operations, users can enter the pick-up mode that establishes the connection between CSD and CFD grid. The main interface will pop up the pick-up operation box, as shown in Fig. 7.17, and then users can use the mouse to pick the CFD grid area associated with the selected structure item. After the operation is completed, the CFD grid block associated with the selected structure item will be displayed in the tree structure. In Fig. 7.18, the results

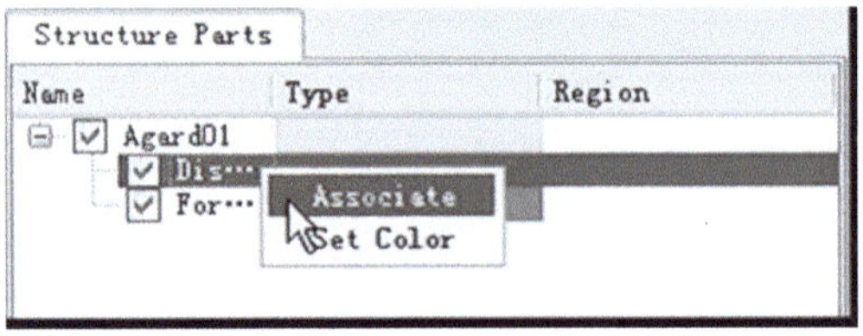

Fig. 7.16: Display and operation of structural information.

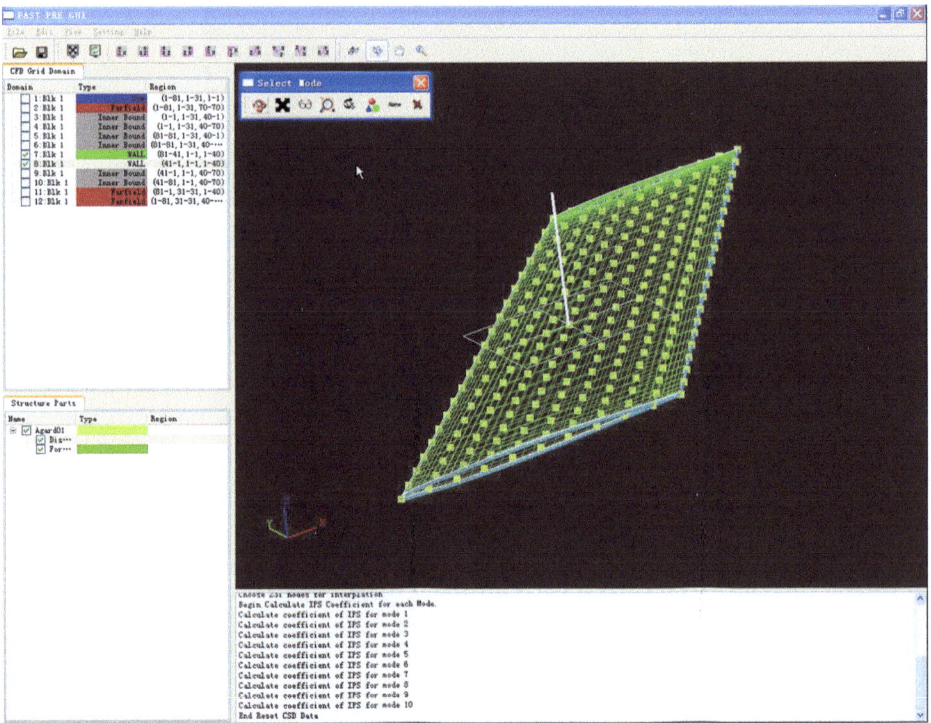

Fig. 7.17: Pick mode.

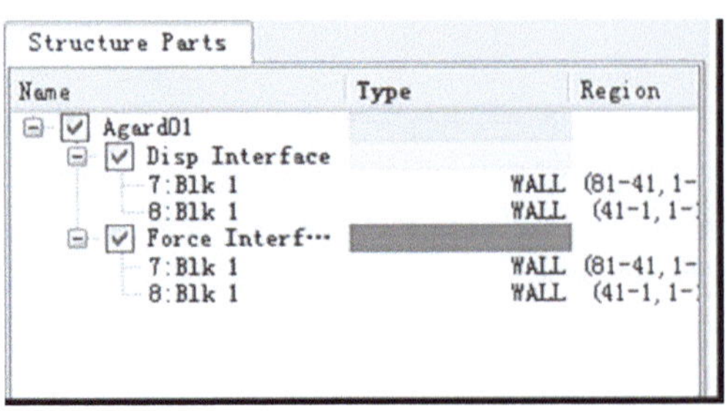

Fig. 7.18: The interpolation interface displayed in the GUI.

are displayed after the selection: two surfaces are selected as the interpolation interface for displacement and force.

After the association of coupling interface, right click on the structure item and enter the setting interface as shown in Fig. 7.18. Users can select the modals in the interface, and set whether the interpolation interface of the force is consistent with the interpolation interface of the displacement. Furthermore, user can input the spacing when filtering the interpolation points. If the distance between two adjacent results is smaller than this threshold, two points will be considered to be one point, that is, only one of them is selected. By setting the normal direction, the direction of interpolation will be along this normal direction, so that the infinite-plate splines (IPS) interpolation can be performed in any desired direction. After the setting is completed, click the OK button. The program will complete the interpolation coefficient according to the modal in background calculation and display the calculation result in the output window of the main interface.

When the setting is completed, user can enter the preview mode by clicking the preview menu under the Edit menu. The setting window of the generalized displacement of each modal is displayed on the interface, as shown in Fig. 7.19. After the desired generalized displacement value is input, click Preview to display the result of the deformation in the display window. Figure 7.19(a), (b) show the mesh deformation as the generalized displacement of the first- and second-order modals of the Agard 445.6 wing is 0.1, respectively. The program also automatically completes the calculation of the mesh deformation in the background. If there is a negative volume or a gap between adjacent grid points, a prompt of warning will be given in the output window.

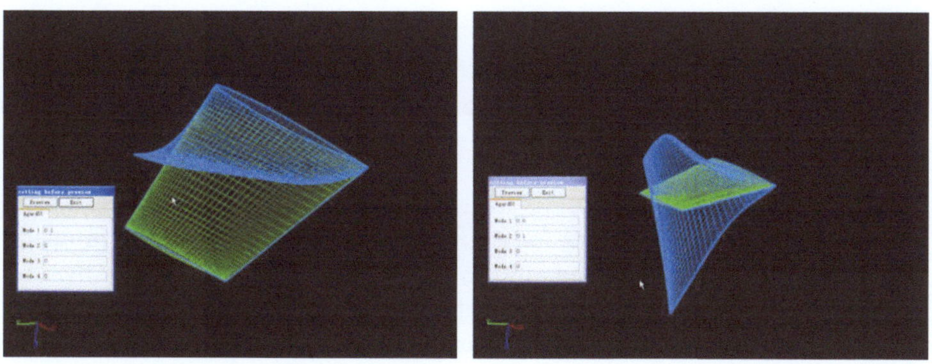

(a) The first -order generalized displacement = 0.1; (b) The second-order generalized displacement = 0.1

Fig. 7.19: Mesh deformation Preview displayed on FAST Pre.

7.3.2 Solver interface

For the convenience of user's setting, the interface is designed for the solver as shown in Fig. 7.20. From top to bottom, there is the scalar bar, the menu bar and the main display region. The main display region can be divided into four parts. The upper right part occupies most of the interface for the display of curves and contour, whilst the lower right part is the output window of the solver solution information. The upper left part shows the structure and CFD model information, and the lower left part is the solution control panel for starting/pausing the solution process.

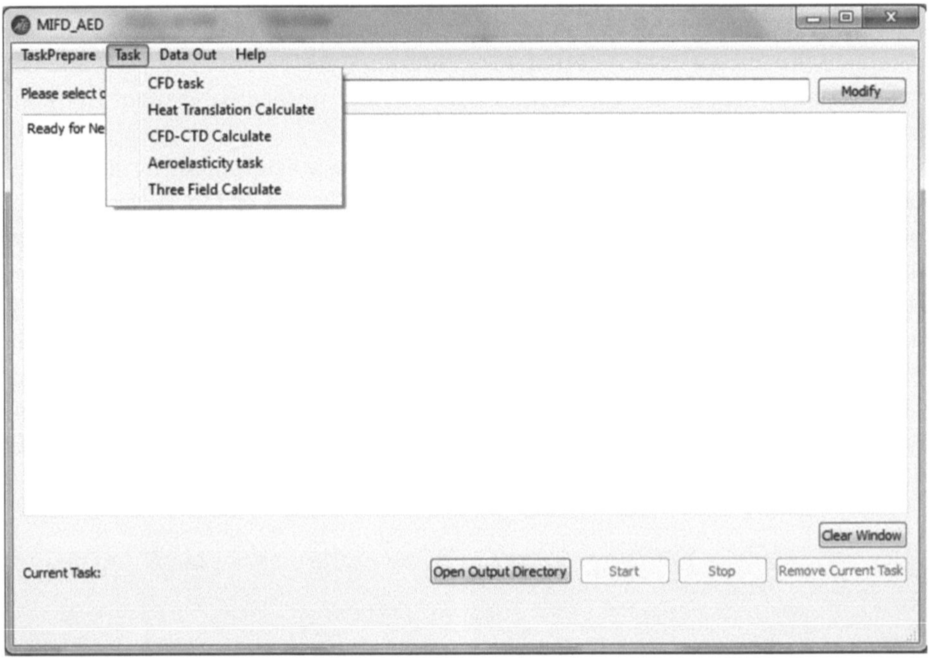

Fig. 7.20: Main interface of solver.

The software developed targets for engineers to solve the aeroelastic problems in engineering conveniently and efficiently. It has the functions such as the flutter boundary prediction and the extraction of dynamic load. There are variety of outputs like contour and curve display model, which can directly display the solution results, which is fully independent on commercial postprocessing software.

7.3.3 Data display and processing

The data display and processing is an important part of the postprocessing work of CFD. CFD postprocessing softwares such as Tecplot and fieldView all have this function, and so are the CFD solvers (such as Fluent and CFD++.). The software we develop also provides the function of data processing and display. Figure 7.21 shows the residual curve of the solver output, the dynamic response during CFD/CSD coupling. OpenGL can be easily implemented for the display of contours and curve. Figs 7.22 and 7.23 show the contours in two-dimensional and three-dimensional cases.

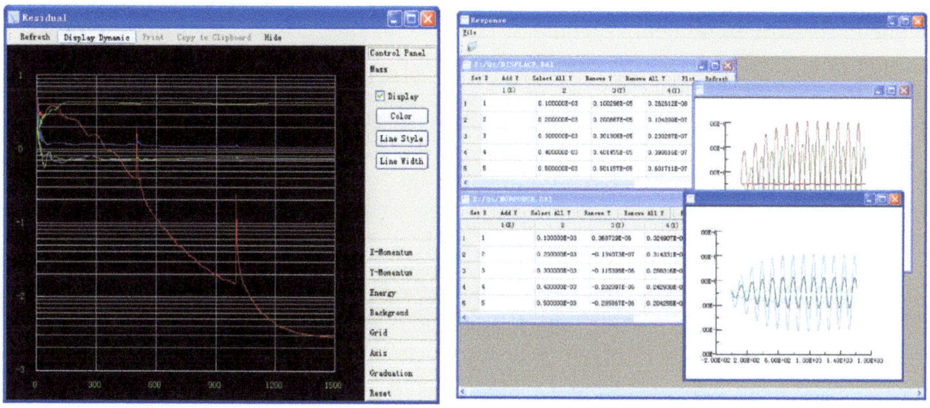

Fig. 7.21: Display of curves and data.

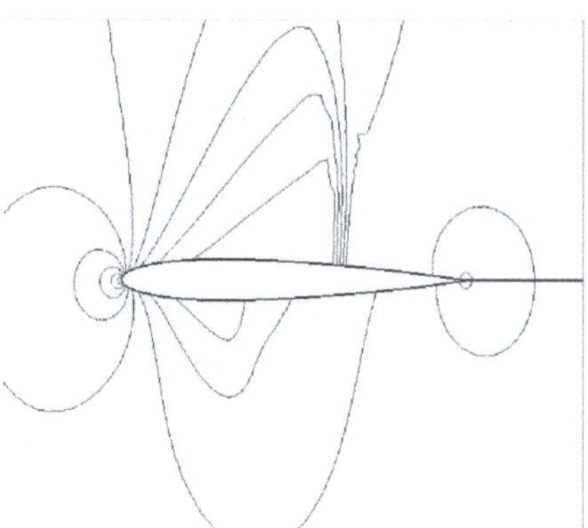

Fig. 7.22: Two-dimensional contour.

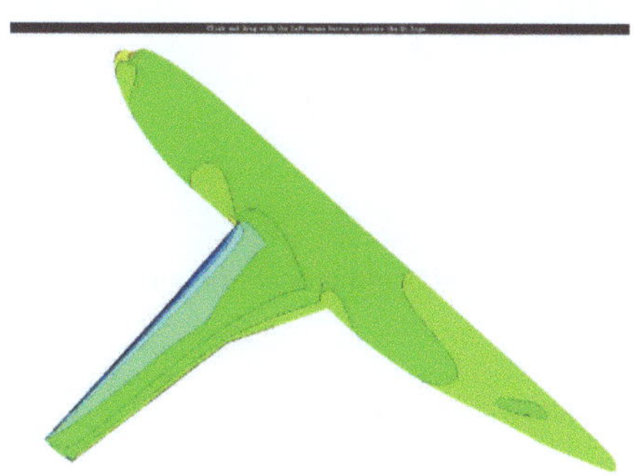

Fig. 7.23: Three-dimensional contour.

7.4 Aeroelastic software application based on CFD/CSD coupling method

7.4.1 Static aeroelastic analysis of large aspect ratio wings

1. Static aeroelastic model of wing

The aeroelastic problem of large aspect ratio wings is generally not just limited to the flutter analysis, but the static aeroelastic problem, that is, the elastic deformation and the effect on the aerodynamic performance as deformation [29, 30]. In this section, a typical wing model of civil airplane is selected. The wing model is composed of beams, ribs, skins and other components of metal materials. The half-span length of the wing is 8.79 m. Figure 7.24 shows the finite element model of the wing, consisting of 48 beam elements, 119 plate elements and 224 shell elements for skin. Table 7.3 shows the frequency and modal shape of the first four modes of the wing. Figure 7.25 shows the shape of the first four modals.

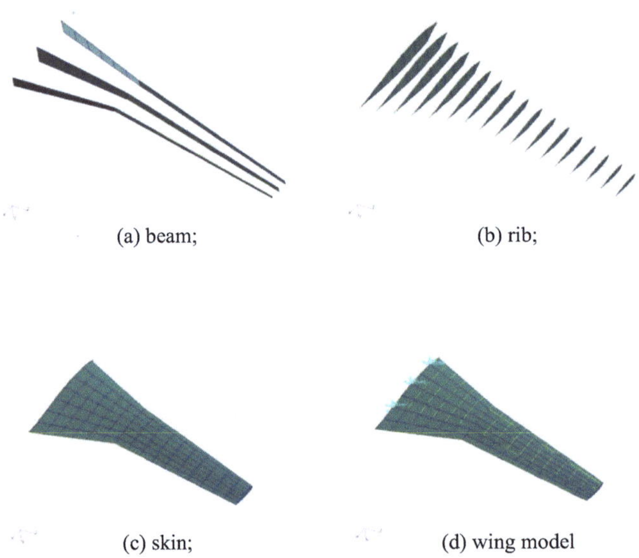

Fig. 7.24: Finite element model of the wing structure.

Tab. 7.3: Wing's modal parameters.

Order	**1**	**2**	**3**	**4**
Frequency ((Hz)	5.22	21.03	42.70	46.96
Modal shape	Bending	Torsion	Torsion	Bending

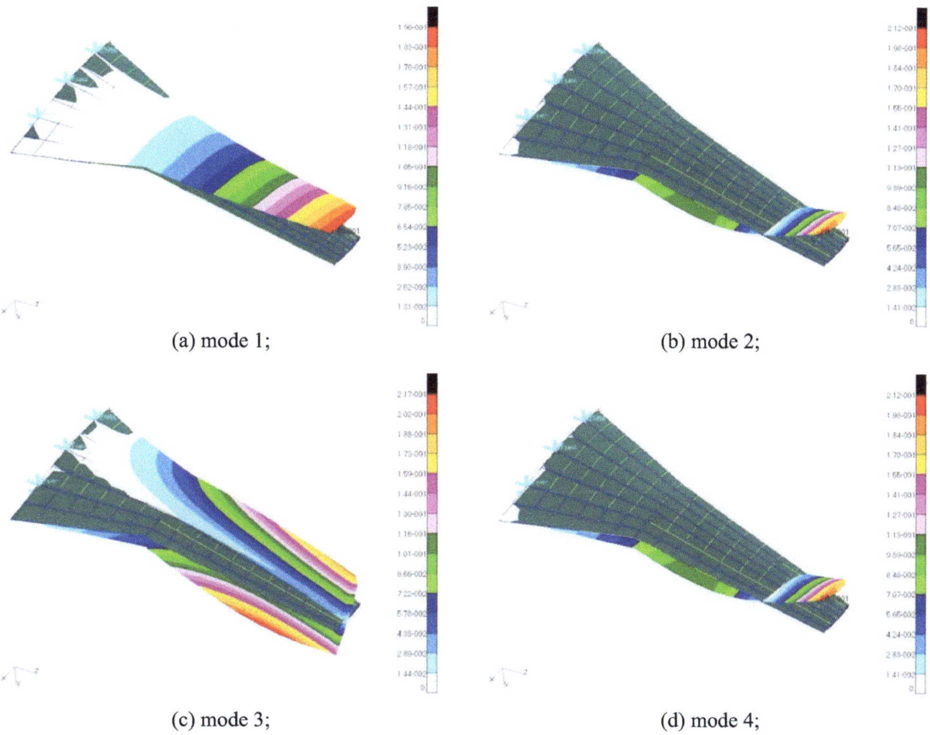

Fig. 7.25: Modal shape of the wing structure: (a) mode 1; (b) mode 2; (c) mode 3; and (d) mode 4.

The aerodynamic mesh of the wing is 81 × 31 × 70, as shown in Fig. 7.26. Figure 7.27 shows the steady pressure coefficient contour at $Ma_\infty = 0.78$, $\alpha = 2.0°$. Figure 7.28 shows the interpolation from the structure shape to the aerodynamic grid using the IPS method. Figure 7.29 shows the mesh deformation of the wing surface, where the deformation at the wing tip exceeds 1.0 m, which proves that the transfinite interpolation method (TFI) method can be used for large deformation.

2. Static aeroelastic analysis based on CFD/CSD coupling

The aeroelastic calculation condition is $Ma_\infty = 0.78$, $\alpha = 2.0°$, and the flight altitudes are $H = 5$, 10, 15, 20 km, respectively. Figure 7.30 shows the aeroelastic response of the wing at different altitudes. The structural response is of well convergence using the CFD/CSD coupling method. It is seen from Fig. 7.30 that the longer convergence time is required with the increase of altitudes. Furthermore, severe oscillation is responding to a small deformation after the convergence. Figure 7.31 shows the deformation and pressure distribution of the wing at 5 and 20 km flight altitudes,

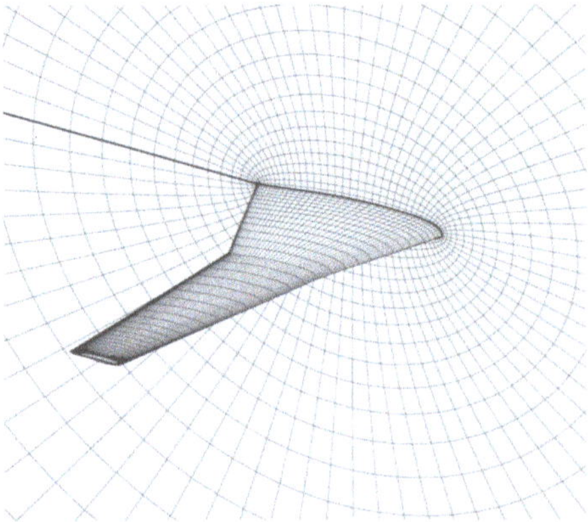

Fig. 7.26: The aerodynamic grid of the wing.

Fig. 7.27: Pressure coefficient contour of wing.

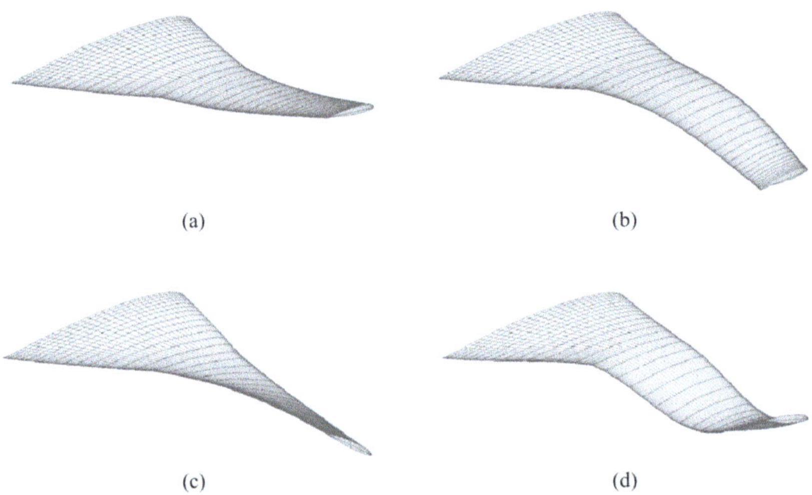

Fig. 7.28: Schematic diagram of structural modal interpolation to aerodynamic grid: (a) mode 1; (b) mode 2; (c) mode 3; and (d) mode 4.

respectively. Table 7.4 shows the lift coefficient C_{Le} after elastic deformation and the ratio of C_{Le} to the one in the rigid case. It is seen that at the same Mach number, the decrease of lift coefficient is smaller with the higher flight altitude or smaller dynamic pressure.

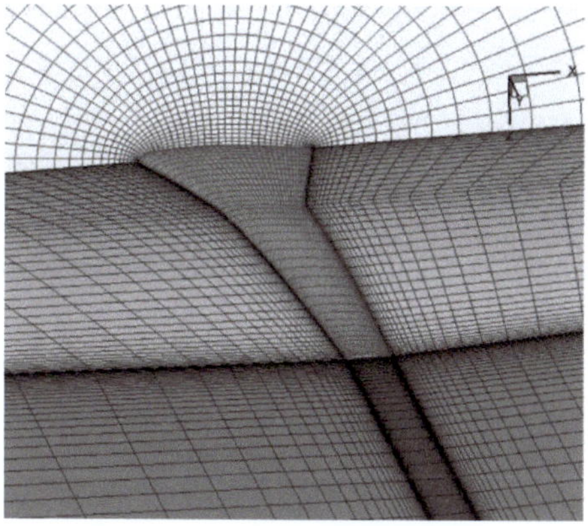

Fig. 7.29: Schematic diagram of mesh deformation.

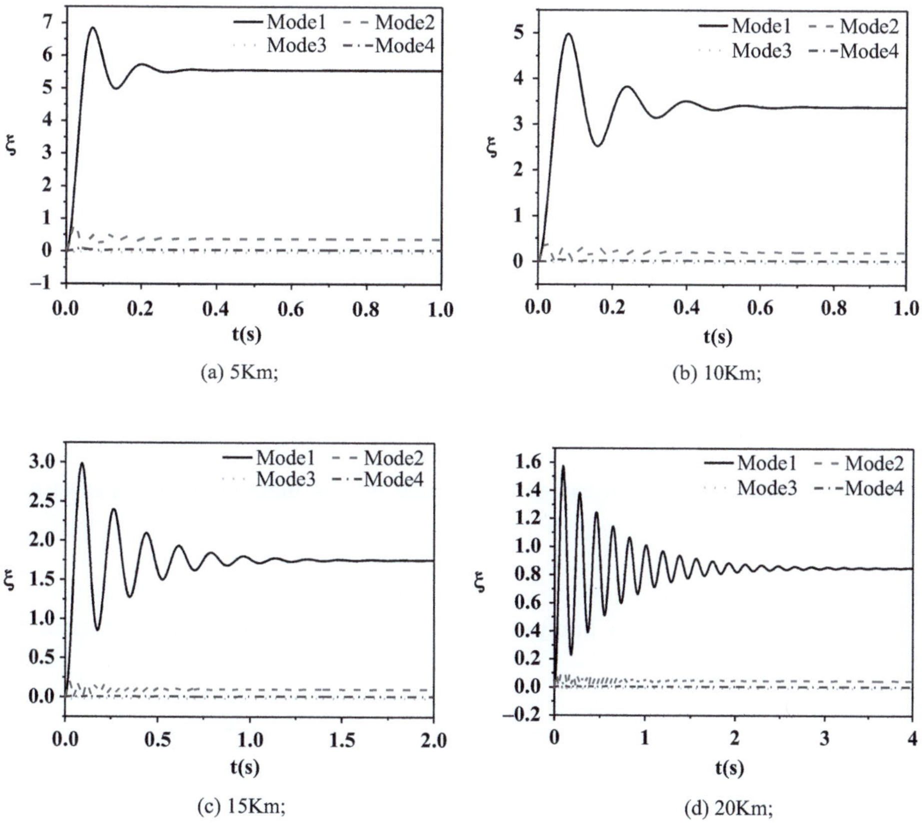

Fig. 7.30: Time history of structural response at different altitudes: (a) 5 km; (b) 10 km; (c) 15 km; and (d) 20 km.

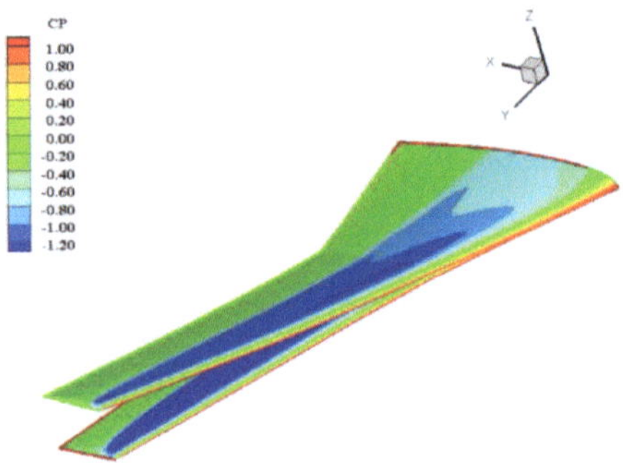

(a) 5km;

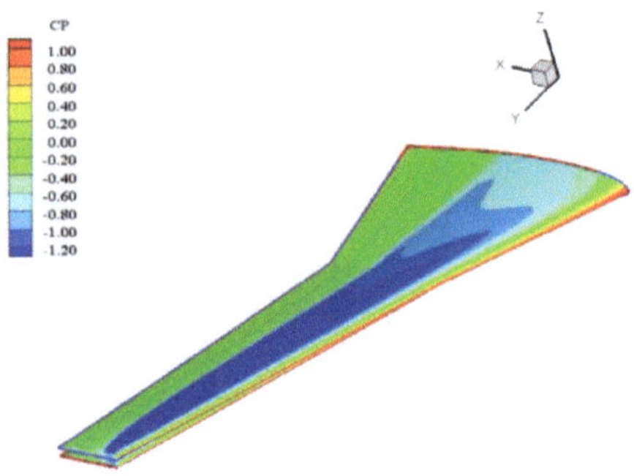

(b) 20km;

Fig. 7.31: Deformation of the wing at different altitudes: (a) 5 km and (b) 20 km.

Tab. 7.4: Comparison of lift loss at different altitudes.

H (km)	Deformation (m)	C_{Le}	C_{Le}/C_{Lr} (%)
5	1.167	0.370	77.24
10	0.707	0.417	87.06
15	0.365	0.448	93.53
20	0.176	0.464	96.87

3. Influence of geometric nonlinearity on static aeroelasticity

In this section, the nonlinear aeroelastic analysis of the wing is carried out. By studying the influence of geometric nonlinearity on the static aeroelasticity of the wing, the necessity of nonlinear geometry role on the aeroelastic analysis is discussed [3, 15, 5, , 8, 21].

Figure 7.32 shows the static deformation of the wing tip. It is seen that the tip deformation exceeds 1.0 m and the remarkable nonlinear phenomenon appears.

The nonlinear aeroelastic analysis is performed with the maximum deformation at $Ma_\infty = 0.78$, $\alpha = 2.0°$ and $H = 5$ km. Consider the nonlinear elastic force of the third-order bending modal, and omit the effect of the nonlinearity of the torsional modal. The number of nonlinear static tests is 19. The nonlinear aeroelastic analysis is performed by CFD/CSD coupling method. Figure 7.33 shows the nonlinear generalized displacement response, while Fig. 7.34 compares the linear and nonlinear responses of the first-order generalized displacement. Figure 7.35 shows the actual deformation between the linear analysis and the nonlinear analysis. Since the wing deformation is basically within the range of linear deformation, the nonlinear result is almost the same as the linear one, with the error between two methods about 1%

for the wing tip deformation. The results indicate that the geometric nonlinearity of the wing is negligible within the deformation case for the aeroelastic analysis.

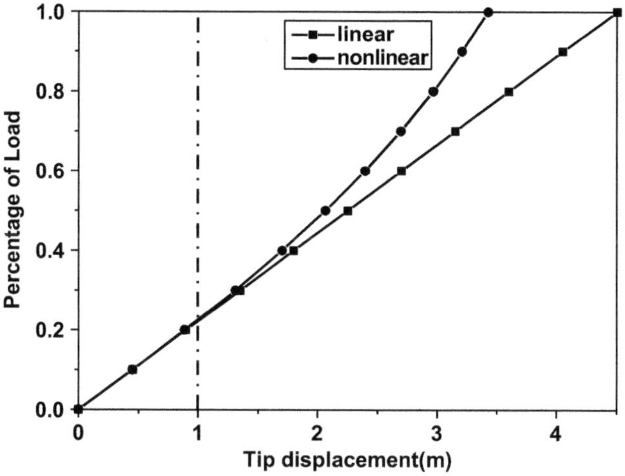

Fig. 7.32: Static analysis of the wing tip node.

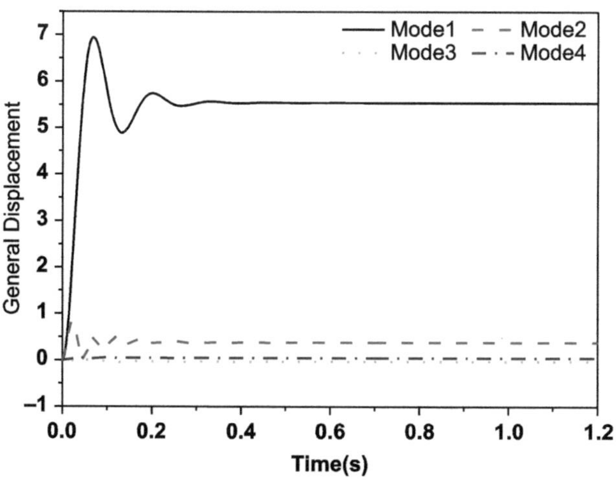

Fig. 7.33: Nonlinear generalized displacement response.

7.4.2 Supersonic flutter analysis of aerodynamic control surface

As an important component of aircraft, the control surface is connected with the aircraft body through the axle with relatively low stiffness. When the aircraft is flying in high speed, the all-movable control surface often encounters severer aeroelastic problem (e.g., flutter) due to the elastic shaft system. In this section, aeroelastic ROM based on Volterra series is used to analyze the flutter of a supersonic all-movable control surface with low aspect ratio [33, 19, 11, 16, 18, 14, 31].

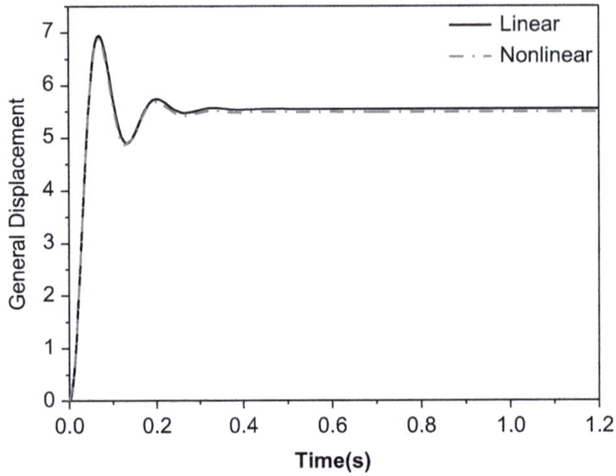

Fig. 7.34: Comparison of first-order generalized displacement.

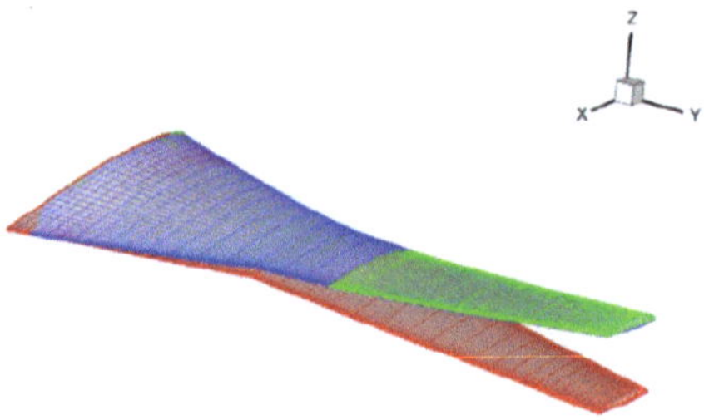

Fig. 7.35: Comparison of actual displacement.

1. Flutter model of control surface

The finite element model of the control surface is shown in Fig. 7.36. The shaft is modeled in beam elements connecting to the control surface by Multi-Point Constraint (MPC). The fixed boundary condition is used for shaft root. The frequency and modal shape of the first four modes are shown in Tab. 7.5. The first two modes are the modals of axis, which are the bending and torsion modal, respectively. The third and fourth modes are for control surface. The deformation modes are the bending and torsion of the control surface, shown in Fig. 7.37.

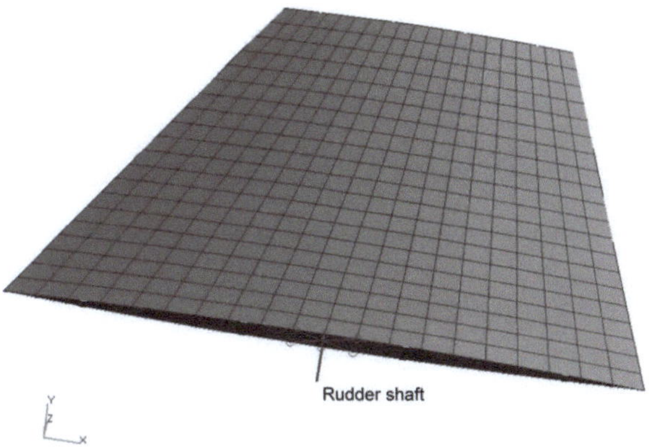

Fig. 7.36: Finite element model of control surface.

Tab. 7.5: Four-order modal frequency and modal shape.

Modal	First order	Second order	Third order	Fourth order
Frequency (Hz)	32.356	52.97	300.87	430.78
Modal shape	Shaft bending	Shaft Torsion	Surface bending	Surface torsion

The aerodynamic grid adopts an O-H topology, and the dimension is $81 \times 31 \times 64$, as shown in Fig. 7.38. The steady aerodynamic calculation is performed first. Figure 7.39 shows the pressure coefficient contour of the surface at $Ma_\infty = 2.0$, $\alpha = 0.0°$. Figure 7.40 is a schematic diagram of interpolating the structure modal shape to the aerodynamic grid using IPS. It is seen that the interpolation works well for the smoothness and coherence of the aerodynamic grid.

2. Flutter analysis based on reduced-order model

Flutter analysis of the control surface is carried out using a reduced-order model (ROM) based on Volterra series. First, the accuracy of the ROM is verified. Figure 7.41 compares the ROM and CFD/CSD coupling results of the first four-order modal force when a forced sinusoidal vibration $\xi_1 = 0.0001\sin(2\pi \times 32.356 \times t)$ with a frequency of 32.356 Hz is applied to the first-order modal at $Ma = 2.0$. The eigensystem realization algorithm/ROM solution agrees well with the CFD/CSD coupling solution in amplitude and phase. Figure 7.42 shows the long-term response of the ROM at $Ma_\infty = 2.0$, and the dynamic pressure for flutter boundary is 236.15 kPa.

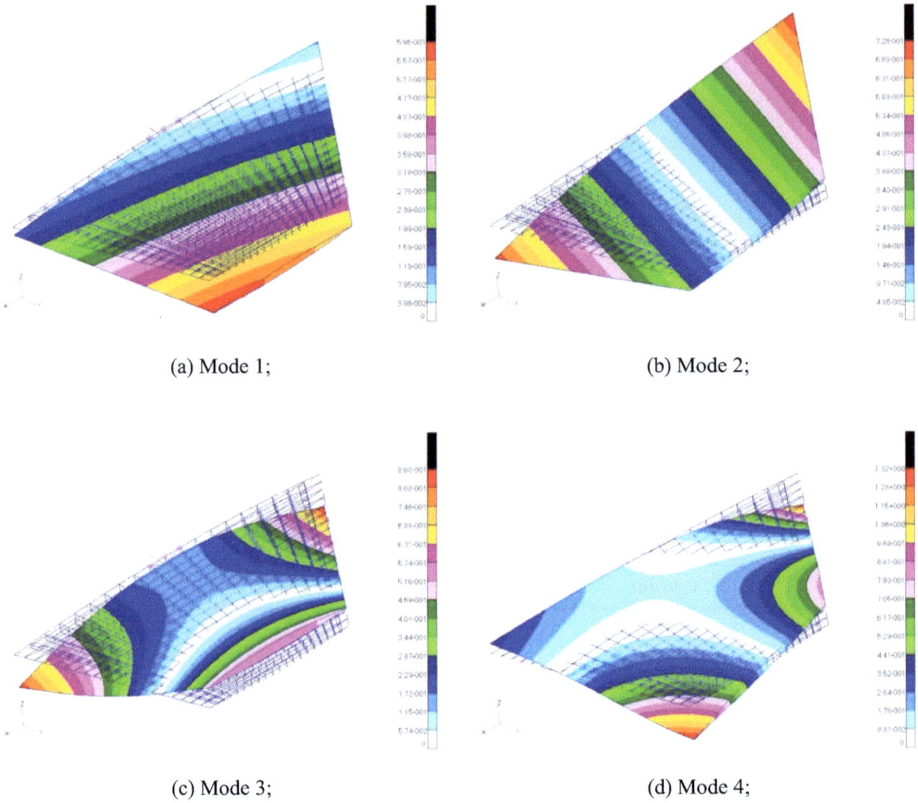

(a) Mode 1; (b) Mode 2;

(c) Mode 3; (d) Mode 4;

Fig. 7.37: Fourth-order mode shape of the control surface: (a) mode 1; (b) mode 2; (c) mode 3; and (d) mode 4.

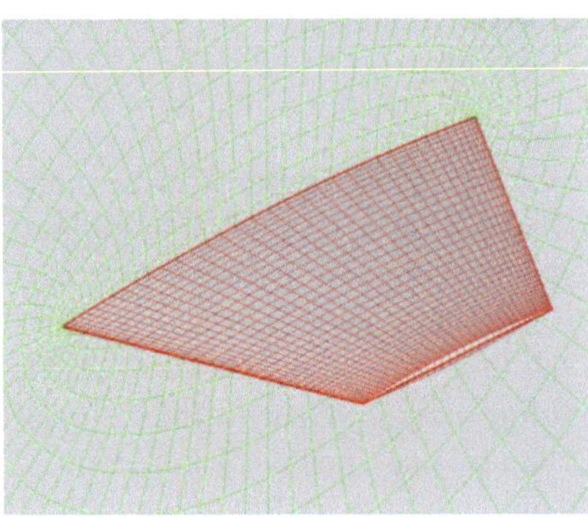

Fig. 7.38: Schematic diagram of the aerodynamic grid of the control surface.

Fig. 7.39: Steady surface contour of the control surface.

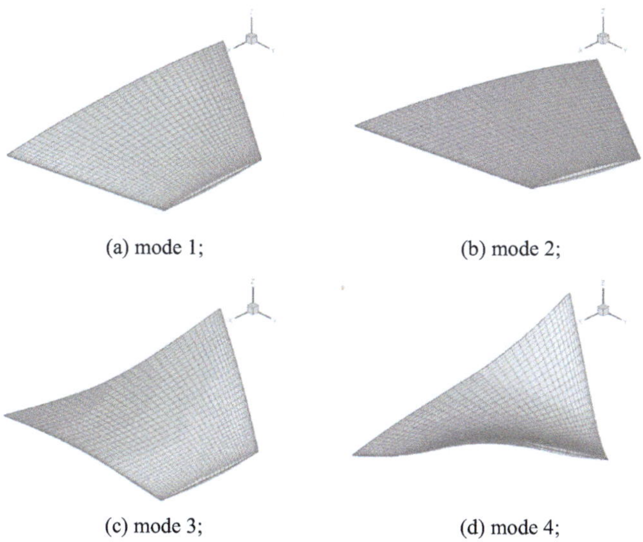

Fig. 7.40: Schematic diagram of modal shape interpolation of the control surface: (a) Mode 1; (b) mode 2; (c) mode 3; and (d) mode 4.

In order to further verify the ROM, CFD/CSD coupling solution is used. Figure 7.43 compares time history of the first two-order generalized displacement in the ROM and CFD/CSD coupling at flutter dynamic pressure 236.15 kPa. The results agree well with each other, which proves that the accuracy and reliability of the ROM.

Table 7.6 shows the flutter dynamic pressure in the range of Mach number from 1.7 to 2.5. In order to obtain the flutter dynamic pressure of the control at the altitude of 2 km, the flutter dynamic pressure at different Mach numbers are compared with the flight dynamic pressures at the altitude of 2 km. Figure 7.44 shows the comparison in the given range of Mach number from 1.7 to 2.5. It can be seen that the Mach number of flutter for the control surface at the altitude of 2 km is about 2.18.

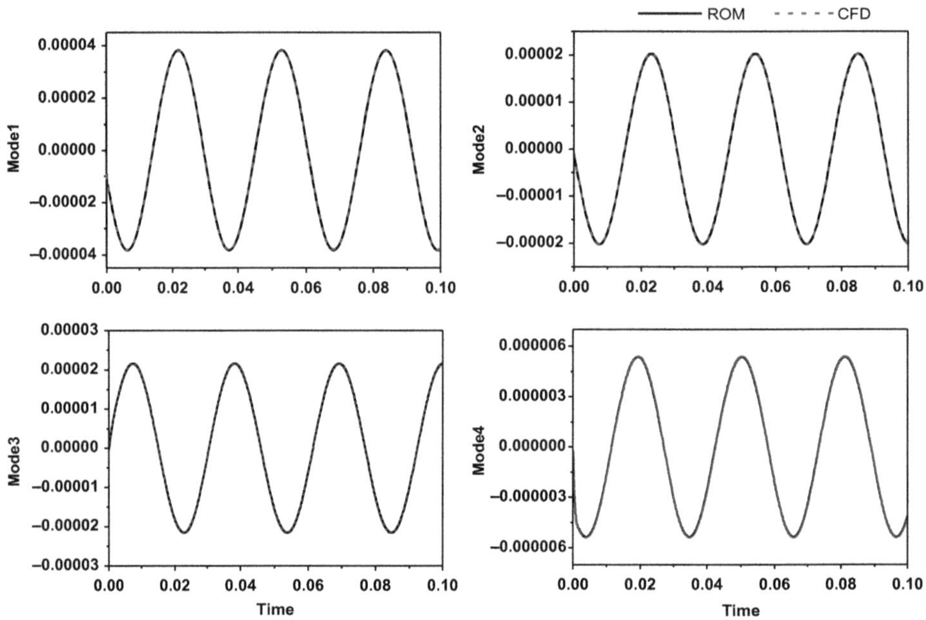

Fig. 7.41: Comparison of first-order modal forces.

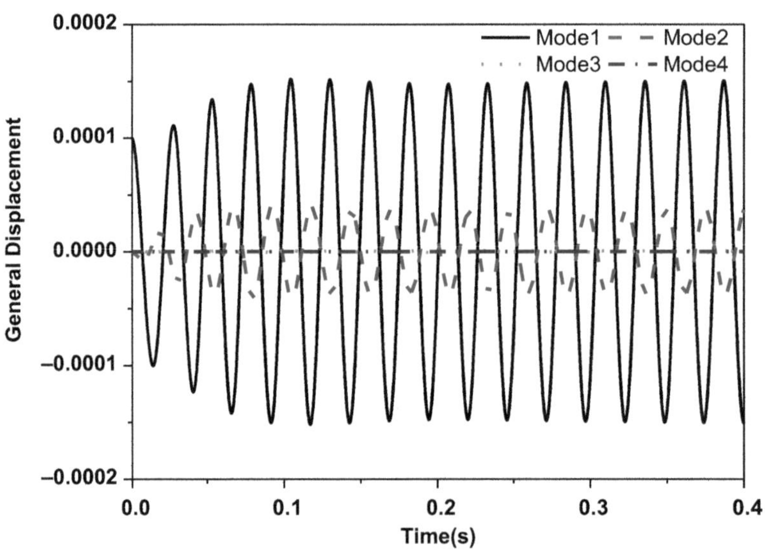

Fig. 7.42: Response of generalized displacement at $Ma_\infty = 2.0$ ($Q_f = 236.15$ kPa).

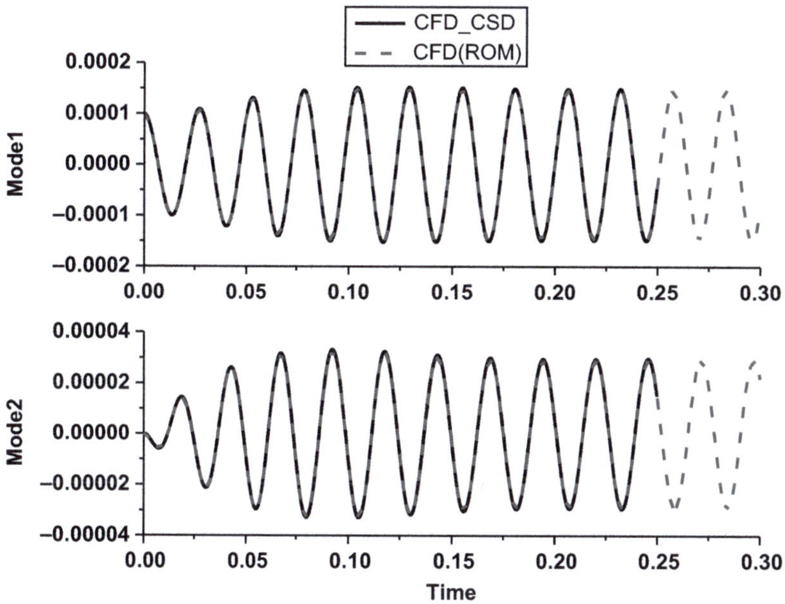

Fig. 7.43: Comparison of generalized displacement of the first two order modals.

Tab. 7.6: Flutter boundary under each Mach number.

Ma_∞	1.7	1.8	1.9	2.0	2.1	2.2	2.3	2.4	2.5
Q_f(kPa)	184.52	202.21	219.51	236.15	252.06	267.55	276.05	294.84	308.18

7.4.3 Dynamic load in aeroelastic analysis

The aeroelastic problem is fundamentally a load problem. The calculation of dynamic load is one of the key technologies for structural design [17, 10], which can provide a reliable basis for structural response calculation, structural dynamic design and fault analysis. Therefore, aeroelastic analysis requires not only the determination of the flutter boundary of the aircraft, but also the dynamic load prediction of the structure during the elastic response.

It is well known that the aeroelastic instability will lead to the destruction of the structure of the aircraft. The phenomenon of torsional instability in static aeroelasticity is relatively simple, that is, the aerodynamic load is overloaded, causing torsional failure of the structure of the lifting surface. However, the flutter phenomenon is a joint coupling among aerodynamic force, inertia force and elastic force, leading to the structural damage. Hence, it is important to understand how these forces are coupling, and the cause of the structure damage.

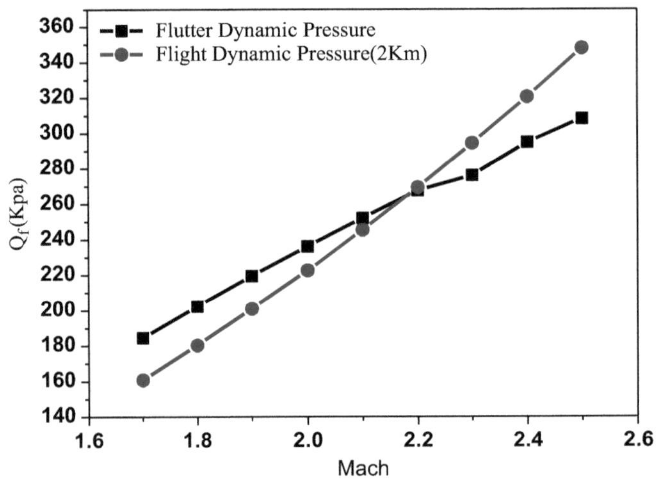

Fig. 7.44: Comparison of the flutter boundary and the dynamic pressure at altitude 2 km.

In this section, two dynamic load analysis methods are proposed for the aeroelastic problems in engineering application. The different functions of aerodynamic and inertial forces in aeroelastic response and the stress changes of the components during elastic deformation are analyzed to provide a deep insight for fault diagnosis and improved design of the aircraft.

1. Dynamic load analysis method

The structural dynamics equation without the damping effect is

$$M\ddot{q} + Kq = Q^F, \tag{7.1}$$

where the three forces are involved in aeroelastic dynamics: inertial force ($M\ddot{q}$), elastic force (Kq) and aerodynamic force (Q^F). Put the inertial force to the RHS of the equation, yielding

$$Kq = Q^F - M\ddot{q}, \tag{7.2}$$

where the left side of the equation is the elastic force of the structure.

Therefore, the elastic force of the structure can be analyzed by calculating the aerodynamic force and the inertial force. The aerodynamic force is solved by the CFD program, and the inertial force is obtained by multiplying the mass by the acceleration in the framework of finite element method. The same method is also applicable to the torque of the component of the structure.

In this section, two dynamic load analysis methods are proposed according to different requirements of engineering application. The first method is to interpolate the mass data of the structure into the CFD/CSD coupling system based on the structural modal superposition method. In order to obtain the forces and moments of the structure, the structural dynamic equation is solved coupling with the aerodynamic computation in CFD solver. This method can be applied to engineering applications concerning the forces and moments of the structure. In Fig. 7.45, yellow circle represents the mass on the structure node.

The second method is to store the unsteady aerodynamic force in the aeroelastic computational process of the structure in time advancing. After interpolating to the surface node of the structure, the transient response of the complete structural model is obtained in CSD software. As shown in Fig. 7.46, the arrows represent schematic diagrams of aerodynamic interpolation to a structural model. Since the method can be used for transient response analysis in a complete structural model, it can provide information such as the stresses on the components.

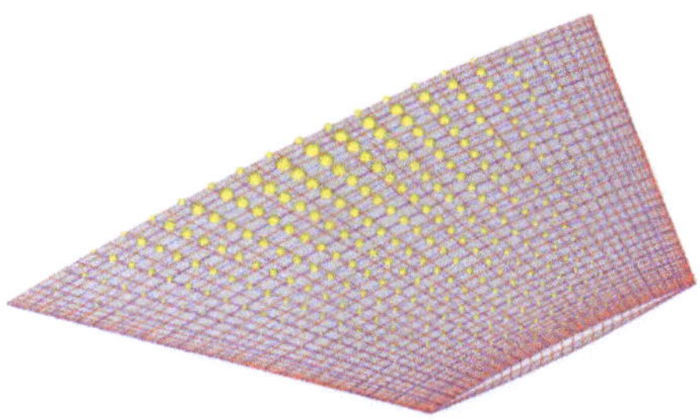

Fig. 7.45: Interpolation from structural mass information to an aerodynamic mesh.

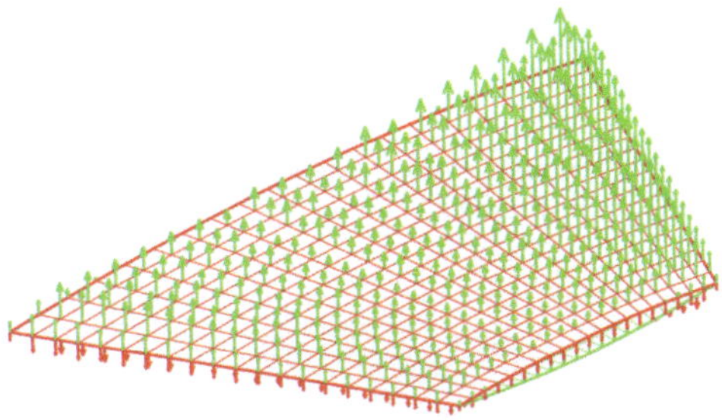

Fig. 7.46: Schematic diagram of interpolating aerodynamic forces onto the surface of the structure.

2. Dynamic load analysis example

The CFD/CSD coupling in time domain is performed for the control surface described in the previous section at $Ma_\infty = 2.18$, $\alpha = 2.0°$, $Q_f = 264.48$ kPa. The deformation and dynamic response of the structure under unsteady aerodynamic forces is analyzed. The transient response of the structural displacement is shown in Fig. 7.47. The dynamic pressure is flutter boundary at the given Mach number, where the structural response is a limit cycle oscillation, which is consistent with the result of the ROM. Due to the nonzero angle of attack. The structure response is asymmetric and of great amplitude. Figure 7.48 shows the deformation and pressure coefficient contour of the control surface at selected time. It is seen that the aeroelastic response of the control surface is mainly caused by the bending and torsion of the control shaft, which is consistent with the flutter analysis results.

The first method is used for dynamic load analysis to obtain aerodynamic forces and inertial forces. Figure 7.49 shows the aerodynamic forces of the control surface in normal direction and the bending moment and torque at the root of the control shaft, and Fig. 7.50 shows the inertial force of the control surface and the bending

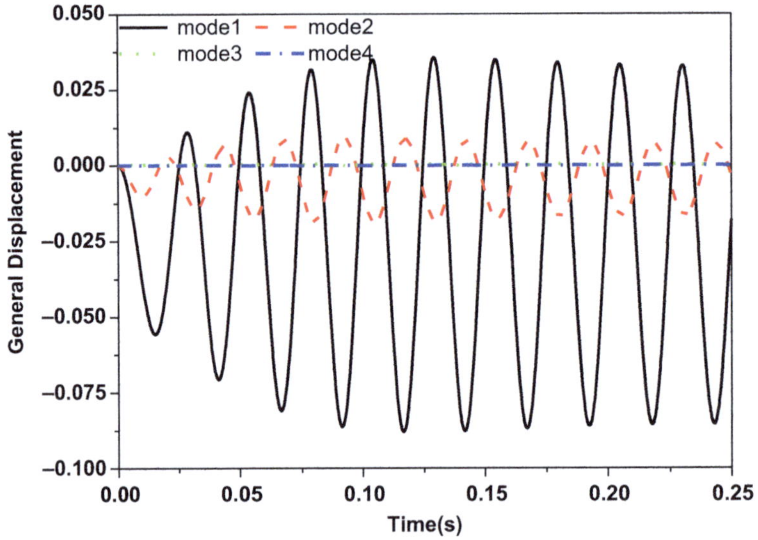

Fig. 7.47: Time history of modal response of control surface at $Ma_\infty = 2.0$, $\alpha = 2.0°$.

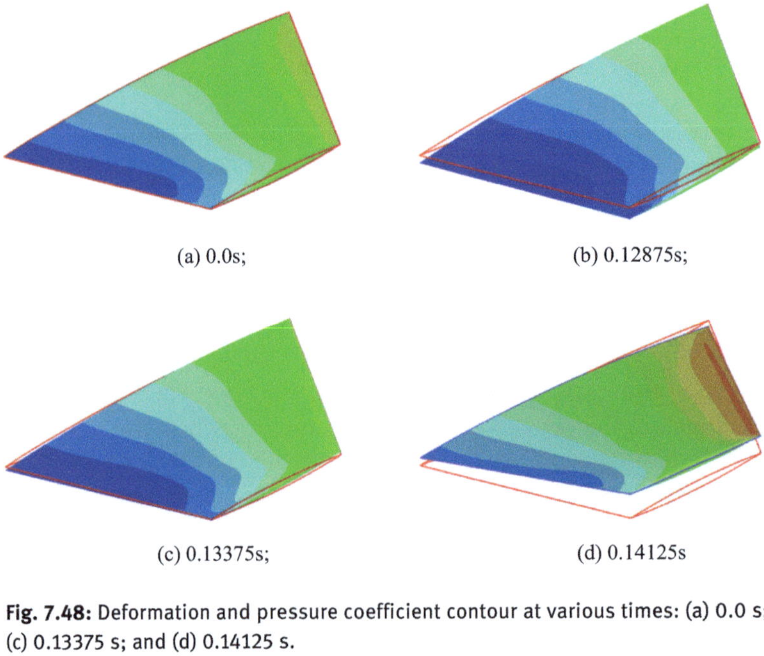

Fig. 7.48: Deformation and pressure coefficient contour at various times: (a) 0.0 s; (b) 0.12875 s; (c) 0.13375 s; and (d) 0.14125 s.

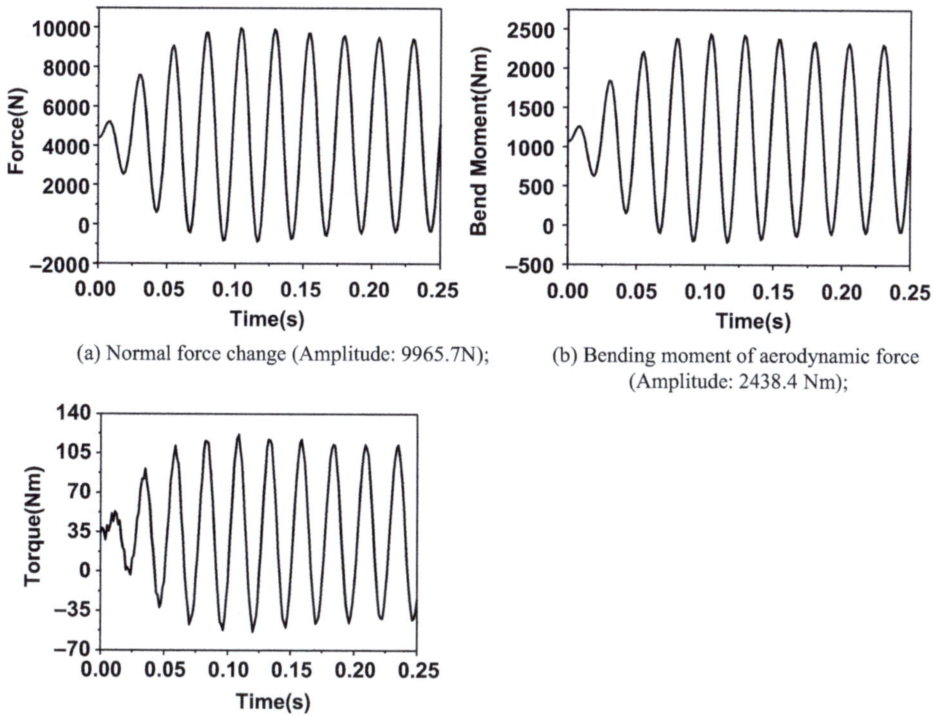

(a) Normal force change (Amplitude: 9965.7N);

(b) Bending moment of aerodynamic force (Amplitude: 2438.4 Nm);

(c) Aerodynamic torque (amplitude: 121.4 Nm)

Fig. 7.49: Time history of aerodynamics and its bending moment and torque: (a) normal force change (amplitude: 9,965.7 N); (b) bending moment of aerodynamic force (amplitude: 2,438.4 Nm); and (c) aerodynamic torque (amplitude: 121.4 Nm).

moment and torque at the root of the control shaft, while Fig. 7.51 shows the total force and total bending moment and total torque. Table 7.7 lists the magnitudes of aerodynamic force, inertial force and the total force on the control surface. It is seen that the aerodynamic force and the inertial force are on the order of magnitude. The total force is bias to the direction of aerodynamic force due to the existence of the angle of attack. Further the torque from inertial force is an order of magnitude higher than the torque from the aerodynamic force. The results can provide a reference of fault diagnosis. If the control shaft is broken due to bending, it can be judged that the aerodynamic force is too large due to the angle of attack. If the control shaft is broken because of twisting, it may be caused by the excessive inertial force induced by the large amplitude during the aeroelastic coupling.

The second method is used to analyze the dynamic load of the control surface. When CFD/CSD coupling analysis is performed, the pressure information is stored every 0.00125 s, which is interpolated to the structural node. Figure 7.52 shows the force on the structural node at 0.10375 s. The force at each time step is stored as a

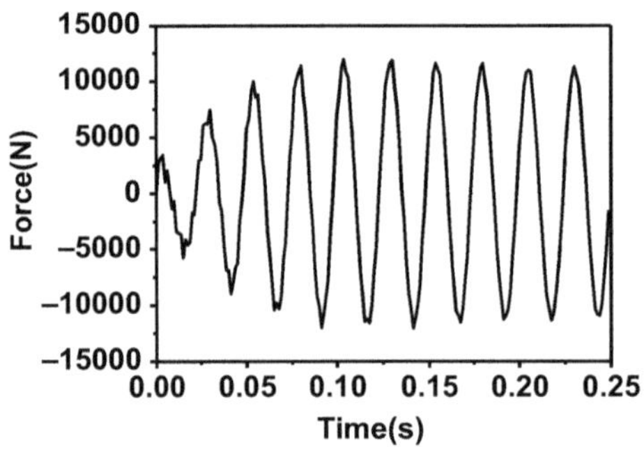

(a) Inertia force (amplitude: 12057.8N);

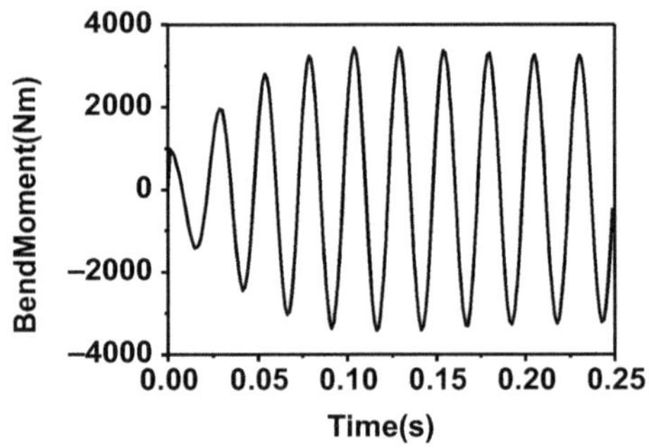

(b) Moment of inertia force (amplitude: −3431.0Nm);

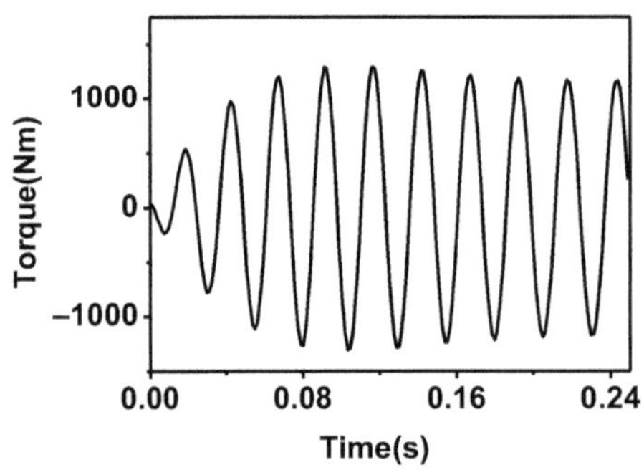

(c) torque of inertia force (amplitude: 1305.4 Nm)

Fig. 7.50: Time history of Inertia force and its bending moment and torque: (a) inertia force (amplitude: 12,057.8 N); (b) moment of inertia force (amplitude: −3,431.0 Nm); and (c) torque of inertia force (amplitude: 1,305.4 Nm).

time-dependent field loads onto the structural model for transient response analysis to obtain the dynamic response of the structure and the dynamic stress. Figure 7.53 shows the time history of displacement of the trailing edge of the control surface. The maximum displacement is about 0.055 m. Figure 7.54 shows the time history of bending stress of the shaft at the root. The maximum stress is about 774 MPa. Figure 7.55 shows the bending stress and the stress on control surface at selected time, where the maximum bending stress occurs at 0.11625 s.

In engineer application, because the unsteady aerodynamic force and inertial force in the aeroelastic coupling cannot be obtained directly, the traditional strength analysis often uses an empirical factor for the estimation of the aerodynamic force as the dynamic loading for the design. Such method cannot describe actual dynamic load conditions obviously. The maximum aerodynamic forces of steady aerodynamic and aeroelastic computation are extracted as a loading onto the structural model for static analysis. The results are compared with the maximum stress value (774 MPa) obtained by CFD/CSD coupling analysis.

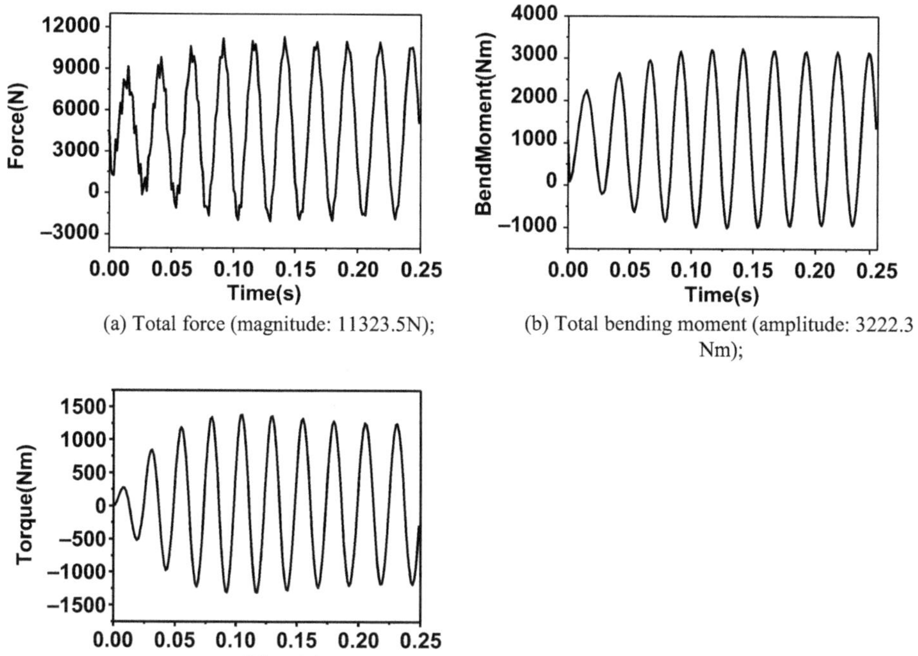

(a) Total force (magnitude: 11323.5N);

(b) Total bending moment (amplitude: 3222.3 Nm);

(c) Total torque (amplitude: 1382.7 Nm)

Fig. 7.51: Time history of total force, total bending moment and total torque: (a) total force (magnitude: 11,323.5 N); (b) total bending moment (amplitude: 3,222.3 Nm); and (c) total torque (amplitude: 1,382.7 Nm).

Tab. 7.7: Comparison of aerodynamic force, inertial force and total force.

	Aerodynamic force	Inertial force	Total force
Force	−884.5 to 9965.7	−12,017.8 to 11,938.8	−2044.8 to 11,323.5
Bending moment	−217.7 to 2438.4	−3415.3 to 3431.0	−1007.7 to 3222.3
Torque	−53.5 to 121.4	−1305.4 to 1291.5	−1309.7 to1382.7

Figure 7.56 shows the results for the case of the steady aerodynamic force onto the model for static analysis. The bending stress at the root of the control shaft is 262 MPa. Further the maximum value of unsteady aerodynamic loading in the aeroelastic response process is used for static analysis of the structure. The root bending stress is 594 MPa as shown in Fig. 7.57. It is seen from the results that the stress for the case of steady aerodynamic force is 33.9% of those from dynamic analysis of

aeroelastic coupling. Even when the static analysis is performed using the maximum value of the unsteady aerodynamic force during the response process, the stress is 76.7% of the reference value. The results indicate that the dynamic stress analysis should use CFD/CSD coupling method to analyze the structural stress during the aeroelastic study.

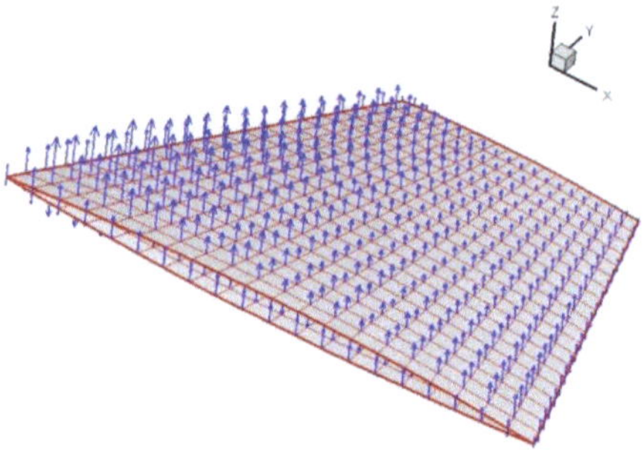

Fig. 7.52: Force vector of control structure at $t = 0.10375$ s.

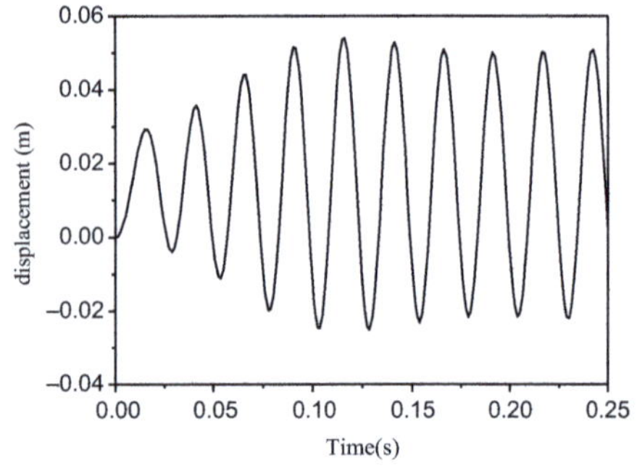

Fig. 7.53: Displacement response of the trailing edge of the control surface.

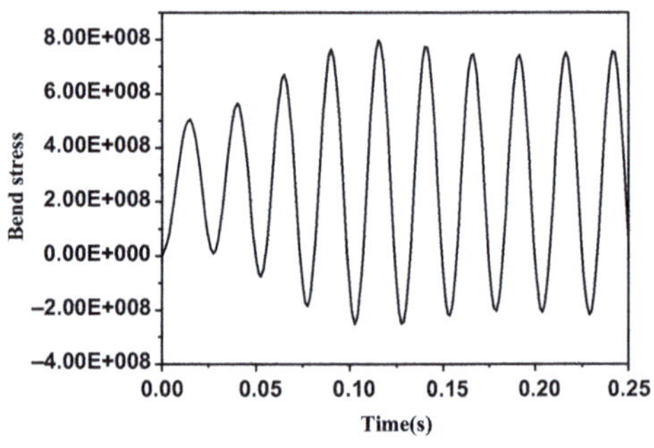

Fig. 7.54: Time history of axial bending stress.

7.4.4 Influence of dynamic perturbation on aeroelasticity

Flutter analysis of aircraft is generally performed under a specified flight condition, regardless of the rigid motion of the aircraft. However, complex perturbations during the flight of the aircraft may result in the rigid motion such as the vibration, the rolling perturbation and the change of the angle of attack caused by the pitching motion. The unsteady aerodynamic force generated by these perturbations will affect the aeroelastic stability of the lifting surface of the control surface. Furthermore, this effect may reduce the flutter dynamic pressure of the aircraft components. Therefore, it is necessary to study the influence of dynamic perturbation on the

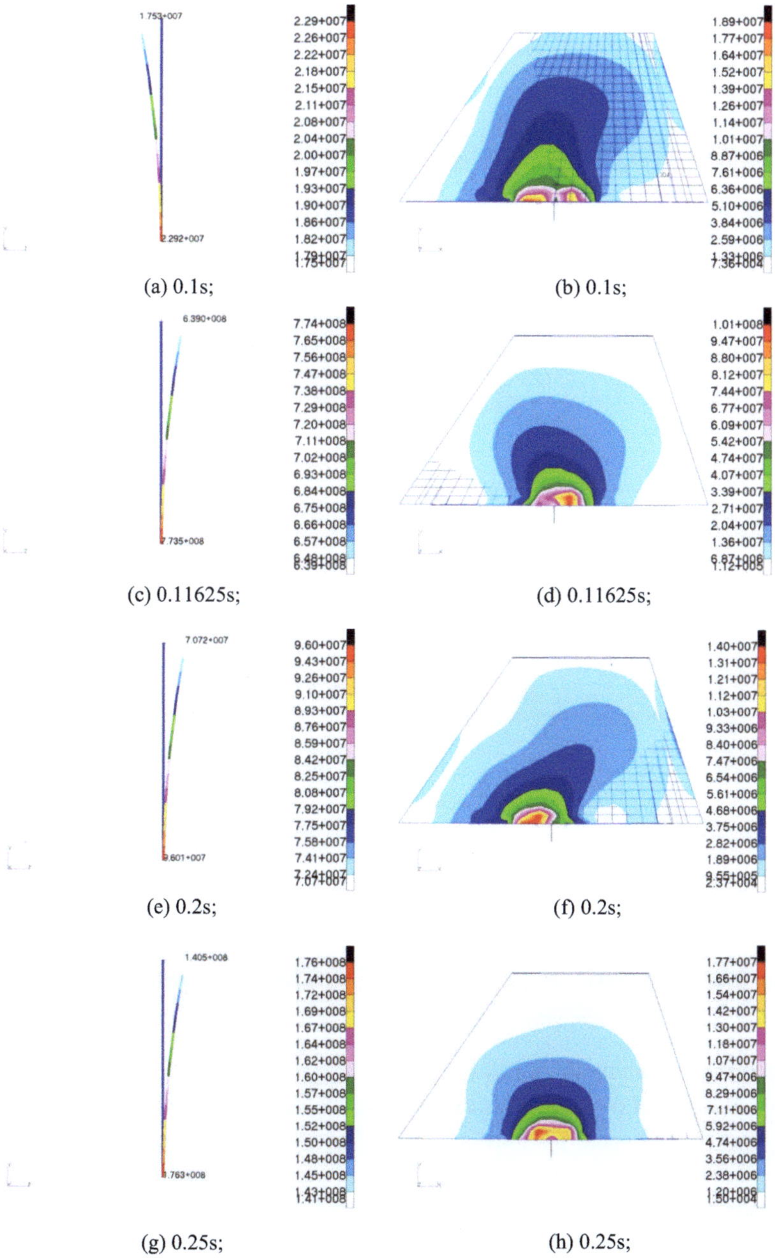

Fig. 7.55: Bending stress of the control shaft and surface at each time. (a) 0.1 s (side view); (b) 0.1 s (top view); (c) 0.11625 s(side view); (d) 0.11625 s(top view); (e) 0.2 s(side view); (f) 0.2 s(top view); (g) 0.25 s(side view); and (h) 0.25 s(top view).

flutter boundary. This section discusses the influence of dynamic perturbation on the aeroelasticity by considering the influence of the rolling perturbation on the flutter boundary of control surface [26, 13].

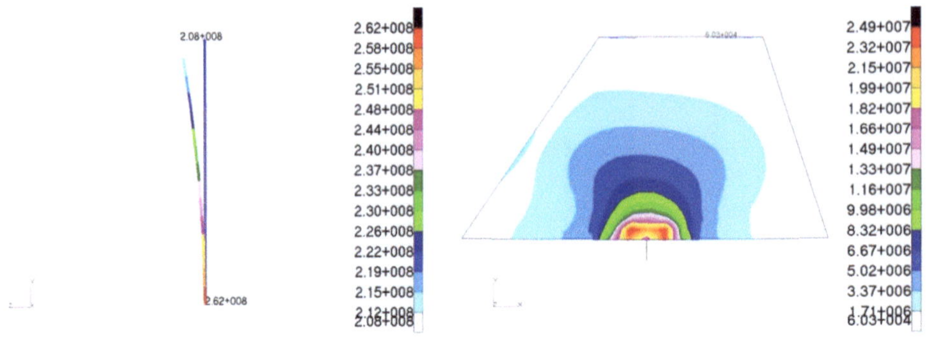

Fig. 7.56: Bending stress of control shaft and surface under constant aerodynamic force.

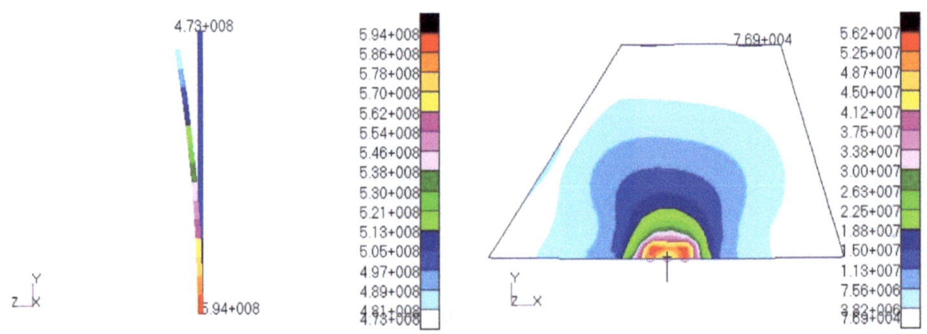

Fig. 7.57: Bending stress of control shaft and surface under maximum aerodynamic force.

1. Dynamic perturbation analysis method

The rolling perturbation of the aircraft is an external excitation to the control surface. It is assumed that this external excitation is not affected by the elastic deformation and acts in its own motion (frequency, amplitude) in a short time. Thus, the motion of the control surface consists of two parts, one part is the rigid motion caused by dynamic perturbation, and the other part is the elastic deformation. Both parts of the motion will affect the unsteady aerodynamic force acting on the control surface.

The rigid motion of dynamic perturbation is applied to the control surface according to its motion equation, and then the aeroelastic analysis is performed using CFD/CSD coupling method. In the aeroelastic computation, the aerodynamic force already takes account of the influence of dynamic perturbation. The roll perturbation is defined by two parameters, the perturbation amplitude and the frequency. The perturbation of the control surface is expressed as follows:

$$\text{Disp} = \frac{A}{(2\pi f)^2} \cdot \frac{D+R}{R} \cdot \sin(2\pi ft), \tag{7.3}$$

where A is the amplitude of the perturbation acceleration at the root of control surface. f is the frequency of the roll perturbation, R is the radius of the missile and D is the distance from the root to the given node.

Figure 7.58 shows the finite element model of the full-movable control surface. The modal analysis is performed on the control surface. The first four modals are selected for aeroelastic analysis. Table 7.8 shows the modals of control surface. Figure 7.59 shows the first four-order modal shapes of the control surface. It can be seen that the first two modals are the bending and the torsion motion, which is characterized by the elastic deformation of the control shaft while the latter two are the bending and torsion of the control surface. The aerodynamic grid is shown in Fig. 7.60. Figure 7.61 presents the interpolation from modal shape onto the aerodynamic grid of the control surface.

Fig. 7.58: Finite element model of control surface.

Tab. 7.8: Modal analysis of control surface.

	1	2	3	4
Frequency (Hz)	50	77.25	248.5	323.95
Modal shape	Shaft bending	Shaft torsion	Surface bending	Surface torsion

2. Influence of dynamic perturbation under different Mach numbers

In this section, the influence of the rolling perturbation on the flutter boundary is studied under different Mach numbers. The maximum overload of the rolling perturbation is 1.5 G, and the rolling frequency is 66 Hz. The range of Mach number is 2.0 to 3.0. The rolling perturbation is considered as a forced motion in the CFD/CSD coupling system. The response of the first four generalized displacements under the aerodynamic load is studied.

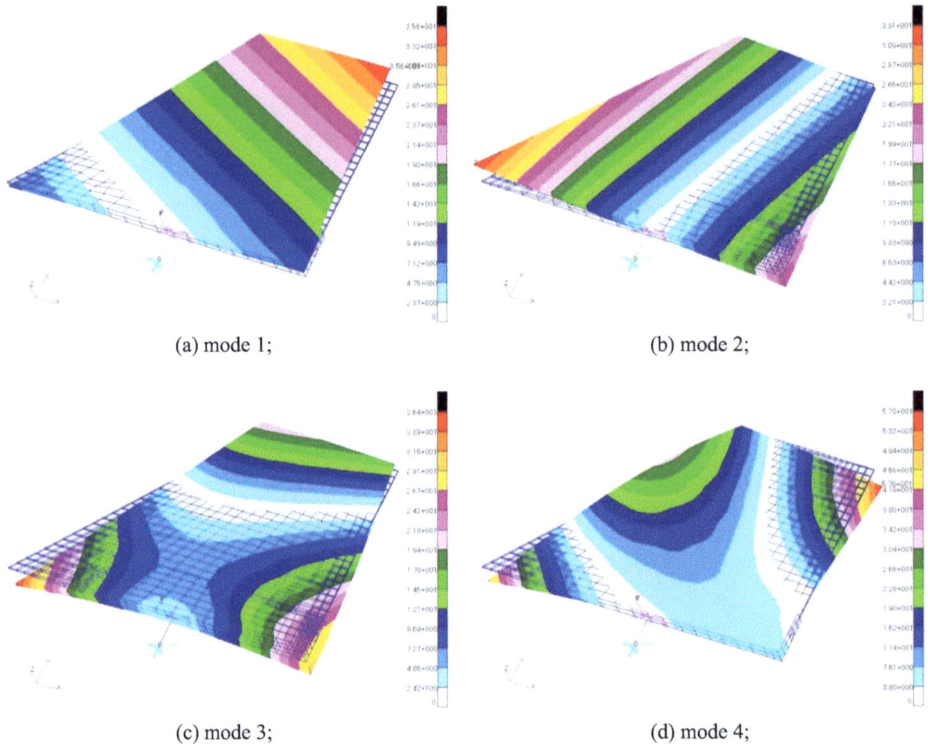

Fig. 7.59: First four-order modal shape of control surface: (a) mode 1; (b) mode 2; (c) mode 3; and (d) mode 4.

The flutter boundary without and with the rolling perturbation are calculated separately. Figure 7.62 shows the flutter response for the undisturbed case at dynamic pressure (Q_f = 339 kPa) at $Ma_\infty = 2.0$ and the perturbation case for dynamic pressure (Q_f = 288.5 kPa). Figure 7.63 shows the flutter response with (Q_f = 340 kPa) and without rolling perturbation for dynamic pressure (Q_f = 404 kPa) at $Ma_\infty = 2.2$. The rolling perturbation causes the decrease of flutter dynamic pressure of the control surface. Furthermore, the structural response includes two distinct frequencies in a beat motion, that is, the flutter frequency and the perturbation frequency. In addition, the rolling perturbation also reduces the flutter frequency of the structure from 61 to 57.6 Hz.

Table 7.9 and Fig. 7.64 compare the flutter dynamic pressure of the control surfacewith and without the rolling perturbation under different Mach numbers. It is seen that the flutter dynamic pressure drops by 14.9–18.3% under rolling perturbation with the increase of Mach number.

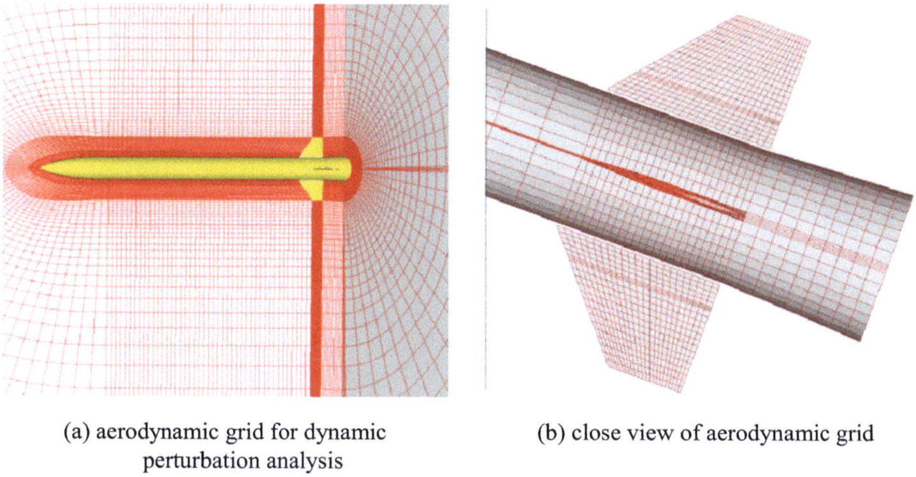

Fig. 7.60: Aerodynamic mesh and close view of aerodynamic grid: (a) aerodynamic grid for dynamic perturbation analysis and (b) close view of aerodynamic grid.

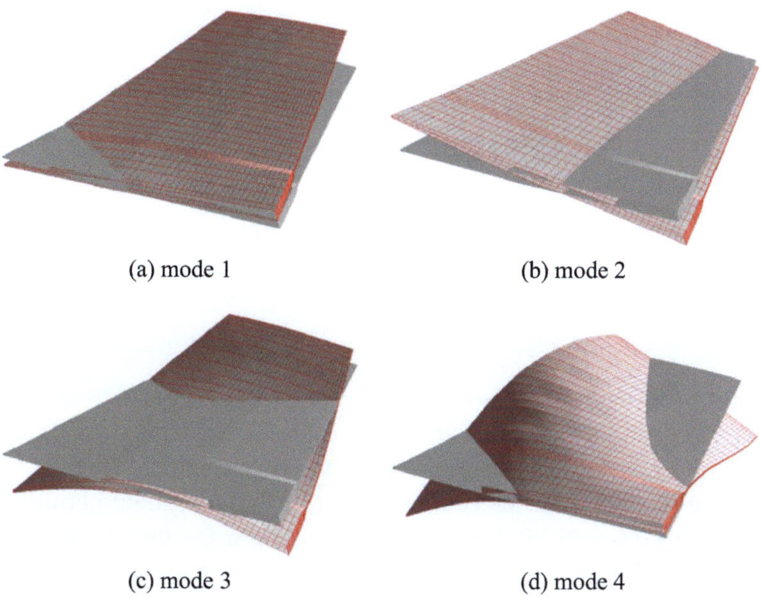

Fig. 7.61: Preview of the modal interpolation: (a) mode 1; (b) mode 2; (c) mode 3;and (d) mode 4.

3. Influence of dynamic perturbation parameters

In this section, the influence of different rolling perturbation parameters on the flutter dynamic pressure is discussed. The computation is performed at Mach = 2.5,

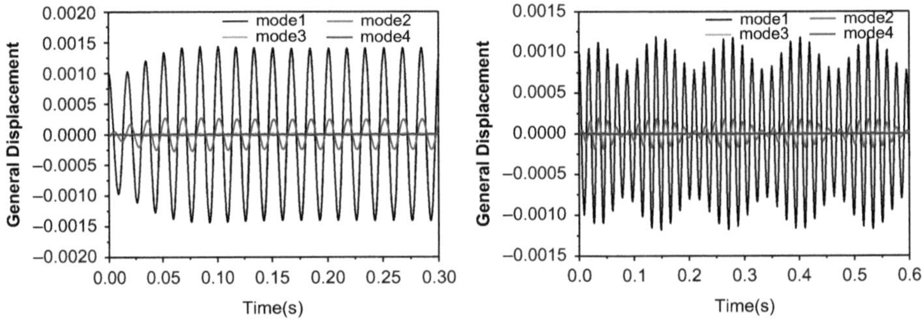

(a)withour perturbation (Q_f=339Kpa)

(b) rolling perturbation (Q_f=288.5Kpa)

Fig. 7.62: Flutter response of the system at $Ma_\infty = 2.0$. (a) Without perturbation (Q_f= 339 kPa) and (b) rolling perturbation (Q_f= 288.5 kPa).

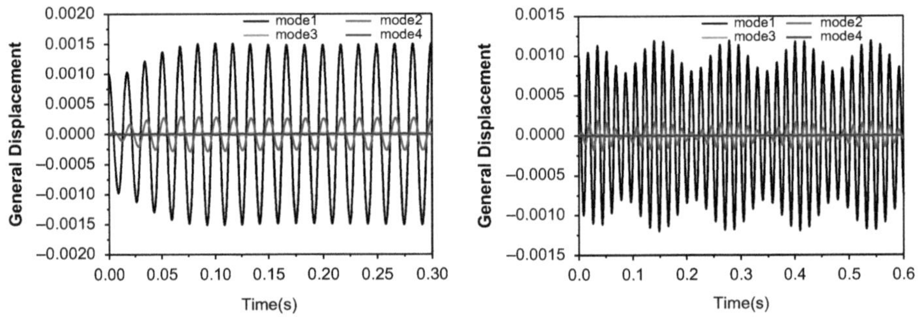

(a) without perturbation (Q_f=387Kpa)

(b) with rolling perturbation (Q_f=327.5Kpa)

Fig. 7.63: Flutter response of the system at $Ma_\infty = 2.2$. (a) Without perturbation (Q_f= 387 kPa) and (b) with rolling perturbation (Q_f= 327.5 kPa).

while the frequency of the control surface is set 66, 60 and 55 Hz, respectively and the overload is 0.5, 1.0 and 1.5 G. Table 7.10 compares the flutter dynamic pressure of the control surface under different motions of rolling perturbations. Figure 7.65 shows the limit cycle oscillation of the control surface at flutter without perturbation at Mach number 2.5 (Q_f= 475 kPa). Figures 7.66–7.68 show the flutter response of the control surface under different overloads as roll perturbations is 66, 60 and 55 Hz, respectively. The results show that the frequency has an impact on the flutter dynamic pressure. When the control flutter frequency (57 Hz) is closer to the frequency of roll perturbation, the flutter dynamic pressure drops greatly. The overload of the perturbation also has an impact on the motion of the control surface. As the overload is increasing, the beating motion of the limit cycle oscillation is pronounced.

Tab. 7.9: Comparison of flutter boundary with and without perturbation.

Ma	2.0	2.2	2.4	2.6	2.8	3.0
Flutter boundary without perturbation (kPa)	339	387	432	477.3	521.5	562
Flutter dynamic pressure with perturbation (kPa)	288.5	327.5	363	397	430	459
Reduction	14.9%	15.4%	16.0%	16.8%	17.5%	18.3%

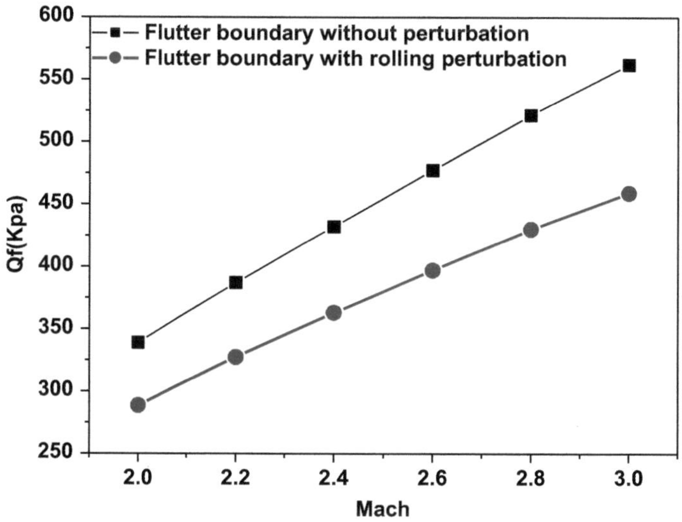

Fig. 7.64: Flutter boundary with and without rolling perturbation.

Tab. 7.10: Flutter dynamic pressure under different frequency and overload (OL) rolling excitation at Mach number 2.5 (flutter dynamic pressure 475 kPa for unperturbation case).

Q_f (kPa) OL / Freq	0.5 G	1.0 G	1.5 G
66 Hz	395	395	395
60 Hz	390	390	390
55 Hz	390	390	390

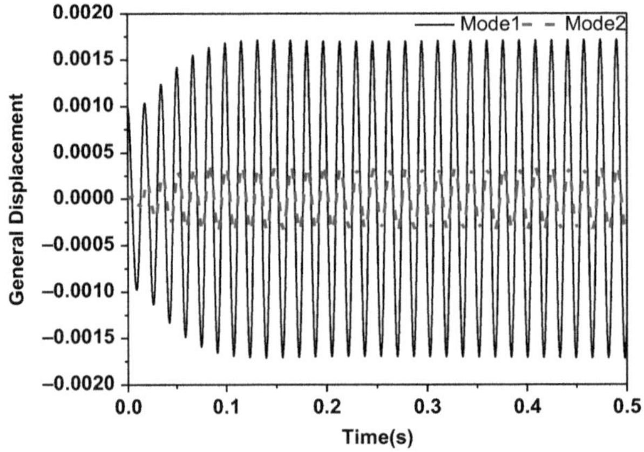

Fig. 7.65: Flutter response of undisturbed case at $Ma_{\infty} = 2.5$ ($Q_f = 475$ kPa).

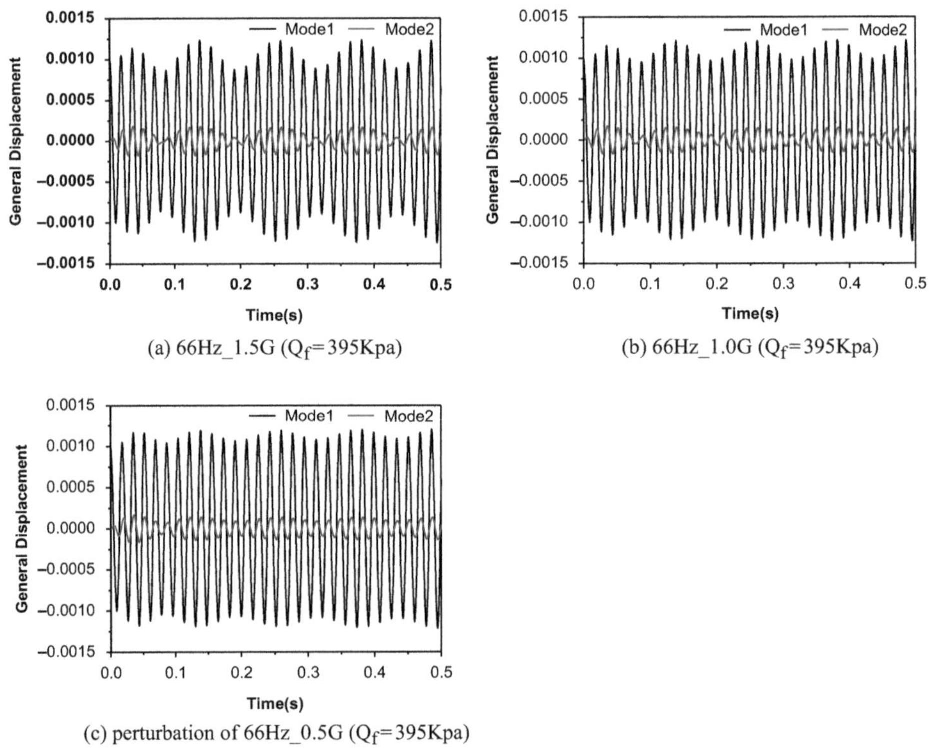

Fig. 7.66: Flutter response of different amplitudes of perturbations with 66 Hz. (a) 66 Hz_1.5 G perturbation ($Q_f = 395$ kPa); (b) 66 Hz_1.0 G perturbation ($Q_f = 395$ kPa); (c) 66 Hz_0.5 G perturbation ($Q_f = 395$ kPa).

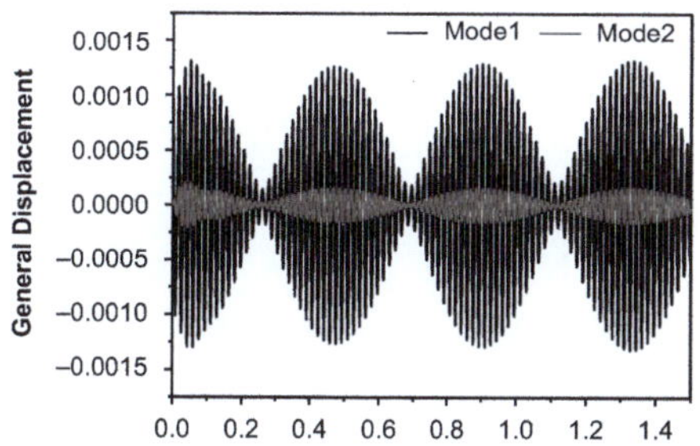

(a) 60hz_1.5G perturbation (Q_f=390Kpa)

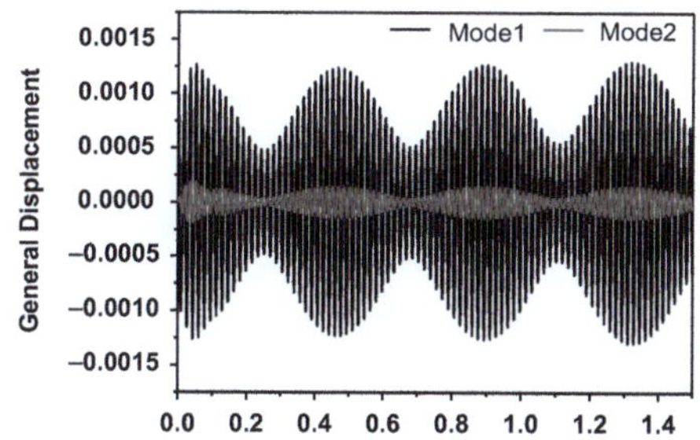

(b) 60Hz_1.0G perturbation (Q_f=390Kpa)

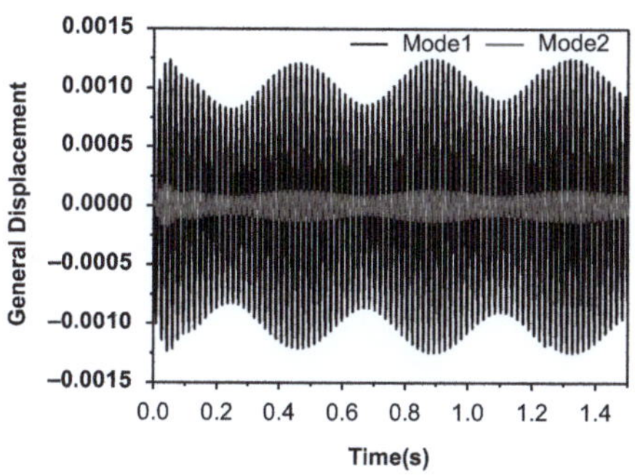

(c) 60Hz_0.5G perturbation (Q_f=390Kpa)

Fig. 7.67: Flutter response of different amplitude of perturbation with 60 Hz. (a) 60 Hz_1.5 G perturbation (Qf = 390 kPa); (b) 60Hz_1.0G perturbation (Qf = 390 kPa); (c) 60Hz_0.5 G perturbation (Qf = 390 kPa).

7.4.5 Aeroelastic analysis of civil aircraft

Current aeroelastic analysis, especially CFD/CSD coupling method, mostly focus on the individual lifting surface components such as the wing and the control surface. The CFD/CSD coupling method requires a multidisciplinary knowledge involving structural modeling, CFD meshing, interpolation and dynamic mesh technique. Aeroelastic analysis technique is developed relatively slowly from individual components to complex shape full aircraft. The principle of aeroelastic modeling for complex shape aircraft is basically the same as that of individual components such as lifting surface. However, it will face many new challenges, including (1) grid blocking technique for aerodynamic grid of complex geometry, (2) block interpolation technique involving multiple elastic components and (3) dynamic mesh technique for complex multiblock mesh. In this chapter, the solution to the earlier three problems is studied by taking an example of the aeroelastic analysis of civil aircraft. The grid blocking strategy for aeroelastic analysis of complex shape aircraft is proposed, and the interpolation and dynamic mesh method in

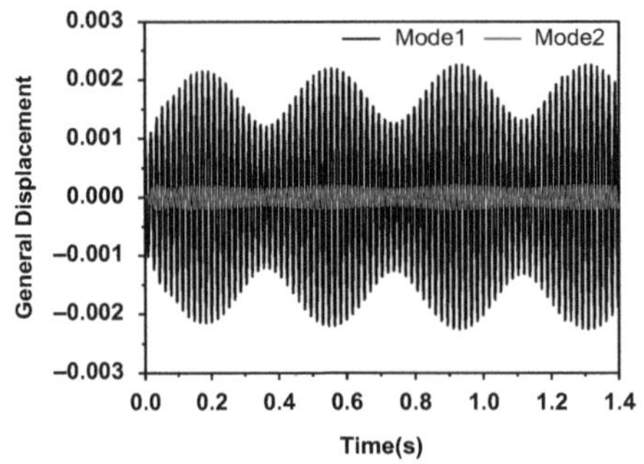

(a) 55Hz_1.5G perturbation (Q_f=390Kpa)

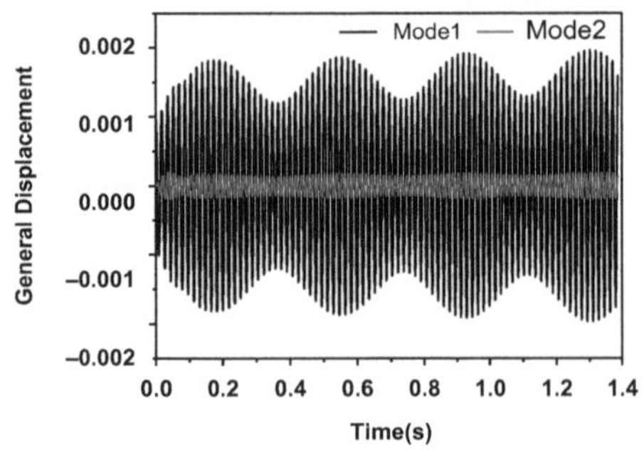

(b) 55Hz_1.0G perturbation (Q_f=390Kpa)

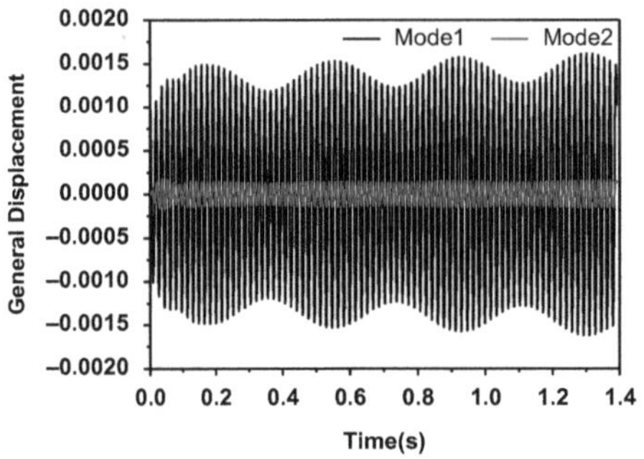

(c) 55Hz_0.5G perturbation (Q_f=390Kpa)

Fig. 7.68: Flutter response under different amplitudes of perturbation with 55 Hz. (a) 55 Hz_1.5 G perturbation (Q_f= 390 kPa); (b) 55 Hz_1.0 G perturbation (Q_f= 390 kPa); (c) 55 Hz_0.5 G perturbation (Q_f= 390 kPa).

the coupling analysis are improved to adapt to the aeroelastic analysis of complex shape aircraft [2, 29, 30, 28, 27, 25, 24, 1, 20].

1. Aeroelastic analysis strategy for civil aircraft

1) Grid blocking strategy

When using CFD/CSD coupling method to analyze the aeroelasticity of complex shape aircraft, the most difficult part is to setup an aerodynamic grid with good quality. If the grid is not established properly, it will bring difficulties to the mesh deformation and may even produce the negative volume of the mesh which directly causes the failure of computation. Therefore, in the mesh generation stage, the influence of the mesh topology on the mesh deformation and the potential mesh deformation motion should be considered before the aeroelastic analysis process. Therefore, the aerodynamic mesh should meet the following requirements: (1) mesh deformation is easy to implement; (2) the mesh can undergo large deformation; (3)

good mesh quality can be maintained after deformation. This section focuses on the blocking strategy of structured grids.

The TFI technique for mesh deformation is characterized by interpolating the deformation displacement of the surface grid node to the far-field grid. Therefore, when preparing a mesh, the none-surface mesh corresponding to the surface mesh should be designed as a far-field or a symmetry boundary, so as to avoid interpolating the deformation of the surface mesh to other surface mesh. The fact is that the least grid block partitioning and the simplest topology can make the mesh deforming block easy to use.

The principle of grid blocking is that the total number of grids is as small as possible, and the grid blocks corresponding to a single elastic component are as simple as possible. The typical civil aircraft with the engine is given in Fig. 7.69 to illustrate the grid-blocking strategy of the aeroelastic analysis of the complex shape aircraft.

Fig. 7.69: Geometry of a typical civil aircraft.

According to the aerodynamic shape of the aircraft, the CFD grid is created. Because the flight attitude of the civil aircraft is relatively unchanged, the sliding angle is set zero in the analysis. The grid for half geometry is used for calculation for the sake of computational efficiency. The influence of mesh topology on mesh deformation should be considered in the initial stage of mesh construction. The basic principle is that the number of meshes should be reduced as small as possible. Figure 7.70 shows the aerodynamic grid of the flow field. The number of grid blocks is 16 and The total points in the number of grids are 1.17 million. Figure 7.71 shows the surface mesh of the aircraft.

The elastic structure of the aircraft are the fuselage, the wing and the tail. The meshing should make the block directly related to an elastic component as simple as possible. Figure 7.72 presents the block directly related to the wing. In order to achieve large mesh deformation, the surface mesh block should connect to the far field mesh.

In order to ensure that the mesh is easy for mesh deformation, it is also necessary to consider the mesh quality. The quality of the aerodynamic mesh is examined by steady computation. Figures 7.73 and 7.74 show the residuals of the computation

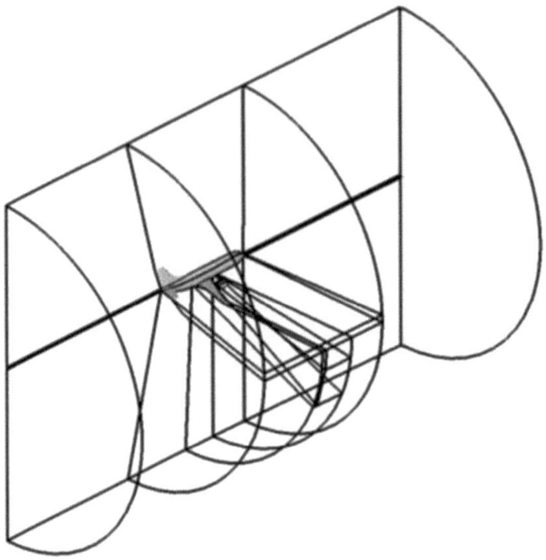

Fig. 7.70: Aerodynamic grid topology.

Fig. 7.71: Aerodynamic surface mesh (semimode symmetry).

and surface pressure contour at 0.6 Mach number and angle of attack two degree. The residual converges quickly in the steady calculation, and the surface flow field is smooth after convergence, indicating that the mesh is of high quality for aerodynamic convergence of CFD/CSD coupling.

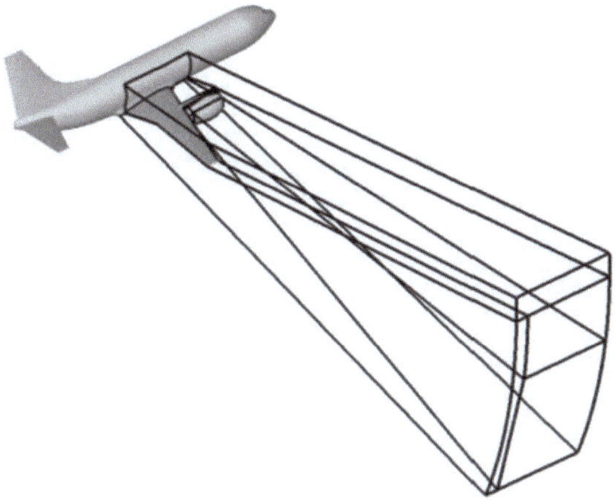

Fig. 7.72: Schematic diagram of a grid block directly related to wing deformation.

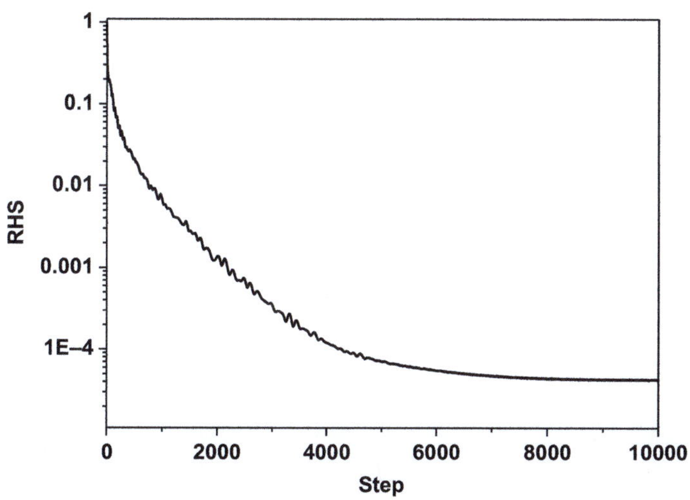

Fig. 7.73: Residual convergence of steady CFD calculation ($Ma_\infty = 0.6$, $\alpha = 2.0°$).

2) Block interpolation strategy

Complex shape aircraft often consists of several main components with distinct geometry and structural modal shape. An interpolation method often fails to meet the interpolation accuracy requirements. Taking the civil aircraft in this section as an example, the aircraft consists of three elastic parts, the fuselage, the wing and the tail. The elastic deformation of these three parts is substantially different, therefore

Fig. 7.74: Surface contour for pressure distribution in steady CFD computation ($Ma_\infty = 0.6$, $\alpha = 2.0°$).

different interpolation strategies should be used for these three parts. Interpolation: the plate-shaped lifting surface components such as the wing and the tail should use IPS interpolation for better interpolation accuracy, while the fuselage should adopt CVT interpolation. For the grid interface between different parts, an error controller needs to be added to eliminate errors caused by different interpolation methods, which may result in poor mesh quality after deformation at the interface. Figures 7.75 and 7.76 show the blocking of the aerodynamic grid and structural nodes when the block is interpolated. Herein, the aircraft structure is divided into three parts for block interpolation.

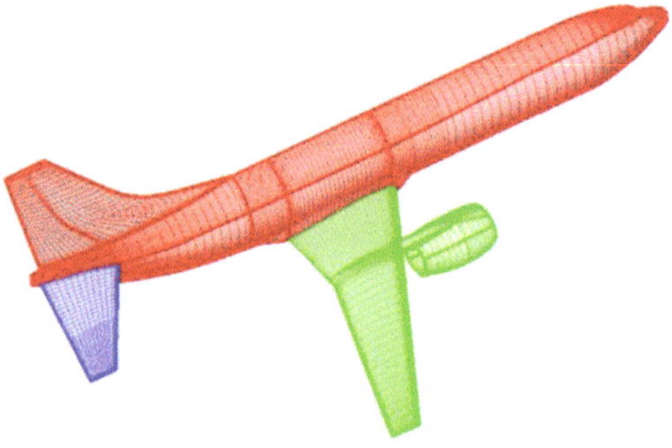

Fig. 7.75: Schematic diagram of block interpolation for aerodynamic node.

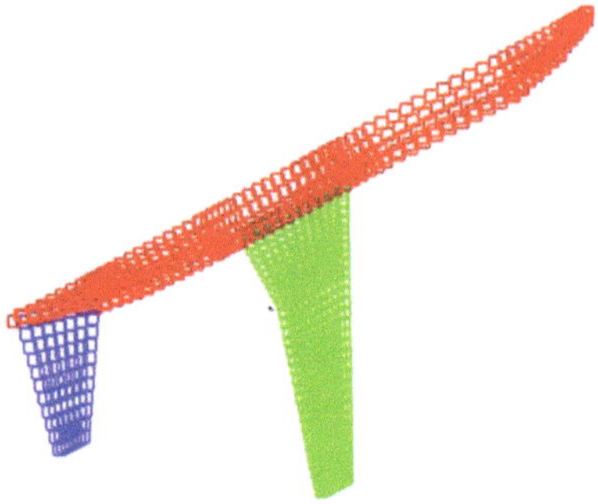

Fig. 7.76: Schematic diagram of the block interpolation for structure node.

3) Moving-grid strategy

After the aerodynamic mesh deformation of the aircraft surface is determined by block interpolation, the deformation of boundary points of the grid block is linearly interpolated to the nonsurface surface boundary by TFI line interpolation so that the displacement of each node on the boundary is obtained. Then, the TFI surface interpolation is used to interpolate the deformation on the boundary line into the mesh surface to obtain the deformation of the nodes in the mesh surface. For the case where there are multiple in-plane elastic components, TFI interpolation method is used same as the case of only a single elastic component, except that the sequence and strategy of grid line interpolation and surface interpolation are more complicated.

Figure 7.77 shows the difference of interpolation sequence in line interpolation and surface interpolation for a single elastic part and two elastic parts. Figure 7.77(a) shows the interpolation sequence of a typical single elastic part. Generally, four line interpolations and three surface interpolations are required. Figure 7.77(b) shows the interpolation sequence of two typical elastic components, which requires seven line interpolation and five surface interpolation. For other forms of multiple elastic components, a reasonable sequence should be selected according to the distribution of components.

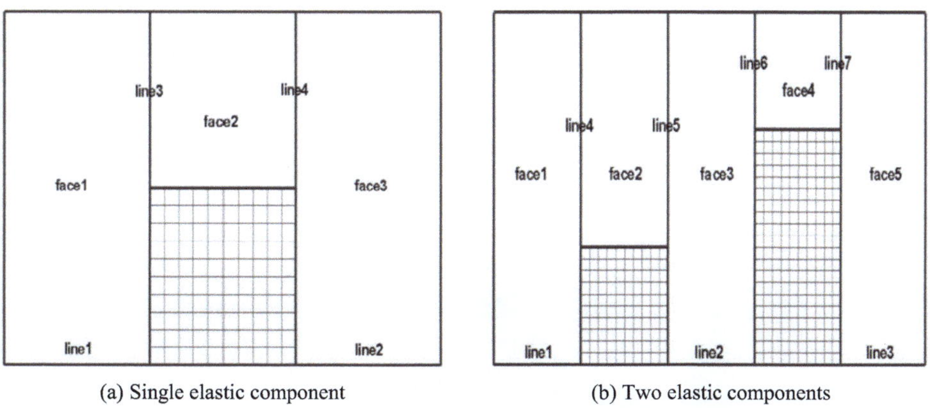

Fig. 7.77: Schematic diagram of grid interpolation sequence for different numbers of elastic components in a block: (a) single elastic component and (b) two elastic components.

Figures 7.78 and 7.79 show the deformation of the surface mesh after a large elastic deformation of the aircraft and the deformation of the mesh block using the TFI dynamic mesh technique. The results show that in the case of large mesh deformation, the surface mesh of the aircraft is smooth with high quality and no negative volume, which can meet the requirement of aeroelastic analysis. It should be noted that the aeroelastic deformation during the actual computation is far less than that shown here.

Fig. 7.78: Preview of surface mesh deformation.

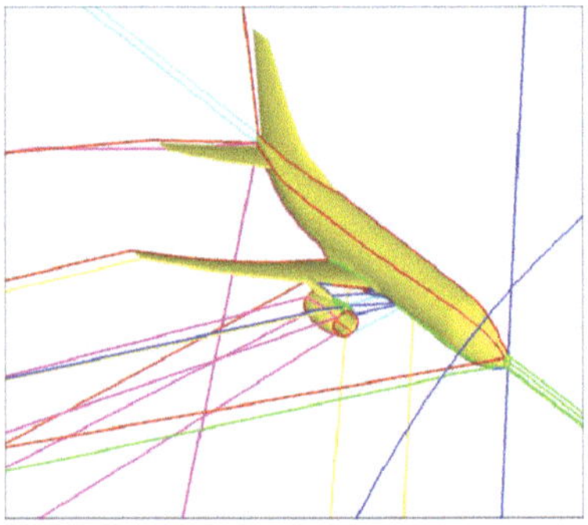

Fig. 7.79: Schematic diagram of the moving grid.

2. Aeroelastic analysis of civil aircraft

1) Structural modeling and analysis

The model of the civil aircraft with double engines in the previous section is considered for the aeroelastic analysis. The finite element model of the aircraft is shown in Fig. 7.80. The modal analysis of the model is carried out. The aeroelastic analysis of the aircraft in the take-off and cruise state is considered, and only the symmetric modals are considered. Table 7.11 and Fig. 7.81 show the frequency and the modal shape of the first four-order symmetric mode of the aircraft, respectively. The first four-order symmetry modals include the first-order bending, second-order bending and first-order torsion of the wing, and the modal of the tail and the fuselage.

Fig. 7.80: Finite element model of civil aircraft structure.

The aircraft is divided into three parts: fuselage, wing and tail. The number of selected structural nodes are 384 nodes of the fuselage, 341 nodes of the wing and 77 nodes of the flat tail, respectively. The CVT interpolation is used for the fuselage while IPS interpolation is for wing and tail. Figure 7.82 shows the interpolation of the four modal shape into the aerodynamic mesh.

Tab. 7.11: Modal selection: select only symmetrical modes.

Modal	7	10	18	20
Frequency (Hz)	2.4	7.0	24.4	25.9
Mode shape	Wing one bend	Wing two bends	Fuselage + wing + tail	Wing twist

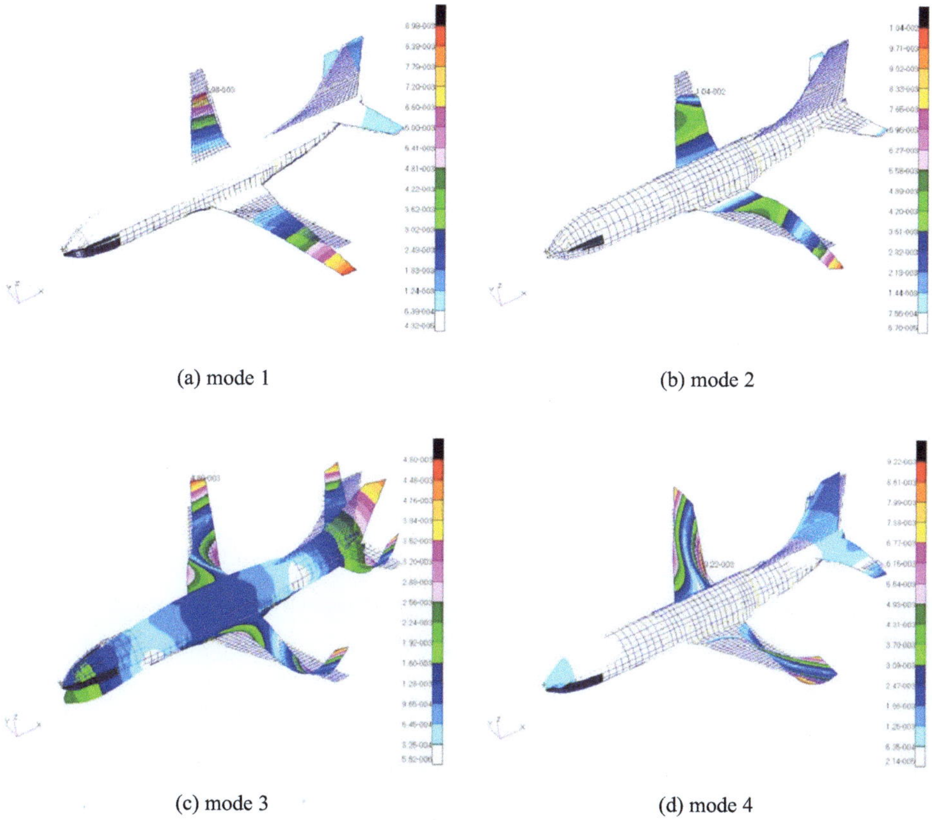

(a) mode 1 (b) mode 2

(c) mode 3 (d) mode 4

Fig. 7.81: Schematic diagram of the first four modes of civil aircraft. (a) Mode 1; (b) mode 2; (c) mode 3; (d) mode 4.

2) Aeroelastic analysis

The aeroelastic effect during the take-off of the aircraft is analyzed first at Mach number = 0.6, angle of attack 10 degree at sea level. Figure 7.83 shows the residual of steady computation. Figure 7.84 shows the surface pressure coefficient distribution of airplane. Figure 7.85 presents the structural response of generalized displacement. Figure 7.86 shows the maximum deformation and surface pressure coefficient after deformation, where the wing tip deformation is about 0.21 m.

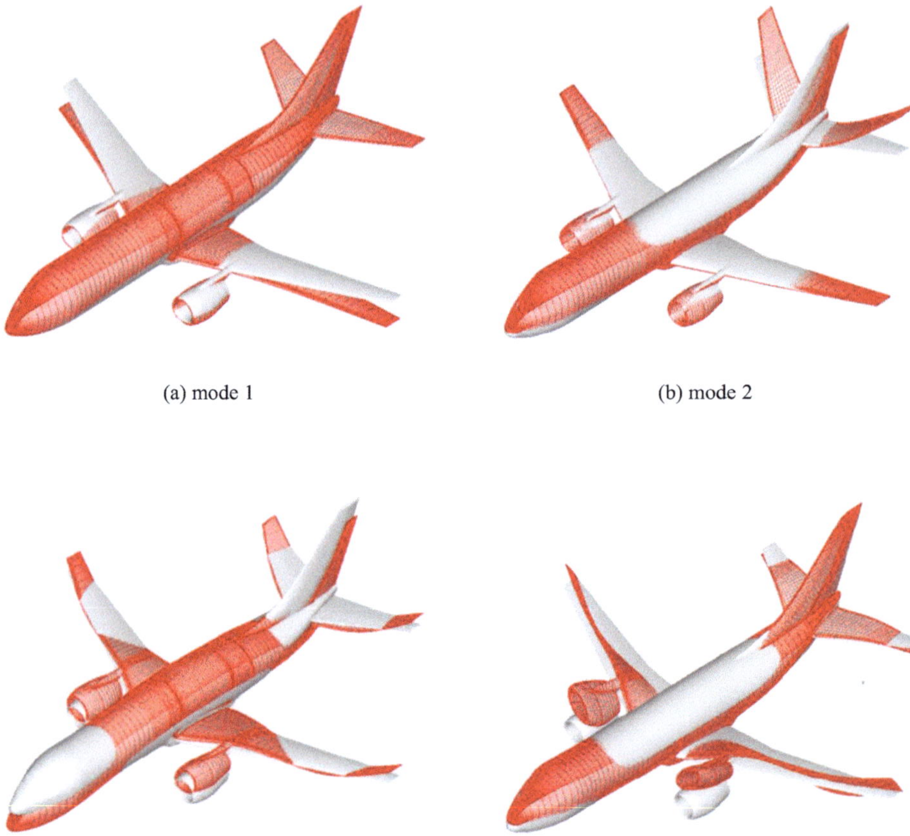

Fig. 7.82: Interpolation preview of civil flight modal: (a) mode 1; (b) mode 2; (c) mode 3; and (d) mode 4.

Furthermore, the influence of aeroelastic deformation on the lift coefficient in the cruise state is analyzed, and the aerodynamic shape is studied to make the lift coefficient in the cruise state equal to the lift coefficient of the rigid shape. The cruising state of the aircraft is set $Ma_\infty = 0.8$, $\alpha = 2.0°$. The lift coefficient of the rigid aircraft is 1.19285. The aeroelastic analysis of the aircraft is carried out and the generalized displacement of the aircraft is (−2.3, −0.07, 0.007, 0.005) after the convergence of aeroelastic computation. The corresponding lift coefficient under elastic deformation is 1.19063, with the reduction of 0.16% compared with that of rigid one.

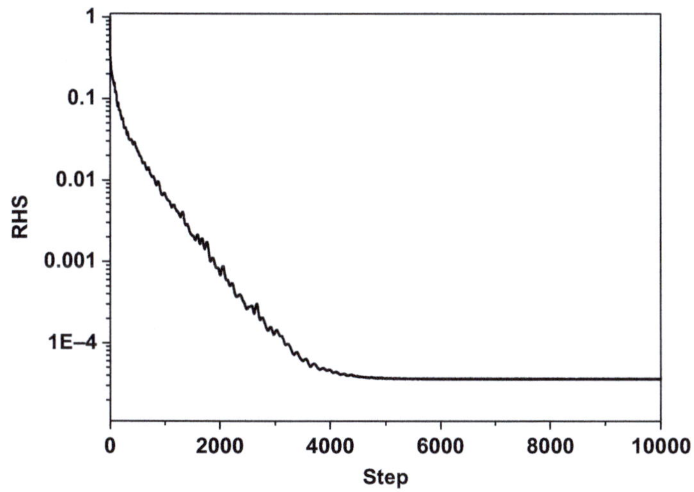

Fig. 7.83: Residual convergence of steady CFD computation ($Ma_\infty = 0.6$, $\alpha = 10.0°$).

Fig. 7.84: The surface contour of pressure coefficient of airplane ($Ma_\infty = 0.6$, $\alpha = 10.0°$).

In the aircraft design, a predeformation in the opposite direction of the elastic deformation can make the lift coefficient of the aircraft under elastic deformation close to that of the rigid airplane, so that the deformation of the aircraft can be compensated by the predeformation. Based on the rigid geometry, the aircraft is implemented a deformation with generalized displacement of (2.3, 0.07, −0.007, −0.005) on the profile of the aircraft. The aeroelastic analysis of the aircraft based on the new shape is performed. Table 7.12 compares the lift coefficient of the rigid shape, aeroelastic

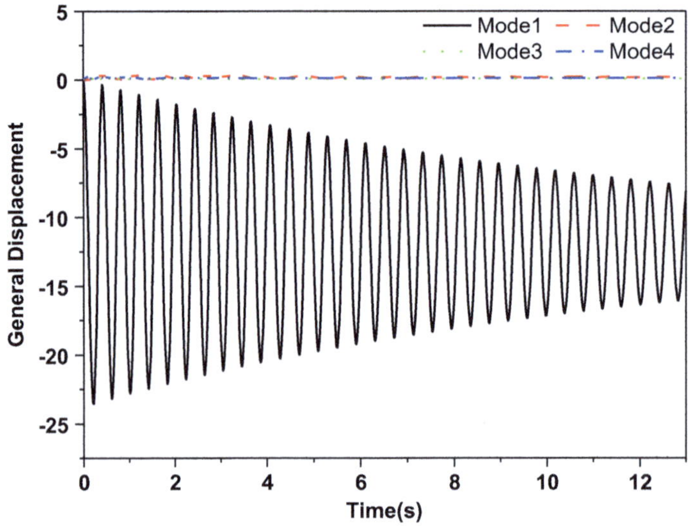

Fig. 7.85: Structure response of civil aircraft.

Fig. 7.86: Maximum aeroelastic deformation of civil aircraft.

Tab. 7.12: Comparison of lift coefficients.

State	Lift coefficient	Lift loss
Rigid shape	1.19258	–
Aeroelastic shape	1.19063	−0.16%
Compensated aeroelastic shape	1.192575	−0.0004%

shape and compensated aeroelastic shape. The lift coefficient of the aircraft is 1.192575, which is close to the lift coefficient of the rigid airplane. It suggests that this design method can well compensate the lift loss caused by the elastic deformation of the aircraft.

Bibliography

[1] Alonso J, Jameson A, editors. Fully-implicit time-marching aeroelastic solutions. 32nd Aerospace Sciences Meeting and Exhibit, 1994.

[2] An X, Xu M, Chen S. A new interface mapping and marching method for aeroelasticity. Journal of astronautics. 2008, 29, 1473–1479. (in Chinese).

[3] An X, Xu M, Chen S. An approximate energy conservation algorithm for shell structure based on Co-rotational (CR) Theory. Journal of Northwestern Polytechnical University. 2011, 2, (in Chinese).

[4] An X, Xu M, Chen S. An overview of CFD/CSD coupled solution for nonlinear aeroelasticity. Advances in Mechanics. 2009, 39(3), 284–298, (in Chinese).

[5] An X, Xu M, Chen S-L. Analysis for second order time accurate CFD/CSD coupled algorithms. Acta Aerodynamica Sinica. 2009, 5, (in Chinese).

[6] An X. CFD/CSD Coupled Solution for Aeroelasticity. Xi'an: Northwestern Polytechnical University; 2006. (in Chinese).

[7] An X. Research on Nonlinear Aeroelasticity Based on CFD/CSD Coupled Solution. Xi'an: Northwestern Polytechnical University; 2009. (in Chinese).

[8] Cai T. Aircraft Aeroelasticity Studies Based on CFD/CSD Coupled Method. Xi'an: Northwestern Polytechnical University; 2009. (in Chinese).

[9] Chen H. Aerothermalelastic Analysis of Lift Component In Hypersonic Regime. Xi'an: Northwestern Polytechnical University; 2013. (in Chinese).

[10] Chen T, Xu M. Effect of thermal load on the dynamic characteristic of panel. Structure & Environment Engineering. 2014, (1), 5, (in Chinese).

[11] Chen T. Investigation of Nonlinear Aeroelastic Modeling Based on Multi-Variable Finite Element Method. Xi'an: Northwestern Polytechnical University; 2014. (in Chinese).

[12] Denegri CM, Johnson MR. Limit cycle oscillation prediction using artificial neural networks. Journal of Guidance, Control, and Dynamics. 2001, 24(5), 887–895.

[13] Dou Y, Xu M, An X, Yao W. Flutter analysis for a fin in hypersonic flow. Engineering Mechanics. 2009, 11, (in Chinese).

[14] Dou Y, Xu M, Cai T, Yao W. Investigation of a two-dimensional panel flutter based on CFD/CSD coupling method. Engineering Mechanics. 2011, 6, (in Chinese).

[15] Dou Y, Xu M. Analysis of energy errors of CFD/CSD coupled algorithms in time domain. Noise and Vibration Control. 2011, (6), 18, (in Chinese).

[16] Dou Y. Investigation of 2-D Panel Flutter With Thermal Effect Based on CFD/CSD Coupling Method. Xi'an: Northwestern Polytechnical University; 2009. (in Chinese).

[17] Guo D, Xu M, Chen S. An effective computation method based on field velocity approach for unsteady flow simulation and obtaining dynamic derivatives. Journal of Northwestern Polytechnical University. 2012, 30(5), (in Chinese).

[18] Liu H, Xu M, Ye M. Investigations of transonic flow field reconstruction and inverse airfoil design based on proper orthogonal decomposition ROM. Acta Aerodynamica Sinica. 2012, 30(4), 539–545, (in Chinese).

[19] Liu H. Investigation of Modeling for Aerothermoelasticity. Xi'an: Northwestern Polytechnical University; 2012. (in Chinese).
[20] Pei X, Xu M, Han B. CFD/CSD coupling aeroelastic-acoustics research in time domain. Journal of Vibration Engineering. 2013, (4), 15, (in Chinese).
[21] Shi Z, Xu M, Chen S. Analysis of moving mesh generation technology. Journal of Air Force Engineering University. 2003, 4(1), 61–64, (in Chinese).
[22] Shi Z. Research on CFD/CSD Coupled Interface Technology. Xi'an: Northwestern Polytechnical University; 2003. (in Chinese).
[23] Slotnick J, Khodadoust A, Alonso J, Darmofal D, Gropp W, Lurie E, et al. CFD Vision 2030 Study: A Path to Revolutionary Computational Aerosciences. NASA. Rapport technique, Nasa; 2014.
[24] Xie L, Xu M, An X, Cai T, Chen W. Research of mesh deforming method based on radial basis functions and nonlinear aeroelastic simulation. Acta Aeronoutica et Astronautica Sinica. 2013, 34(7), 1501–1511, (in Chinese).
[25] Xie L, Xu M, Zhang B, An X. Space points reduction in grid deforming method based on radial basis functions. Journal of Vibration and Shock. 2013, 32(10), 141–145, (in Chinese).
[26] Xu M, Chen G, Chen S, Wei X. Design method of active control law based on reduced order model of unsteady aerodynamics for aeroelasticity. Xibei Gongye Daxue Xuebao/Journal of Northwestern Polytechnical University(China). 2004, 22(6), 748–752, (in Chinese).
[27] Xu M, Chen S. Study of date exchange method for coupling computational CFD/CSD . Chinese Journal of Applied Mechanics. 2004, 21(2), 33–36, (in Chinese).
[28] Xu M, Shi Z, Chen S. A suitable method for transferring information between CFD and CSD grids. Journal-Northwestern Polytechnical University. 2003, 21(5), 535–538, (in Chinese).
[29] Yao W, Xu M, Chen Z. An efficient method for flow field inverse design. Engineering Mechanics. 2011, (2), 8, (in Chinese).
[30] Yao W, Xu M, Chen Z. Efficient method for aeroelastic virtual simulation and modeling. Journal of System Simulation. 2011, (6), 2, (in Chinese).
[31] Zeng X, Xu M. Integrated guidance and control of supersonic cruise missile for dive-attacking. Computer Simulation. 2011, 11, (in Chinese).
[32] Zhang N, Xu M. Based on aerodynamic/aeroacoustic/structure coupling simulation study. Structure & Environment Engineering. 2012, 1, (in Chinese).
[33] Zhou G. Numerical Simulation Research for Nonlinear Aeroelasticity. Xi'an: Northwestern Polytechnical University; 2012.

Index

https://doi.org/10.1515/9783110576689-008

Bei Fragen zur Produktsicherheit wenden Sie sich bitte an:
If you have any questions regarding product safety,
please contact:

Walter de Gruyter GmbH
Genthiner Straße 13
10785 Berlin
productsafety@degruyterbrill.com